The Delaware
Naturalist Handbook

The Delaware Naturalist Handbook

EDITED BY McKAY JENKINS AND SUSAN BARTON

UNIVERSITY OF DELAWARE PRESS
NEWARK, DELAWARE

DISTRIBUTED BY THE UNIVERSITY OF VIRGINIA PRESS

Copyright © 2020 by University of Delaware Press
All rights reserved

Printed in the United States of America on acid-free paper

First published 2020

ISBN 978-1-64453-199-0 (paperbound)
ISBN 978-1-64453-200-3 (e-book)

9 8 7 6 5 4 3 2 1

Library of Congress Cataloging-in-Publication Data
is available for this title.

Book design by Robert L. Wiser, Silver Spring, Maryland

Cover. A great blue heron surrounded by shorebirds at Bombay Hook National Wildlife Refuge.
Frontispiece. Wintering gulls and seaducks at the Indian River Inlet.
(Photos courtesy of Ian Stewart.)

TABLE OF CONTENTS

Acknowledgments...7

INTRODUCTION   Becoming a Delaware Master Naturalist...9
McKAY JENKINS

CHAPTER ONE   Human Impact on the Land in Delaware: A History...11
McKAY JENKINS

CHAPTER TWO   Geology...37
TOM McKENNA

CHAPTER THREE   Watershed Ecology...59
GERALD McADAMS KAUFFMAN

CHAPTER FOUR   Botany and Soil...83
SUSAN BARTON AND JOCELYN WARDRUP

CHAPTER FIVE   Field Sketching and Nature Photography...109
JULES BRUCK, SUSAN BARTON, AND JON COX

CHAPTER SIX   Plant Identification and Taxonomy...135
SUSAN BARTON AND ANNA WIK

CHAPTER SEVEN   Insects: The Little Things That Run the World...163
DOUG TALLAMY

CHAPTER EIGHT   Amphibians and Reptiles...185
JIM WHITE AND AMY WHITE

CHAPTER NINE   Birds...205
IAN STEWART

CHAPTER TEN   Citizen Science and the Scientific Method...227
TARA TRAMMELL

CHAPTER ELEVEN   Weather and Climate Change...243
JENN VOLK

CHAPTER TWELVE   Invasive Species and Habitat Management...267
DOUG TALLAMY

CHAPTER THIRTEEN   Environmental Justice...291
VICTOR W. PEREZ

CONCLUSION   Sustainable Landscapes...311
ANNA WIK AND SUSAN BARTON

Contributors...337

ACKNOWLEDGMENTS

WE WISH TO THANK Angelia Seyfferth and Amy Shober from the University of Delaware's Department of Plant and Soil Sciences for their generous support through the Center of Food Systems and Sustainability; Julie McGee and Tracy Jentzch for their support with grants from the university's Interdisciplinary Humanities Research Center; Dan Rich and Lynnette Overby for their support with a grant from the university's Community Engagement Initiative; John Ernest, chair of the university's English Department, and Kaylee Olney, the English Department's business manager, for administrative support; Julia Oestreich from the University of Delaware Press; and Blake Moore at the university's Cooperative Extension Service for his help constructing the infrastructure for the state's Master Naturalist Certification program. Special gratitude goes to Chris Williams, Michael Chajes, and Kacey Stewart for their consistent support, and especially to Jason Wardrup, a tireless and gifted colleague from the university's Institute for Public Administration, without whose contributions this project would never have gotten off the ground.

We would also like to thank Helen Fischel and Joe Sebastiani from the Delaware Nature Society for their early and consistent support; the Hagley Museum for the use of images from their photographic archives; and Dennis Coker of the Delaware Lenape Tribe, for his close reading of our summary of local Native American history. Finally, we thank the members of the Alliance of Natural Resource Outreach and Service Programs, and especially the Maryland Master Naturalist Program's Kelly Vogelpohl, Wanda MacLachlan, and Joy Shindler Rafey, for their help envisioning Delaware's program.

Jon Cox wishes to thank Susan Barton and Jules Bruck for their wonderful collaboration and for working on multiple iterations of the text to capture a cohesive voice.

Jerry Kauffman wishes to thank the Delaware Department of Natural Resourcesand Environmental Control (DNREC); the National Oceanic and Atmospheric Administration (NOAA); the U.S. Geological Survey (USGS); the U.S. Department of Agriculture (USDA); the National Weather Service (NWS); and the University of Delaware Water Resources Center.

Tom McKenna wishes to thank John Chris Kraft, Professor Emeritus, from the University of Delaware's Department of Geology; William S. Schenck; and Ray Troll, who created the beautiful geologic time scale featured in his chapter.

Victor Perez wishes to offer his admiration and respect to the individuals and families throughout Delaware working toward a just and equitable place to "live, work, and play."

Ian Stewart wishes to acknowledge Joe Sebastiani and Jim White from the Delaware Nature Society for reading his manuscript.

Doug Tallamy wishes to thank Kimberley Shropshire, his research assistant at the University of Delaware, and Drs. Desiree Narango, Adam Mitchell, and Ashley Kennedy, his recent PhD students, for their help on all aspects of the research reported in his chapters.

Tara Trammell wishes to thank the University of Delaware Research Foundation.

Jennifer Volk wishes to thank the Office of the Delaware State Climatologist; the Delaware Department of Natural Resources and Environmental Control's Division of Climate, Coastal, & Energy; and the Northeast Climate Hub.

INTRODUCTION

Becoming a Delaware Master Naturalist

McKAY JENKINS

So you want to become a Delaware Master Naturalist. Congratulations, and welcome to the team!

As you may know, or as you will soon surely discover, Delaware is a state of great and varied natural beauty, and also a state that needs considerable help in returning to full ecological health and resilience. We hope that in reading this book, you will join the growing number of people exploring our state's many wonders; committing themselves to deepening their understanding of environmental systems; and adding their hearts, minds, and bodies to our state's ecological restoration efforts.

In this volume, you will learn from experts in a wide variety of fields who have made the observation, study, and rehabilitation of the natural world their life's work and their life's passion. These professional scientists and veteran environmental educators offer their wisdom on everything from Delaware's geological foundations and its changing meteorology to the complex ecological systems formed by the state's forests, rivers, and soils. You will learn about the region's myriad species of birds, insects, reptiles, and amphibians, and how skilled naturalists can train themselves to notice, identify, and photograph (or draw) what they see.

As anyone who has scoped shorebirds at Bombay Hook National Wildlife Refuge, or canoed the Brandywine River, or watched hawks migrating over Cape Henlopen State Park knows, Delaware is a world-class place to study and appreciate the natural world. Hearing a wood thrush sing the forest to sleep in White Clay Creek State Park, watching the arrival of horseshoe crabs along the Atlantic Coast, listening to spring peepers emerge in a vernal pool along the Delaware River—these are among the many magical events that make Delaware a wonderfully rich place to explore.

But as in the rest of the mid-Atlantic, the country, and the world at large, Delaware's ecosystems are under considerable pressure, most of it caused by human impact. This volume—and the Master Naturalist certification program it supports—have been intentionally constructed with a sense of urgency as well as a sense of awe. The breadth of the material covered here provides some indication of just how interwoven natural processes and human history have become. An environmental historian can look at the Brandywine River and see four hundred years of human history—not just pretty farms and forests, but also damaging mill dams, industrial pollution, and the presence of pharmaceutical drugs in Wilmington's drinking water. An entomologist, trained to understand complex food webs between soils, native plants, insects, and birds, can look at a suburban lawn and know that the grass growing there, although it might look "neat and tidy," is actually about as ecologically valuable as a patch of cement. A forest ecologist can not just look at Delaware's woodlands and see gorgeous, hundred-foot-tall trees but can also recognize that these trees represent a tiny fraction of the trees that once grew here, and that the ones that remain are imperiled by a staggering infestation of invasive vines. Experts in fields as diverse as sustainable landscaping and environmental justice can look at a cornfield or a chicken farm in Sussex County and see profitable industries layered on top of land burdened by a history of slavery, migrant labor, soil depletion, and water pollution.

To put it simply, everything is connected to everything. In this light, you will undoubtedly notice some overlap in these chapters: it is impossible for a watershed ecologist not to talk about geology beneath the soil (or about agricultural pollution in the soil), just as it is impossible for a bird biologist (or a scientist interested in insects) not to mention our urgent need to plant more native plants. As scientists and environmental educators, our hope is that by reading this book, and (if you are able) by taking the Master Naturalist training course for which this book serves as the curriculum, you will deepen your understanding of historical and ecological systems that have made Delaware's landscape what it is today. By understanding our past, and closely observing our present, we hope you will join a growing team of experts and volunteers alike working to restore our land and water to an equilibrium that has been tipping out of balance not just for a few decades but for hundreds of years. Especially given distressing global trends—notably climate change, rising sea levels, deforestation, soil depletion, and a radically growing number of endangered species—the time could not be more pressing for more of our citizens to invest themselves in caring for our home. It is our hope that by learning about our state's environment, you will deepen your sense of enjoyment of the state's many natural wonders, and also commit yourself to restoring our state to optimal ecological health. There is much that needs to be done, and much that we can do ourselves. We hope this book offers a step in that direction.

CHAPTER ONE

Human Impact on the Land in Delaware: A History

McKAY JENKINS

A COUPLE OF WEEKS AGO, one of my students at the University of Delaware told me he had recently seen a regal bird standing in the shallows of White Clay Creek. He didn't know much about birds, he said, but he'd seen a lot of nature shows on television. He guessed the bird was a blue flamingo. I said flamingos were pink, Delaware was a long way from Florida, and that the bird he saw was probably a great blue heron. Another student said the problem with studying the environment was that nature, in today's culture, is essentially irrelevant. Trees would get more respect, he said, if they were functionally useful, like cell towers. I asked how he felt about the oxygen he was breathing, and decided it was time to take the class outside.

For the past twenty-five years, I have asked my undergraduates to spend at least an hour a week wandering the forest and streambank inside White Clay Creek State Park, part of which sits just alongside campus. For fifteen or sixteen weeks each semester, students learn to tune into the rhythms they see unfolding before them: the coming and going of the seasons; the arrival and disappearance of migratory songbirds; the rising and falling of the creek itself. I tell them I want their field journals to be as precise as possible, to chronicle not just "trees" but "white oaks"; not just "birds" but "belted kingfishers"; not just "foliage" choking the forest canopy but invasive Asian bittersweet, autumn olive, and English ivy.

But I also want my students to unplug from their electronic devices, and get to know their local landscape with their bodies as well as their brains. I want them to remember, or reawaken, their capacity to *see*, to *sense*, to *feel* their presence on the land. I want them to experience what the Japanese call *Shinrin-yoku*, or "forest bathing"—the immersive, therapeutic experience of being surrounded by trees. At a time when more than a third of all college students report having diagnosable mental illness, such experiences are in distressingly short supply.[1]

Beyond nudging students to see what is right in front of them, I also want them to recognize what they *don't* see, in the forest, but also back on campus and in their own (mostly) suburban neighborhoods. I want them to recognize that the White Clay Creek watershed, as ecologically imperfect as it is, represents not just a precious remnant of the region's past, but also a reminder of how ecologically compromised the rest of the state (and indeed the mid-Atlantic region itself) has become. I want students to apply their observation skills to the landscape inside *and* outside the forest, to look out the windows of their cars and *see* the Phragmites growing alongside the highway, or the Japanese honeysuckle draping the native trees alongside the supermarket, or the stormwater pouring off the parking lot of their local shopping mall. I want them to *see* the industrial soybean and chicken farms, to *see* the beach resorts built atop former wetlands, to *see* the subdivisions that have replaced forests with chemically sanitized lawns.

Most of my students arrive in class thinking that Delaware's "environment" is nothing more than suburban subdivisions, shopping malls, and superhighways—a landscape utterly given over to providing people with goods and services. In this, of course, they are not wrong: Delaware in the early twenty-first century is a highly developed, engineered, even manicured landscape, a state slashed through by interstate highways, intensive urban industry, extensive suburban housing development, and global-scale industrial agriculture.

This state of affairs, I tell them, did not happen overnight. Our contemporary culture, often critiqued as a generation of digital distraction, is also the logical, even predictable result of four hundred years of environmental extraction. Finnish lumberjacks (and many others) extracted trees in great numbers from the mid-Atlantic beginning as early as the seventeenth century. Dutch engineers extracted crops from wetlands they channeled, diked, and filled. Plantation owners used slave labor to extract tobacco and corn from their soil. Industrialists dammed the Brandywine and the Christina, extracting river power to make everything from bread flour to dynamite. More recently, Delaware's land has been used to build endless subdivisions, and to support extractive industries that harvest everything from soybeans to chickens and create the petrochemicals used in vehicles, cosmetics, and cookware.

This extractive, consumption-driven approach to our land and water is not universal, and may indeed be a product of our own sense of dislocation. Native American writers like Robin Wall Kimmerer remind us that, in many non-European traditions, human beings are considered the youngest of all species, dependent on the wisdom and generosity of animals and plants to live fully and well. "After all these generations since Columbus, some of the wisest of Native elders still puzzle over the people who came to our shores," Kimmerer writes:

They look at the toll on the land and say, "The problem with these new people is that they don't have both feet on the shore. One is still on the boat. They don't seem to know whether they're staying or not." This same observation is heard from some contemporary scholars who see in the social pathologies and relentlessly materialist culture the fruit of homelessness, a rootless past. America has been called the home of second chances. For the sake of the peoples and the land, the urgent work of the Second Man may be to set aside the ways of the colonist and become indigenous to places. But can Americans, as a nation of immigrants, learn to live here as if we were staying? With both feet on the shore?[2]

This has long made me wonder: what would it take for us—all of us—to return to a deeper, more intimate, more reciprocal relationship with our land and waters? What would it mean to more fully *know* our environment, so that we might learn to *love*, and then to *heal* it?

As you will read in the following pages of this essay, and the other chapters collected in this book, four centuries of dramatic human impact, layered on top of billions of years of geological (and millions of years of biological) change, have constructed the baseline onto which we must map our sense of place, and our sense of environmental responsibility. Even if many of us feel fundamentally disconnected from natural systems, the evidence of system *failure*—climate change, species extinctions, toxic pollution, water contamination—is becoming harder and harder to ignore. Many of my students are aware of this. Even those who don't self-identify as environmentally aware exhibit at least a low-grade sense of cynicism, or dread, about the state of the world. I consider it part of my job both to increase their environmental awareness and to help them keep their chins up. There's too much restoration work to be done.

For starters, there is nothing to be gained by romanticizing (or bemoaning the loss of) prehistoric wilderness. The environmental conditions that present themselves today are the result of several centuries of incomplete awareness or willful blindness. They have also resulted from a willingness to sacrifice long-term ecological health for short-term material well-being. Indicators are arising everywhere that we no longer have the luxury of ignorance or passivity. This is not just an issue in Delaware, obviously, and given the state's inextricable place in the globalized economy, I have tried to put the state's challenges in the context of some of the issues facing the country and the world.

Our work today is to see the world as it is, to understand how it evolved that way, and to figure out how we can regenerate and restore systems that previous generations cared for less well than they might have, and to which our own generation is paying far too little attention. Some of this wisdom will surely come from science, and some of it must inevitably come from ways of being that are both far more subtle and far more pragmatic in their approach to human health and ecological systems. Never before have we had more information at our fingertips, and never before has our energy and imagination been more urgently needed. As Dennis Coker, principal chief of Delaware's Lenape Tribe, recently told me, it is possible to return to the view that the rivers and trees that surround us, along with the plants and animals with whom we share the world, are not our "natural resources," but our *relatives*.

We encounter today a Delaware ecological landscape that is beautiful, compromised, and in great need of our attention and care. This essay will try to lay out how this has come to be. My hope is that by understanding something of the path we have been traveling, we might come to know our place more intimately, and see that now is the time to invest ourselves more fully in knowing, caring for, and restoring Delaware's land and water.

THE BASICS

Let's start with the basics. At 1,954 square miles (ninety-six miles long and between just nine and thirty-five miles across), the state of Delaware is the second smallest state in the country. It is bounded to the north by Pennsylvania; to the east by the Delaware River, Delaware Bay, and Atlantic Ocean; and to the west and south by Maryland. The state's boundaries include a pair of oddities that have long had both political and ecological implications. The Mason-Dixon Line runs north-south along the border between Delaware and Maryland, and effectively cuts Delaware off from the upper reaches of the Chesapeake Bay. The odd arc at the state's northwest edge, which describes a twelve-mile radius from a center point in the town of New Castle, was intended to separate Delaware and Pennsylvania, but it also divided a natural watershed that includes drinking water supplies flowing through the Brandywine, the Red Clay, and the White Clay Creeks. Today, officials hoping to improve water quality in Delaware have to convince policymakers in Pennsylvania to plant more trees, as there is no better protector of water than a healthy forest.

The state's average profile is just sixty feet above sea level, giving it the lowest average altitude of any state. Delaware has among the country's highest percentages of both wetlands and floodplain. Its highest point, marked by the Ebright Azimuth, is just 442 feet above sea level; only Florida has a lower high point. Add to this the state's 381 miles of shoreline, and the implications of Delaware's geography, especially in an era of rising sea levels, are substantial. Water levels near Lewes have already risen a foot over the last century, meaning the state (and the mid-Atlantic) is facing one of the world's most rapidly changing coastlines.

Of the state's roughly 1.25 million square acres of land, about 490,000 are currently in farmland, some 340,000 square acres of which are committed to the growing of industrial-scale corn and soybean crops. A great deal of this grain is grown to feed the roughly 260 million chickens currently in residence in the state. In Delaware, there are about 270 times as many chickens as people.

As small as it is, the state still has some 13,500 lane-miles of highway, notably the colossus of Interstate 95, which was completed in the late 1960s. And even with 110,000 acres of state parks and recreation areas, only about seven percent of Delaware's land is publicly owned. This is partially because only 2.3 percent, mostly in the form of the National Wildlife Refuges at Bombay Hook and Prime Hook, is owned by the federal government. By comparison, Pennsylvania has more than 17 percent of its land, about 4.8 million acres, under state or federal protection. Maryland has protected about 11 percent of its land, about 683,000 acres.[3]

If Delaware's industrial cities and large-scale farms often get the most attention from environmental groups, since the end of World War II, the most intensive change in Delaware's landscape has actually been the explosion of suburban housing. Farms in densely populated areas no longer grow food, the saying goes; they grow houses. The pace of suburban development in Delaware has been scorching: about half of the state's developed land, more than 121,000 acres, was developed between 1982 and 2007. In this, Delaware looks a lot like the rest of the country: in the last six decades, the United States has lost four million family farms, much of them to suburban development. Nationwide, every year between 1982 and 1997, some two million acres of land were lost to development. Add to this the nearly 44,000 square miles of blacktop and 62,500 square miles of ecologically sterile lawns we maintain around our homes, and you begin to understand the national scale of the suburban ecological impact.

What this has meant for Delaware (and the country) is a radical reduction in forest and cropland. The state's once contiguous forest is now roughly one-quarter of its original size, and nearly 50 percent of the remaining woodlots are smaller than ten acres. The impact of this human footprint has been profound. In Delaware, as of 2002, 40 percent of our native plants were either threatened or already gone, and 41 percent of native forest-dwelling bird species were rare or absent. We have lost 78 percent of our mussel species, 34 percent of our dragonflies, 20 percent of our fish species, and 31 percent of our reptiles. The causes for these dire numbers are not just the construction of too many new houses, but also corollary problems: predation by domestic cats, collisions with cars and office building windows, and the wide use of lawn chemicals. The decline of monarch butterflies, which may be teetering on the edge of extinction, has been caused in large part by the widespread use of agricultural herbicides, which (even if unintentionally) kill milkweed, the monarch's only larval food source.[4]

What all this large-scale ecological degradation, mostly happening on privately owned land, means is that ecological restoration cannot just be done by "professionals" working in public parks. It will have to happen on private land—in our neighborhoods and schools, along our roadways, and on our farms. And it will have to be done by all of us. In other words, everyone in the state bears some responsibility for caring for our land. And what that means is that we all need to understand our common predicament. For me, that means taking not just environmental science students but everyone—English majors and history majors and math majors—out into the woods to put their hands on the bark of a sycamore tree, to hear the eerie call of a wood thrush, to trail their fingers in the water sliding beneath their canoes. I want them to dig up invasive multiflora rose, and taste the leaves of invasive garlic mustard, and chop thigh-thick Asian bittersweet out of native maple trees. And I want them to understand how the forests came to be so compromised, and how they might help heal them.

So let's start at the beginning.

THE STATE'S FIRST INHABITANTS, AND THE PRESSURES ARRIVING FROM EUROPE

Like the rest of North America's aboriginal people, Delaware's first inhabitants are thought (by Western anthropologists) to have been descendants of immigrants who arrived here in several ways: some crossed the Bering Strait from Asia to Alaska around 10,000–9,000 B.C.; others are thought to have crossed the Atlantic Ocean's ice edges from Europe sometime between 25,000 and 13,000 years ago.[5] Indigenous communities often take exception to this narrative, and have very different stories of how their people came to be on the land. Lenape Chief Coker notes that mastodon skulls and indigenous spearpoints have been discovered forty miles off the Atlantic Coast, where (when sea levels were far lower than they are today) the continental shelf was an expansive (and animal-rich) grassland. The spearpoints, traced to a mine twenty miles up the Susquehanna River, have been dated to twenty-three thousand years ago. "The Creator put us here," Chief Coker told me. "We've always been here."

Regardless of the origin story, the land early peoples discovered between the Delaware and Chesapeake Bays would have been similar in topography to the land today with one primary difference. With so much of the world's water still locked up in colossal glaciers like the Laurentide Ice Sheet, which once covered North America as far south as Southern New York, the continental coastline would indeed have extended much further into the ocean. Wildlife would have been abundant, including white-tailed deer, caribou, and a wide variety of small mammals, birds, and fish. As the glaciers began melting, and ocean levels began rising, once-brackish estuaries would likely have become overwhelmed by rising levels of salt water, leaving shellfish like crabs, clams, mussels, and oysters in relatively short supply. Coastal trees would have suffered under increasingly salty soils.

Like other issues related to our current climate crisis, this part of the state's past may prove to be prologue.[6]

The region's first human residents—from a few hundred to a few thousand—were a loosely associated people living between New York and the

Chesapeake Bay. European colonists initially labeled these groups the "river people" and then the "Delaware"; the people called themselves the Lenape. Often banding in groups of twenty-five to thirty, they took up residence on the west bank of the Delaware River from Pennsylvania's Lehigh Valley in the north to Delaware's Leipsic River Valley in the south. Other scholars argue that the Lenape's range was far greater, extending from the mouth of the Delaware north to the Lower Hudson Valley. Smaller regional bands of Native Americans included the Nanticokes, living along Broad Creek in Sussex County, and a small group known as the Siconese (also spelled Sekonese or Ciconicin) northwest of Cape Henlopen. Descendants of these peoples continue to live in the state.

The solutions that individual Native American groups found to the basic problems of getting food, clothing, and shelter "were rooted in a shared cultural tradition that had been evolving in the Chesapeake Bay region for several millennia," historians Helen Rountree and Thomas E. Davidson write. The state's inland forests at the time were rich with oak, beech, chestnut, tulip poplar, hickory, walnut, and elm, with stands of pines closer to the coast and south of the St. Jones River. The state's northern hardwood stands produced great quantities of "mast"—tubers, nuts, berries, roots, and seeds—that supported a diversity of wildlife and provided copious human food supplies.[7]

As the post-glacial ocean levels stabilized, the state's many estuaries began once again producing a vast array of fish and shellfish. Especially in Northern Delaware, the Lenape practiced "horticulture" rather than formal "agriculture," cultivating edible plants but also relying on seasonal hunting and gathering of the forests' cornucopia of mast and edible plants. Indigenous understanding of the subtleties of the land were deep, intimate, and intuitive, and even if the land was "used," it was also, for the most part, renewed. The relationship between people and the land was less extractive than reciprocal. "Native science extends to include spirituality, community, creativity, and technologies that sustain environments and support essential aspects of human life," Native American scholar Gregory Cajete has written. "This gives rise to the view that creation is a continuous process but certain regularities that are foundational to our continuing existence must be maintained and renewed. If these foundational patterns are not maintained and renewed, we will go the way of the dinosaurs."[8]

Recent research around the world maintains that the diet of hunter gatherers was, and in some parts of the world remains, far more diverse and nutritionally dense than the diet of agricultural communities. "Corn got here late because we didn't need it!" Chief Coker told me. This is especially worth contemplating given our current investment in industrial agriculture, which has built a global food system out of a tiny handful of grains like corn, wheat, and soybeans, favors the production of high-calorie but low-nutrition foods, and has resulted in global epidemics of obesity and diabetes.[9]

Technology—notably the bow and arrow—made hunting deer, turkey, and (occasionally) elk far more efficient. So too did the annual burning of forests to clear the forest's understory, which opened up hunters' lines of sight, created foraging meadows for deer and other prey animals, and allowed more sunlight to reach native berries as well as cultivated food crops. So broad was the Native American use of fire along the mid-Atlantic Coast that in 1632, a Dutch mariner wrote that the land "could be smelt before it is seen."[10]

Contrary to popular mythology, indigenous people across North America, and on other continents, had the capacity to dramatically impact or even eliminate populations of animals long before European colonization. In North America, Native Americans likely reduced the continent's megafauna—such as giant sloths, mastadons, and the wooly mammoth—and surely decreased populations of more familiar animals like deer, beaver, and elk. The mastodon disappeared from the continent about thirteen thousand years ago, part of "a wave of disappearances that has come to be known as the megafauna extinction," Pulitzer Prize-winning environmental journalist Elizabeth Kolbert has written. "This wave coincided with the spread of modern humans and, increasingly, is understood to have been a result of it."[11]

By the time Europeans began arriving in the early seventeenth century, then, Delaware's landscape had already felt the firm hand of humans. But the human capacity for altering natural systems grew dramatically from the beginning of colonization. This was caused by a change in technology, notably guns, and it was caused by a change in culture, notably European-style forest, farming, and economic practices. Small-scale indigenous economies were quickly overwhelmed by extractive, market-driven economies, especially by the exploding colonial market for such things as timber, deerskins, and beaver pelts. This colonial market would itself explode into an international export market that sent untold millions of tree and animal products to voracious European markets for everything from ship masts to bird-feather and beaver-skin hats. Up and down the East Coast, once dense populations of hardwood trees and native animals were reduced to extreme scarcity, and ecological systems began to feel pressure that only intensified over the next four centuries. And the people themselves were wiped out by diseases, notably smallpox, spread (sometimes intentionally) by Europeans. "Every time there was an outbreak, we would lose 50 percent of the people living in a village," Chief Coker told me. "After a half-dozen outbreaks, we'd lost all our elders, all our wisdom keepers, and all our young. That was a devastating time."

EARLY SETTLERS

It is worth remembering that by the time Europeans arrived in North America, their home continent had already been utterly deforested, denuded of wildlife, and planted to meadows and pastureland for many hundreds of years. As early as the eleventh century, Europe's woodlands had already been reduced to less than 15 percent of the continent's landscape. Dense "forests," like "wild animals," existed more in European folklore than they did in the actual landscape. Adding to this, European Christianity inherited (and then reinforced) agricultural ideas of land use that encouraged the subjugation of the natural world, often in the overt service of centralizing and scaling up political and economic power. Mankind's relationship to the natural world was fundamentally hierarchical: man had been made in God's image; the water and land, and all that lived there, had been placed on the earth for man's use.

"The fear of you and the dread of you shall be upon every beast of the field, and upon every bird of the air, upon everything that creeps on the ground, and all the fish of the sea; into your hands they are delivered," Genesis declares. "I give you everything." As environmental historian Roderick Nash has reported, the biblical Hebrew words outlining man's relationship to the world include *kabash*, translated as "subdue," and *radah*, translated as "have dominion over." Man was given "dominion" over the earth, and was installed in a hierarchical relationship to all other creatures. Along with many centuries of agricultural practice, this vision solidified the European notion of mankind as not just superior to but in control of all other beings. "The image is that of a conqueror placing his foot on the neck of a defeated enemy, exerting absolute domination," Nash writes. "Both Hebraic words are also used to identify the process of enslavement."[12]

The triumph of European Christianity (and Christian-supported colonialism, imperialism, and capitalism) over indigenous ways of knowing and being on the land "was the greatest psychic revolution in the history of our culture," Lynn White wrote in his famous 1967 essay "The Historical Roots of Our Ecological Crisis." By destroying indigenous cultures, Christianity "made it possible to exploit nature in a mood of indifference to the feelings of natural objects." In recent years, a growing number of "ecotheologians" and writers from Vine Deloria to Wendell Berry have articulated just how damaging the marriage of religion and exploitation have been. Many have also argued that a parallel "green Christianity," supported by a host of biblical passages, has always been present, urging a far more integrative and less exploitive relationship between humans and the earth.[13]

In Delaware, the earliest Dutch, Swedish, and English explorers found forests—of oak, beech, and hickory—of a darkness and density they had not experienced in thousands of years. Along the coast they discovered a constellation of fertile estuaries, swamps, and wetlands that supported a great bounty of fish, shellfish, and waterfowl—and, of course, mosquitoes. The forests, swamps, and skies were dense with creatures unfamiliar to the newcomers: passenger pigeons, bald eagles, fireflies, cicadas. Whales patrolled Delaware Bay, which, like the Chesapeake, was also home to vast oyster beds. The new arrivals quickly got to work changing the Delaware topography to reflect the European landscapes they had left behind. They cut down trees, drained swamps, dammed creeks, constructed roads, planted row crops like wheat and barley, and imported their own domestic pigs, sheep, goats, and cattle.[14]

Out in the forests, lumberjacks from Finland, a country that supplied perhaps 30 percent of the Delaware Valley's earliest settlers, practiced the same destructive slash-and-burn forestry that had gotten them booted out of Swedish forests. The Swedish Forest Law of 1647 condemned those "who come into the forests with a ravenous appetite, without explicit permission," and instructed authorities "to capture and, as with other noxious animals, strive to get rid of them." Dutch immigrants used their engineering knowledge to dike and drain coastal wetlands into farmable land, fencing them in by turning 140-foot-tall cypress trees into rot-resistant fenceposts. As William Williams put it in his essential environmental history *Man and Nature in Delaware*, it was the state's fate "to begin its European era under the influence of the most successful forest-conquering culture in history, but one that used fire ceaselessly, chopped down trees indiscriminately, practiced agriculture negligently, and destroyed wildlife enthusiastically."[15]

In the river corridors, Europeans saw not animals but "natural resources" they could harvest and sell. By 1663, European fur traders in the Delaware Valley were purchasing some ten thousand pelts a year from Native hunters; within forty years, so many beaver had been extracted from the landscape that the beaver trade collapsed. Whalers were so effective in hunting Delaware Bay that they ran out of local quarry. And the Delaware market represented just a tiny fraction of a continent-wide plundering. Between 1769 and 1868, the Hudson's Bay Company sold at London auction, among other items, the pelts of 891,091 fox; 1,052,051 lynx; 68,994 wolverine; 288,016 bear; 467,549 wolf; 1,507,240 mink; 94,326 swan; 275,032 badger; 4,708,702 beaver; and 1,240,511 marten. During the same period, two other companies, the North West Company and the Canada Company, were trading in similar numbers of animals.[16]

By the end of the seventeenth century, then, Europeans had killed or driven off most of the state's Native American occupants, privatized much of their former lands, radically reduced local populations of animals, and set in motion cultural and economic forces that would dramatically alter ecological systems across the state, and across the continent. Such a vision, of course, was in marked contrast to spiritual systems created by indigenous peoples in North America (and many other places around the world) that considered humans to fit into a subtle, interdependent, and reciprocal web of being that included not only animals but also plants, rivers, mountains, and weather.

As Europeans fully settled into the Delaware landscape, they began engineering systems to resemble and feed an economic model that would expand inexorably for the next several hundred years. This engineering included the implementation of social systems (like slavery), ecological systems (deforestation, damming, and ditch-digging), and transportation systems (canals, roads, and railroads) to maximize yields, and distribution systems to expand the capacity for exporting goods to distant markets. Very little attention was paid to the tremendous damage these systems did to the integrity of human or ecological communities.

SLAVERY, INDUSTRY, AND CARVING UP THE LAND

One of the first things early settlers had to figure out was how to work vast stretches of land in a region with relatively few people. One answer came in the form of slavery. "Although most people think of slavery in terms of big cotton plantations in the Deep South, it was the area surrounding the Chesapeake Bay—Maryland, Delaware, and Virginia—where slavery originated and where it flourished for almost two hundred years prior to the development of big cotton plantations in the South," sociologist Margaret Anderson has written.[17] Slavery in Delaware, as elsewhere, "grew like a cancer, at first slowly, almost imperceptibly, then inexorably, as colonists eager for material gain imported hundreds of thousands of Africans to toil in their fields," historian Peter Kolchin has written. "During the eighteenth century, slavery became entrenched as a pervasive—and in many colonies central—component of the social order, the dark underside of the American dream."[18]

In 1721, an estimated two thousand to five thousand slaves lived in Pennsylvania and the three lower counties on the Delaware (New Castle, Kent, and Sussex); perhaps five hundred of these people were kept in the three lower counties. By 1790, the three Delaware counties alone held some nine thousand slaves and four thousand free Blacks, most of whom worked as farm laborers or domestic servants. Wealthy planters in Kent County also used skilled slaves as tanners, shoemakers, carpenters, and tailors, and as labor in iron foundries.

By 1829, the number of slaves in Delaware had decreased by half, and by 1860, the number was down to less than eighteen hundred. These numbers should not leave an impression that the decline was a marker of good will, historian James Newton writes. It was, rather, a matter of money:

The usual explanation given is humanitarianism and religious feeling, abolitionist efforts, and runaways.... In reality, Delaware farmers found it cheaper to hire free black labor than to keep slaves. Furthermore, Delaware, the most northern of the slave states, had no great crop of tobacco or cotton to be looked after during all seasons of the year. The land was wearing out, and state law forbade the sale of slaves out of state. Thus, slave owners could not benefit from breeding slaves as in a state like Virginia.[19]

The tortuous arrival of freedom to Delaware's African Americans "may well be the most enigmatic such episode in the history of emancipation in the Americas," historian Patience Essah has written. The state was split between those in the more populous north, who espoused the Quaker-led emancipation values of Pennsylvania, and those in the agricultural south, who resisted the movement to free slaves. The Underground Railroad passed through Wilmington, and when slaves finally got across the Christina River, they were free. Although the state had banned the international slave trade as early as the 1780s and vigorously debated the abolition of slavery entirely, it did not ratify the Thirteenth Amendment until 1901—thirty-six years after the end of the Civil War and the effective end of slavery nationwide.[20]

CUTTING THE FORESTS, DRAINING THE SWAMPS

If the enslavement of human beings left a lasting cultural wound in Delaware, deforestation, market hunting, and industrial engineering left an indelible imprint on the land. By 1841, lumbermen had so fractured the state's habitat, and hunters had so thoroughly scoured the state for white-tailed deer, that the General Assembly banned deer hunting. Bird populations did not enjoy similar protections; by the early nineteenth century, the ten thousand turkeys that are thought to have lived in the state before colonization were virtually extinct.[21]

By 1910, 75 percent of Delaware's forests had been cut down, and most of its remaining trees were less than sixty years old. Virtually every creek in the state had either been redirected, diked, or silted over by farm runoff. Market hunters, who had already burned through the state's deer and turkey populations, turned to bird hunting to fill the state's bellies and provide the feathers to help outfit society's upper-class women with fancy hats. Steamboats and railcars opened up vast new consumer markets for meat from ducks and geese; hatmakers had a bottomless demand for the plumage of great blue herons, egrets, ibises, and swans. The forest-dwelling ruffed grouse was gone from Northern Delaware by 1870, and from the entire state by the early twentieth century. The passenger pigeon, once the most populous bird in the world (with an estimated population of four billion), was hunted into extinction by the mid-nineteenth century. Hunters used giant punt guns to take down entire flocks of birds with a single blast of birdshot.

It wasn't until 1911 that state conservationists pressured state legislators to ban the commercial shipping of birds killed in Delaware. Diamondback terrapins, once so common they were fed to slaves, were virtually eliminated by the late 1930s. Trappers pulled 100,000 muskrat pelts from Delaware estuaries in the late nineteenth century. Delaware's last bear was shot in Sussex County in 1906.[22]

Along the coast, engineers continued their three hundred-year quest to tame the swamps. By the 1930s, the Civilian Conservation Corps (CCC) had been ordered by state and federal authorities to dig a grid of ditches around Delaware's ocean resorts to protect one of the state's growing economic engines from mosquitoes. The CCC dug some two thousand miles of ditches across the salt marshes in Kent and Sussex Counties, typically in two-man teams, one using a long, heavy spade, the other a "potato hook." Over time, some forty-four thousand acres of wetlands were channeled, deforested, dried out and (some years later) sprayed with DDT. Once majestic cypress swamps were reduced to a fraction of their former range, in many cases replaced by pines. Oysters, once abundant, were hammered in the nineteenth century: first, by overharvesting (especially by iron-toothed dredging machines); then by industrial pollution; and then, in the 1950s, by a parasite called MSX. Shad—anadromous fish that move from freshwater streams in the spring, out to the oceans in fall, and then return several years later to spawn—were once ubiquitous in streams throughout the state, but were decimated first by the damming of their streams and later by overfishing and pollution.

INDUSTRIALIZATION

By the mid-eighteenth century, Pennsylvania's Chester and Lancaster Counties had emerged as the region's most fertile agriculture areas. Early entrepreneurs like William Shipley began bringing grain overland to the north bank of the Christina River and then shipping it up the Delaware River to Philadelphia. In 1739, the growing port town on the north bank—with water deep enough to allow ships drawing fully fourteen feet (among the largest of the era)—was officially named Wilmington. It wasn't long before all this grain moving into Wilmington gave rise to new industries, especially with the construction of not only grain mills, but also paper and gunpowder mills powered by the Christina and the Brandywine. These two rivers combined to make Wilmington one of the most powerful centers of eighteenth-century grain production in the country.[23]

The pressures of a growing population fell heavily on the region's already damaged and fragmented forests. Most settlers lived in wooden homes heated with wood fires. Many worked in iron foundries that turned tens of thousands of acres of trees into charcoal. In 1802, Éleuthère Irénée du Pont established a gunpowder mill near Wilmington that laid the foundation for what would become Delaware's huge chemical industry. Delaware manufacturers began turning out vulcanized fiber, textiles, paper, medical supplies, metal products, machinery, and machine tools.

All this commerce required and depended on transportation networks to move products around and out of the state. Millers, merchants, and bankers in Philadelphia were keen to expand south, toward markets in Baltimore and south along the Chesapeake. With trees disappearing, rivers and streams had long since become clogged with sediment, and were thus less useful as transportation corridors. By the early nineteenth century, the region's structural engineers were well into the practice of cutting canals through the land. After twenty-five years of on-again, off-again construction, the Chesapeake and

Delaware Canal (C&D) finally opened in 1829. Stretching 13.6 miles, the canal connected the Delaware River, about six miles south of the town of New Castle, to Back Creek, a tributary of Maryland's Elk River.

As economically beneficial as the new canal became, the construction of the C&D—like that of Wilmington's many dams (and the mill ponds and millrace canals that fed them)—also entailed broad-scale destruction of forests, wetlands, and traditional fish migrations, and turned once wild rivers into domesticated commercial waterways. Shad could no longer make their annual migrations (as, 150 years later, they could no longer get past the four hydroelectric dams on Pennsylvania's Susquehanna). Heavy rainfall, once absorbed by marshes and swamps, now flooded cities and towns, and ruined roadways. As late as the early nineteenth century, Delaware's roads were still in sorry, muddy, rutted shape. But by 1808, the General Assembly began contracting with "turnpike" companies to start constructing a network of roadways out of wood, stone, gravel, and clay, topped with macadam, and graded to permit proper drainage. By 1825, a web of roadways connected Philadelphia and Lancaster to Wilmington, New Castle, Newport, Christiana Bridge, and Newark through to the Northern Chesapeake and Baltimore.[24]

The 1837 completion of the Philadelphia, Wilmington and Baltimore Railroad rapidly changed Wilmington's economy from that of a grain-and-flour port to one of a national center of manufacturing. Within a few short decades, the Christina River had become lined with coal-fired factories, which turned Wilmington into a hub for tanning, and for the making of railcars, iron ships, cotton textiles, fertilizer, and gunpowder. The city's skies were now blanketed by soot belching from coal-burning factories and wood smoke from domestic fireplaces. The city streets were fouled by garbage and animal excrement, especially from the six thousand horses that annually contributed enough manure to cover an acre of land fully seventy-two-feet high. Runoff from this waste, along with spillover from the city's outhouses and even the outflow from eroding cemeteries, left the city's rivers at a polluted, typhus-contaminated low. In 1877, a physician named L. P. Bush wrote that pollutants flowing out of Wilmington's factories were "quadrupled by those which are swept into the Brandywine by every heat rainfall from the hillsides and valleys, consisting of earthy and excrementous substances."[25]

This only compounded an existing problem: up the Delaware River, Philadelphia by the early twentieth century was dumping 200,000 tons of raw sewage into the river every year, and military contractors were pumping industrial waste on top of it. By mid-century, Philadelphia was already known as one of the foulest freshwater ports in the country, and all that waste flowed downstream, where it met first Camden's and then Wilmington's own polluted effluent. It wasn't until after World War II that public outcries finally led to major upgrades in urban sewage treatment. By the late twentieth century, Northern Delaware was both more densely industrial and more densely populous than almost anywhere in the United States.

FIGURE 1.1. Rockford and Kentmere cotton mills, Wilmington, 1937. (Joseph Bancroft & Sons photographs. Dallin Aerial Survey, Hagley Museum.)

FIGURE 1.2. Chesapeake and Delaware Canal, 1937. (Dallin Aerial Survey, Hagley Museum.)

FIGURE 1.3. The Pyrites Company, Wilmington, 1936. (Hagley Museum.)

Between World War II and 2000, the state's population had jumped from 266,505 to 783,600. More than one hundred manufacturing plants, chemical plants, and oil refineries lined Northern Delaware's riverbanks. New Castle County's population was fifteen times denser than the national average and, on its own, produced far more garbage, human waste, and polluted air from cars and homes than did forty-three other states.

Delaware was home to some of the country's worst toxic waste dumps, and the state's 718,000 registered automobiles, combined with untold millions of cars passing though on the state's interstate highways, routinely left the state's air quality below federal health standards. Benzene, mercury, vinyl chloride—it was all in the air at levels that were among the worst in the country. In 1990, Delaware companies released 12,455,947 pounds of toxic chemicals, half of it into the state's air. In 2001, the American Lung Association gave all three Delaware counties an air quality grade of "F." A national insurance company in 1989 ranked Delaware as the country's most unhealthy state. The state's relatively affluent consumers threw away 860,000 tons of garbage in 1989 alone.[26]

Beyond this general degradation, such heavy industrial pollution frequently, and disproportionately, impacts minorities and the poor. "When decisions are made about where to locate heavily polluting industries, they often end up sited in low-income communities of color where people are so busy trying to survive that they have little time to protest the building of a plant next door," journalist Steve Lerner writes. "The health impact of this patently unwise zoning formula is predictable: residents along the fenceline with heavy industry often experience elevated rates of respiratory disease, cancer, reproductive disorders, birth defects, learning disabilities, psychiatric disorders, eye problems, headaches, nosebleeds, skin rashes, and early death. In effect, the health of these Americans is sacrificed, or, more precisely, their health is not protected to the same degree as citizens who can afford to live in exclusively residential neighborhoods."[27]

THE GROWTH—AND ECOLOGICAL CHALLENGES— OF DELAWARE FARMING

Down on the farm, Delaware's overworked soils continued to suffer from centuries of nutrient depletion. By the mid-nineteenth century, farmers were so desperate to improve their soils that they began importing bird guano from Peru. Local soil amendments—especially "marl," composed of minerals and soil mixed with ancient shells, along with crushed shells of horseshoe crabs and oyster shells—eventually proved easier to come by. It wasn't until the late 1880s that scientists discovered the wonders of cover cropping with legumes, which fixed nitrogen in the soil by pulling it straight out of the air. Red clover became so beneficial for soil improvement that it earned the nickname "mortgage-lifter."

As the decades unfolded, the problem of too little fertilizer became a problem of too much. Two developments led to significant challenges that remain today: first, the post-World War I invention of synthetic nitrogen, which allowed farmers to purchase and spread as much petrochemical fertilizer as they could afford; and second, the explosion of the chicken industry. Up until the early 1900s, chickens had been raised primarily for their ability to lay eggs, not for their meat. Then one day in 1923, as legend has it, a woman named Cecile Steele from Ocean View ordered fifty chicks for her backyard flock, but due to a miscalculation, the hatchery sent her five hundred chicks instead. "She kept them, she raised them, she sold them sixteen weeks later for sixty-three cents per pound for meat," the state's former Agricultural Secretary Ed Kee once recounted. "The word spread rapidly, and that was the birth of the Delaware and Delmarva poultry industry."[28] Delaware's annual broiler production jumped from two million birds in 1928 to sixty million just sixteen years later, leading to a parallel explosion in the need for chicken feed, mostly in the form of soybeans and corn. Even by 1943, the state's chickens ate enough feed every year to fill a freight train 110 miles long. If the state was going to grow its chicken industry, it was going to have to grow corn and soybeans. And it has.

The chicken industry is now controlled by a handful of large factory farms, or "integrators," including Perdue, Tyson, and Mountaire, which can produce astonishing numbers of birds with relatively few farmers (some of whom keep forty thousand birds in a single enclosure). In 2017, the Delmarva chicken industry raised 605 million chickens and fed them 87 million bushels of corn, 36 million bushels of soybeans, and 1.6 million bushels of wheat.

There are two pressing environmental challenges posed by all these birds. The first is the hundreds of thousands of acres of monoculture grains required to feed them and the agricultural chemicals used on those crops. Beyond the ecological cost of converting forests, meadows, and wetlands to monoculture grain crops, this farmland is heavily chemical-dependent, and public anxiety about the potential harm of these chemicals is beginning to mount. In 2015, the cancer research arm of the World Health Organization declared glyphosate (marketed by Monsanto as Roundup) to be a "probable human carcinogen." In 2018 and early 2019, major court decisions awarded tens of millions of dollars to people claiming glyphosate had contributed to their cancer. Roundup is the world's most popular herbicide, and though there remains considerable debate about its toxicity to humans, its wide use on farms and suburban lawns alike is devastating to biodiversity. Using weed killers along roadsides, agricultural fencerows, and suburban lawns is a primary cause of the collapsing of populations of native bees and monarch butterflies, to say nothing of the decline of numberless insects scientists have not had the time to count.[29]

The second issue is the colossal amount of waste that chickens produce. Large poultry companies have concocted a business model in which they own the birds and the feed, and contract with local farmers to raise the birds

FIGURE 1.4. Darling property farm near Wilmington, 1937. (Dallin Aerial Survey, Hagley Museum.)

to market weight. The chicken manure—remember, we're talking about hundreds of millions of birds—is left to the farmers to deal with.

At first, these mountains of manure seemed like a gift, helping farmers replenish the depleted nitrogen and phosphorus in their soils. The "chicken litter" that farmers use for fertilizer, made from manure mixed with wood shavings, is desirable primarily for its nitrogen and organic content. The trouble is its high levels of phosphorus, a mineral that is added to chicken feed in order to strengthen their eggs and bones so they will survive mechanical processing without breaking. Chickens are very inefficient processors of phosphorus, so most of what they ingest comes out in their excrement. Over decades of this practice, the soil on many farms has become saturated with phosphorus, and even if some if this mineral binds with soil, there is growing scientific evidence that saturated soils leach phosphorus into nearby creeks and rivers.

Downstream, the manure fertilizes aquatic plants, notably algae, which, under the influence of these excess nutrients, flourish and die in spectacular numbers, creating a brand new environmental catastrophe: "cultural eutrophication," in which decaying algae suck all the oxygen out of an aquatic system and leave behind hypoxic "dead zones" in which nothing else can live. In 2019, for example, Mountaire Farms was ordered to clean up one million gallons of wastewater that had spilled from a chicken plant near Millsboro that processes some two million chickens per week. The company has been found to have violated wastewater permits a number of times, notably for releasing high levels of nutrients and bacteria in its wastewater. Add such events to the assault by farm chemicals, untreated urban sewage, and the fertilizer runoff of suburban lawns, and you begin to understand why Delaware's estuaries and inland bays have become so vulnerable.[30]

THE SUBURBAN EXPLOSION

And those are the farms that survived suburban sprawl. Since World War II, real estate development in Delaware—and around the country—has exploded in once-rural, unincorporated areas outside of cities and towns. Especially during periodic economic booms, developers have bought vast tracts of farmland (often from farmers who can no longer afford to keep their land in agriculture) and turned them into huge developments far from existing water and sewer lines, schools, or even substantial roads. Environmental historian Adam Rome has called the postwar building boom "an environmental catastrophe on the scale of the Dust Bowl." For the first time, builders began putting hundreds of thousands of homes in environmentally sensitive areas, including wetlands, steep hillsides, and floodplains. Rome writes, "Builders also began to use new earth-moving equipment to level hills, fill creeks, and clear vegetation from vast tracts. The result was more frequent flooding, costly soil erosion, and drastic changes in wildlife populations."[31]

By 2010, 71 percent of Delawareans lived outside of cities or towns. What this movement from cities to suburbs has meant is that many more acres of forests and cropland have been turned into houses and shopping malls. To combat this, state land use planners have continually tried to convince policymakers to spend money on schools, sewer lines, police stations, hospitals, and public transportation in places where people already live—Wilmington, for example—rather than letting builders put more roads in rural parts of the state. "It is easy to see that the cost of providing these services is greatly affected by our pattern of land use," Delaware's Office of State Planning Coordination notes. "In general, the more spread out we are, the more costly it is for taxpayers and the more of our environmental resources will be consumed."[32]

The trouble with all this building outside of existing population centers is that it has continued to chop up state forests that have already been under intense pressure for four hundred years. By 1995, the average patch of forest in the state was just ten acres, far too small to support a functioning, diverse ecosystem within. In 2015, a state Ecological Extinction Task Force issued a formal report urging policymakers to encourage landowners to understand the impact they have on local watersheds and local species of plants and animals. The average woodlot is just "too small and too fragmented to sustain biodiversity into the future," Doug Tallamy, a member of the task force, wrote.

Some 93 percent of state land is privately owned, including 77 percent of its forests, so stewardship largely depends on the wisdom and good graces of the land's owners. Yet, in Delaware's countless suburbs, fully 92 percent of the landscape is ecologically useless lawn, 79 percent of the plants are introduced species, and only 10 percent of the trees that could be planted have been planted. "Lawns do not support pollinators, do not support natural enemies of pest species, do not sequester much carbon, do not support the food webs that support animals, and they degrade our watersheds," Tallamy wrote. "Invasive plants alter soil conditions and nutrient cycling, change habitat structure, and compete with native plants for natural resources. There are over 3,300 species of (non-native) plants in North America and we are introducing more every year. There are now 432 species of North American birds at risk of extinction (more than a third of all species)."

The extinction report urged the state to encourage the planting of native plants, to the removal of invasive species, and to the replacement of non-native trees with native trees. It encouraged governments to lead by example, by putting these practices into place on public lands as well. "A critical part of demonstrating to the public the importance of native species to our local ecosystems is by our government taking the lead and providing native species landscape and management on our public property," the report noted. Local and state governments should be encouraged to pass legislation supporting the protection and restoration of native species, and perhaps to limit the sale of invasive plants, many of which can still be found in nurseries throughout the state. In 2017, an analysis of plants sold by nurseries in New Castle County found that more than 83 percent were non-native and less than 17 percent were native.[33]

Some public officials appear to have taken notice. In 2018, New Castle County Executive Matthew Meyer signed an executive order mandating that the county would only use native plants when developing or restoring public lands.[34] This is a promising sign in a state where the public response to structural, industrial, and agricultural challenges has ranged from frequently mute to occasionally inspired. In the 1960s, Governor Elbert Carvel pushed back against real estate developers and kept the state's beaches in public hands. Governor Russell Peterson followed up by preventing the energy industry from building massive oil shipping ports along the state's coastline. Historian Carol Hoffecker has called Peterson's signature Coastal Zone Act Delaware's "greatest and most comprehensive legislative achievement toward maintaining a livable environment." In 1995, Governor Tom Carper established the Office of State Planning Coordination to provide some oversight of the rapid suburbanization of the state. That same year, the state began funding the Delaware Agricultural Lands Preservation program that had been created five years before. As of April 14, 2015, the program had spent over two hundred million dollars to preserve over 116,223 acres of crops and forestland. More recent statewide efforts—some effective, some chronically underfunded—have sought to further preserve forests and farms, and to prepare the state for rising sea levels and other effects of climate change.[35]

As we move deeper into the twenty-first century, Delaware remains confronted by a host of emergent environmental challenges, including: rising sea levels; forest fragmentation and the attendant drop in plant and bird species; and the growth of invasive plants throughout the state's forests and wetlands. The state's deer population, once in disastrous decline, has now rebounded to the point that populations are threatening woodland understories everywhere. State planners remain anxious to convince homeowners and policymakers alike to regenerate urban forests, support farmers markets (and the local, small-scale farming economy generally), and live in denser, more concentrated dwellings with bike lanes, urban parks, and retrofitted green buildings and green infrastructure.

Indeed, even at this late stage of Delaware history, when so many of us feel the pinch of a compromised landscape, natural systems—from water and soils to plants to the weather itself—are calling for us to pay closer attention to what we are doing, and to take better care. In my teaching, I have taken this quite literally: it is no longer enough, I've decided, to "study" environmental problems. We have encountered a time when an intellectual interest in such matters is not sufficient. I now routinely take my Environmental Humanities students canoeing on local rivers to show them the damage done by hydroelectric dams. I take them into forests to physically remove invasive species. And, on a day in early March 2019, I brought fifty of my students to spend the

FIGURE 1.5. Looking down east of Blackshire Road, Wawaset Park, 1916. (Wawaset Park photographs, Hagley Museum.)

FIGURE 1.6. Suburban lawns near White Clay Creek State Park, 2013. (Photo courtesy of Doug Tallamy.)

day working on five hundred acres of abandoned farmland that had recently been acquired by the Bombay Hook National Wildlife Refuge.

The land had been through a number of iterations over the last four hundred years: it had been cleared for timber, planted with tobacco and worked by slaves, then replanted with industrial corn. But it also benefited from quieter, more reciprocal work as well. For four years, starting in 1938, the Civilian Conservation Corps Company 3269-C—the state's only segregated, all African American unit—was told to create a twelve hundred-acre waterfowl refuge. Part of the work was repairing some of the centuries-old damage that had been done to the local forests. CCC workers created a tree nursery holding nearly forty-two thousand seedlings, then added fifteen thousand more a year later. Early plantings included Norway maples, honey locust, green ash, black locust, persimmon, hackberry, yellow and red pine, and horse chestnut seedlings.[36]

Now, more than ninety years later, the U.S. Fish and Wildlife Service is returning farmland to a version of its precolonial state. My students helped plant (in a single day) some seven hundred native hardwood trees—swamp white oaks, pin oaks, red maples, sycamores—that will help the land regain ecological stability. To my mind, such work—learning about, and caring for, an embattled local ecosystem in an environmentally embattled state—might well serve as a lesson (and perhaps a requirement) for all college students. And for the rest of us as well.

NOTES

1. "Among College Students, Mental Health Diagnosis and Treatment are Up, Stigma is Down." American Psychiatric Association, November 5, 2018. See https://www.psychiatry.org/newsroom/news-releases/among-college-students-mental-health-diagnosis-and-treatment-are-up-stigma-is-down.
2. Robin Wall Kimmerer, *Braiding Sweetgrass* (Minneapolis: Milkweed Editions, 2015).
3. See "2015 Delaware Strategies for State Policies and Spending," prepared by Delaware Office of State Planning Coordination: http://www.stateplanning.delaware.gov/strategies/documents/2015-state-strategies.pdf. See also USDA/NRCS National Resource Inventory, https://www.nrcs.usda.gov/wps/portal/nrcs/main/de/technical/dma/nri/. For corn and soybean data, see USDA's 2018 State Agriculture Overview for Delaware: https://www.nass.usda.gov/Quick_Stats/Ag_Overview/stateOverview.php?state=DELAWARE. See also "Environmental Policy in Pennsylvania," *Ballotpedia*, https://ballotpedia.org/Environmental_policy_in_Pennsylvania.
4. Douglas Tallamy, *Bringing Nature Home* (Portland, OR: Timber Press, 2007), 29, 34–26. See also "Final Report of the State Ecological Extinction Task Force," December 1, 2017, http://www.dnrec.delaware.gov/Admin/Documents/de-eetf-final-report.pdf.
5. Dennis J. Stanford and Bruce A. Bradley, *Across Atlantic Ice: The Origins of America's Clovis Culture* (Berkeley: University of California Press, 2012), 14.
6. Jay F. Custer, *Delaware Prehistoric Archaeology: An Ecological Approach* (Newark: University of Delaware Press, 1984), 39–47.
7. Helen C. Rountree and Thomas E. Davidson, *Eastern Shore Indians* (Charlottesville: University of Virginia Press, 1997), 1.

8. Gregory Cajete, *Native Science: Natural Laws of Interdependence* (Santa Fe: Clear Light Publishers, 2000), xi, 3.
9. Marshal Joseph Becker, "Lenape Population at the Time of European Contact: Estimating Native Numbers in the Lower Delaware Valley," *Proceedings of the American Philosophical Society*, CXXXIII, no. 2 (1989) 116–17. See also Custer, *Delaware Prehistoric*, 89–93. See also Zach Elfers, "The Way of the Groundnut," February 11, 2019, www.nomadseed.com.
10. Shepard Kretch, *The Ecological Indian: Myth and History* (New York: Norton, 1999), 101.
11. Elizabeth Kolbert, *The Sixth Extinction: An Unnatural History* (New York: Picador, 2014).
12. Roderick Nash, *The Rights of Nature: A History of Environmental Ethics* (Madison: University of Wisconsin Press, 1989).
13. Lynn White, "The Historical Roots of Our Ecological Crisis," *Science*, Vol. 155, #3767, March 10, 1967.
14. Delaware Department of Agriculture, "Delaware Agricultural History." See https://agriculture.delaware.gov/agricultural-history/.
15. Terry G. Jordan and Matti E. Kaups, *The American Backwoods Frontier* (Baltimore: The Johns Hopkins University Press, 1989), 57, 96; William Williams, *Man and Nature in Delaware* (Dover: Delaware Heritage Press, 2008), 95.
16. Rountree and Davidson, *Eastern Shore Indians*, 35–37. See also Barry Lopez, *Arctic Dreams* (New York: Bantam Books, 1986), 337.
17. Margaret Anderson, "Discovering the Past/Considering the Future: Lessons from the Eastern Shore," in Carol Marks, ed., *A History of African Americans of Delaware and Maryland's Eastern Shore* (Wilmington: The Delaware Heritage Commission, 1998), 103. See https://archives.delaware.gov/wp-content/uploads/sites/156/2018/08/A-History-of-African-Americans-of-Delaware-and-Marylands-Eastern-Shore.pdf.
18. Peter Kolchin, *American Slavery, 1619–1877* (New York: Hill and Wang, 2003), 4.
19. James Newton, "Black Americans in Delaware: An Overview," in Marks, ed., *A History of African Americans of Delaware and Maryland's Eastern Shore*.
20. Patience Essah, *A House Divided: Slavery and Emancipation in Delaware, 1638–1865*, (Charlottesville: University of Virginia Press, 1996), 1–3.
21. Jordan and Kaups, *The American Backwoods Frontier*, 96. See also Williams, *Man and Nature in Delaware*, 45, 61, 66, 155.
22. Williams, *Man and Nature in Delaware*, 128.
23. J. Thomas Scharf, *History of Delaware*, 1609–1888 (Philadelphia, L. J. Richards, 1888), 632; Carol Hoffecker, *Brandywine Village: The Story of a Milling Company* (Wilmington, DE: Old Brandywine Village, 1974), 16–19; Carol Hoffecker, *Wilmington, Delaware: Portrait of an Industrial City, 1830–1910* (Charlottesville: University of Virginia Press, 1974), 17.
24. Williams, *Man and Nature in Delaware*, 53–54, 76, 81–82, 85.
25. Hoffecker, *Wilmington, Delaware: Portrait of an Industrial City, 1830–1910*, 19; Williams, *Man and Nature in Delaware*, 115; Lewis Potter Bush, *Some Vital Statistics of the City of Wilmington* (Washington, DC: Library of Congress, 1877), 8.
26. Williams, *Man and Nature in Delaware*, 169–74; 192.
27. Steve Lerner, *Sacrifice Zones: The Front Lines of Toxic Chemical Exposure in the United States* (Cambridge, MA: The MIT Press, 2010), 6.
28. Mark Eichmann, "Delaware's Growing Poultry Industry," *Radio Times*, WHYY, August 11, 2014. See https://whyy.org/articles/delawares-growing-poultry-industry/.
29. Holly Yan and Madeline Holcombe, "Jurors Say Roundup Contributed to a 2nd Man's Cancer. Now Thousands More Cases Against Monsanto Await," CNN, March 20, 2019. See https://www.cnn.com/2019/03/20/health/monsanto-verdict-federal/index.html.

30. Maddy Lauria, "Southern Delaware Chicken Plant Spilled Up To 1 Million Gallons of Wastewater: State," *Delaware News Journal*, February 13, 2019. See https://www.delawareonline.com/story/news/local/2019/02/13/delaware-chicken-plant-facing-lawsuits-spills-wastewater/2864817002/.

31. Adam Rome, *The Bulldozer in the Countryside: Suburban Sprawl and the Rise of American Environmentalism* (Cambridge: Cambridge University Press, 2001), 3.

32. "2015 Delaware Strategies for State Policies and Spending," prepared by Delaware Office of State Planning Coordination. See: http://www.stateplanning.delaware.gov/strategies/documents/2015-state-strategies.pdf.

33. "Final Report of the State Ecological Extinction Task Force," December 1, 2017. See http://www.dnrec.delaware.gov/Admin/Documents/de-eetf-final-report.pdf. See also Nicole Alvarez and Jennifer Parrish, "Analysis of the Sale of Native Species in Retail and Wholesale Outlets in New Castle County, Delaware," Appendix D, "Final Report of the State Ecological Extinction Task Force."

34. See New Castle County Executive Order 2018–10: https://www.nccde.org/DocumentCenter/View/26671/Exec-Order-2018-10-Native-Species.

35. Carol Hoffecker, *Democracy in Delaware: The Story of the First State's General Assembly* (Wilmington, DE: Cedar Tree Books, 2004), 223.

36. Robert Mayer, "Creation of a Legacy: The Story of the Civilian Conservation Corps at Bombay Hook, 1938–1942," National Fish and Wildlife Service. See https://www.fws.gov/uploadedFiles/CREATIONOFALEGACY.pdf.

CHAPTER TWO

Geology

TOM McKENNA

INTRODUCTION: LEARNING TO THINK IN DEEP TIME

To understand geology, we must first grasp what geologists call "deep time," a concept so abstract that it requires—for most of us, anyway—a good metaphor. Imagine that the full geologic time scale, from the formation of the Earth, 4.6 billion years ago, to the present day, has been compressed into a single calendar year. At this scale, one million years is represented by just two hours. When do Homo sapiens arrive on the scene? On December 31, the last day of the year, at just twenty-two minutes before midnight.

Or perhaps a spatial analogy will help. Say you are driving south on Route 1 in Delaware, from Interstate 95 to Fenwick Island at the Maryland border. As you drive, you are moving through geologic time, from the formation of the Earth (at I-95) to the present day (at the Maryland border). At this scale, you are moving through geologic time at a rate of forty-three million years per mile.

You start at Interstate 95 in the hellfire of the Hadeon Eon, and enter the Archeon Eon at Boyds Corner Road in New Castle County, just south of the toll plaza. By the time you enter the last eon in the Precambrian, the Proterozoic Eon, you are already at the south end of the Dover Air Force Base in Kent County. After driving for about 91 miles of the 101-mile trip, you are approaching the Indian River Inlet south of Dewey Beach in Sussex County. The only drivers you have seen on the road so far are simple-minded, marine organisms. As you get closer to the inlet bridge, you enter the Cambrian Period, and suddenly there is an explosion of complex life forms.

On the bridge, you enter the Ordovician Period. Volcanoes loom offshore, and you see the recently uplifted Appalachian Mountains, as high as the current Himalayas. Near the Tower Shores community, you drive through the Silurian and Devonian Periods and see fish all around you. On through the Mississippian and Pennsylvanian Periods, you pass through lush, tropical vegetation and enter the Permian Period just north of the Delaware National Guard base. You are riding on the supercontinent of *Pangaea*. At South Maplewood Avenue, in

Bethany Beach, there is a major highway mishap: the remnants of the greatest extinction in Earth's history—at the end of the Permian Period—block the road. You pass through the end of the Paleozoic Era and into the Mesozoic Era at the start of the Triassic Period.

You continue on, sharing the road with a growing number of dinosaurs in the Jurassic and Cretaceous Periods. Suddenly, at Fenwick Island State Park, there is another major accident, and much of life, once again, goes extinct. The dinosaurs disappear here at the end of the Cretaceous and the start of the Paleocene Epoch and the Cenozoic Era. You continue through the Cenozoic, and, as life begins to regenerate, more and more mammals begin crowding the highway. You reach the Pleistocene Epoch in Fenwick Island, only one block from the Maryland border. Here, you start driving in and out of ice ages. You have still not yet seen any people! In that last block, your ancestors evolve, and start honking their horns. Welcome to the present epoch, the Holocene.

Geologists are students of time and space. They begin their exploration of past landscapes by examining what they can see on or under the ground in the present day: the visible textures, compositions, and structures of the small bits of rock that crop out here and there, following the simple law that the present is both the key to the past and a clue to predicting the future. They know that a million years is a small bit of geologic time, and they can envision geologic processes occurring at grand spatial scales, like entire continents colliding to create mountain chains. Then they ask questions to figure out how a place came to be over geologic time. Did these rocks form above or below ground? How did they get to their location—by uplift, gravity, water, or wind? Were they deformed? Due to what stresses?

From these inquiries, students of geology can begin to paint a mental picture of the relevant geologic processes. Though to normal human perception many geological processes may appear to be slow, they nonetheless result in the geologic wonders we observe today. The bed of the Colorado River, for example, is eroding at about the thickness of a fingernail (0.25 millimeters) per year. But multiply that by six million years and you get the Grand Canyon!

The Himalayas reach an elevation over 5.5 miles at Mount Everest, yet marine fossils found at the summit indicate that the rocks were formed at sea level (zero feet). These mountains, like many mountains, resulted from the collision of two continents (in this case India and Asia) at the incremental rate of only two inches per year, but this slow collision still results in a present-day uplift rate of about 0.4 inches per year. Over one million years, this has pushed the Himalayas up 5.5 miles! (Uplift has been ongoing in the Himalayas for about forty-five million years, and is still happening, but the erosion on the land surface is occurring at about the same rate as the uplift so there is little net change in elevation).

Geology is an integrating science, using deductive reasoning that requires knowledge and the application of physics, chemistry, biology, mathematics, and other disciplines. In its early stages, geology was mostly an observational and descriptive science, and it remains focused on the study of earth, the rocks and other materials that make it up, the processes acting on and resulting in the formation and structure of those materials, the fluids that occur at the surface and in the pores of the rocks, and how materials evolve through geologic time.

Geology is also part of larger (and pressing) environmental conversations, including debates over the exploration and sustainable development of natural resources including energy resources (like oil, gas, coal, uranium), mineral resources (like metals and fertilizers), and water resources (like groundwater). Geologists address natural hazards (like volcanoes, earthquakes, landslides, and floods), man-made hazards (like land subsidence and mass movement near construction sites), and environmental contamination (related to the byproducts of mining, industrial, and agricultural industries). They also bring the viewpoint of geologic time and the evolution of the Earth and its atmosphere to the climate change debate and attempt to anticipate what could take place in the future.

It is worth stating right up front that geology is a field with a lot of terms to learn. Many terms are abstract, either because they refer to ancient time periods, or because they refer to physical structures that are invisible to everyday human observation (like trenches at the bottom of the sea, or molten rock miles below the surface of the earth). Other terms, some of which refer to common ideas, may seem unnecessarily complex: a beach, for example, is a type of *sedimentary depositional environment*. But as difficult as some of these terms may be to absorb, trust that they (like the geological formations they describe) have evolved over time, and will fit together sensibly once you become conversant with them.

THE GEOLOGIC TIME SCALE

At the onset of the contemporary field of geology in the eighteenth century, scientists observed sedimentary rocks in the field and tried to organize them into some type of chronological order. If rocks were found in the same outcrop, or were discovered close by one another, scientists could correlate rock layers based on physical appearance and/or the similarity between sequences of two or more layers. The trouble is this often cannot be done accurately over long distances because sedimentary rocks may change their physical character across space even if they were deposited at the same time. For instance, think of present-day sediments deposited at the same time in different places: the sands on Dewey Beach are physically different from muddy sediments in the middle of nearby Rehoboth Bay.

Over time, scientists solved this problem when they discovered that fossil assemblages in rocks vary in a systematic way. Combined with the understanding that a sedimentary rock layer in a sequence is necessarily younger than the one beneath it (and older than the one above it), and a set of principles of relative time dating, scientists started constructing the *geologic time scale*.[1]

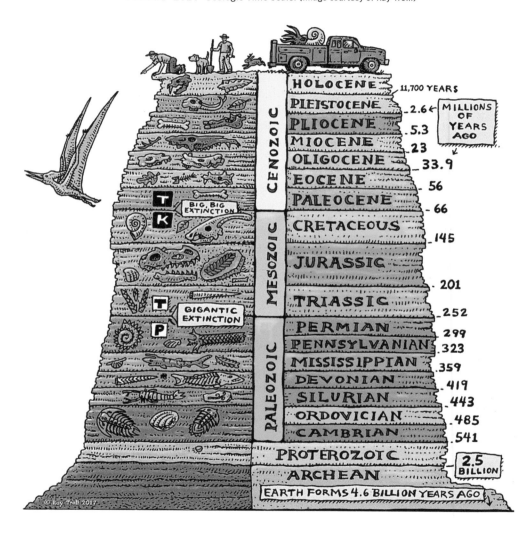

FIGURE 2.1. Geologic Time Scale. (Image courtesy of Ray Troll.)

Using this scale, we divide Earth's history into discrete intervals of time going back to the formation of our planet 4.6 billion years ago. There is nothing magical or complicated about the scale, but again, it is critical to understand "deep time" to think like a geologist. Geologic time is divided along a *relative time scale*, with the subdivisions representing different lengths of time. The scale is broken down into *eons, eras, periods, epochs*, and *ages*. From youngest to oldest, the geologic *eons* are the *Phanerozoic, Proterozoic, Archean*, and *Hadean* (the latter three make up the *Precambrian*, considered an informal term). The current period of the Earth's existence, the *Phanerozoic Eon*, is broken down into three *eras*: the *Paleozoic* (representing 53 percent of the period), the *Mesozoic* (34 percent), and the *Cenozoic* (12 percent). Most of the subdivisions are based on the occurrence (or absence) of particular fossils, the position of fossils relative to fossils in other rock layers, and some simple rules for relative dating, which are discussed below. In many cases, divisional boundaries can be determined by identifying macroscopic fossils. Others require the examination of *microfossils* under the microscope.

While there was single-celled life in the Precambrian, there are insufficient fossils to subdivide time. Within the Cenozoic Era, the most commonly used subdivision is the *epoch*, from youngest to oldest: *Holocene, Pleistocene, Pliocene, Miocene, Oligocene, Eocene*, and *Paleocene*. In the Mesozoic and Paleozoic Eras, the most frequently used subdivision is the *period*. The periods in the Mesozoic, from youngest to oldest, are the *Cretaceous, Jurassic*, and *Triassic*. The periods in the Paleozoic are *Permian, Pennsylvanian, Mississippian, Devonian, Silurian, Ordovician*, and *Cambrian*.

To augment the relative time scale, geologists (working in the subfield known as *geochronology*) have used *radiometric dating* techniques to develop an *absolute time scale*, typically expressed in "millions of years before present." As work by geologists continues, the correlations of the absolute to relative scales continues to evolve, with new ages assigned to the relative succession of rocks in the geologic time scale. A detailed time scale for North America that includes both relative and absolute time is available online from the Geological Society of America.[2]

THE EARTH'S INTERIOR, AND PLATE TECTONICS

The outermost layer of the earth, the layer we live on, is called the *crust*.[3] It is often compared to the skin of an apple, since it comprises less than 1 percent of Earth's volume. It is less dense than—and actually "floats" on—the underlying *mantle*. The crust is about three to six miles thick under the oceans (where it is called the *oceanic crust*) and thirty-one miles thick under the continents (the *continental crust*). Oceanic crust is composed primarily of darker, denser rocks like basalt, diabase, and gabbro that contain the minerals olivine, pyroxene, amphibole, and biotite. Continental crust is composed mostly of less dense, lighter-colored rocks, like granite, that contain the minerals feldspar and quartz. The upper crust is relatively brittle and breaks (like pottery) when geologic forces are applied. Earthquakes typically occur in this brittle upper crust. The deeper crust is under higher pressures and temperatures, making it more *ductile* (like putty). Ductile materials move and deform plastically under geologic forces. The *mantle* makes up 84 percent of Earth by volume and is composed mainly of a rock called peridotite. It is denser than the crust but less dense than the underlying core. Most of the mantle is solid rock but it is ductile enough to be able to flow and slowly deform. The *core* of the Earth (16 percent of Earth by volume) is mostly metallic iron with some nickel. The *outer core* is liquid. The *inner core* is under so much pressure that it is a solid.

In addition to the compositional classification above, the Earth's layers have been subdivided based on how each layer responds to stresses. The *lithosphere* is the outermost physical layer of the Earth and consists of the crust and upper mantle. The *oceanic lithosphere* ranges in thickness from 3 to 84 miles and is relatively rigid. The *continental lithosphere* is thicker, at 56 to 120 miles, and more ductile. The *asthenosphere* is the part of the mantle that lies below the lithosphere. It is mechanically weak and moves like a wet putty due to the forces of internal, circulating *convection currents* created by the Earth's internal heat. The lithosphere is broken into large, thin, relatively rigid *tectonic plates* driven about the globe by the convection currents in the asthenosphere.

The movement of the plates, called *plate tectonics*, is a unifying theory for understanding the dynamic evolution of the Earth through geologic time. It incorporates the older idea of *continental drift*, postulated by Alfred Wegener in 1912. Others had recognized the jigsaw puzzle shapes of the continents we see on a globe, but in 1912, Wegener presented a theory that continents drifted around the earth in geologic time. From 1915 through 1929, in several editions of his book *The Origin of Continents and Oceans*, Wegener compiled an impressive amount of evidence to support his theory. However, he lacked a theory for the mechanism for this drifting, and his ideas gained few supporters.

The idea was debated, but was not generally accepted, until the 1950s and 1960s. The breakthrough came when scientists used new technology developed during World War II to map the ocean floor. What they found was astounding. They knew that the Earth's magnetic field had switched directions many times in the geologic past, with the North Pole wandering to different locations in the Northern Hemisphere and sometimes even jumping to the Southern Hemisphere. Those working in the field of *paleomagnetism* knew that when igneous rocks form, the existing magnetic field imprints on them. The scientists sailed on a research ship, measured the magnetic field in rocks on the floor of the Atlantic Ocean and found that the magnetic patterns were symmetrical on both sides of a *ridge* in the middle of ocean. It appeared that the crust was spreading out evenly from the ridge. Other scientists produced new maps of the ocean floor depth. They found a system of mountain ridges 1.1 to 1.9 miles tall. In the center of those ridges, the ocean floor was splitting apart and erupting with molten lava. Here at last was a mechanism that explained the spreading crust and continental drift. The energy for this mechanism was provided by the Earth's internal heat.

Scientists also found that the deepest parts of the oceans were found in narrow *trenches* adjacent to lines of active volcanoes. Others documented the locations of earthquakes that almost always occurred along the ridges and trenches. The earthquakes at the trenches occurred at similar depths, suggesting a similar geologic process in action. Scientists recognized that these areas were *tectonically active*; they were places where crust was being created at ridges and deformed at trenches. With new drilling and sampling technology, other scientists were able to collect samples of the ocean floor at different distances from the ridges. In another stunning discovery, scientists using radiometric dating found that the age of the rocks systematically increased as they got farther from a ridge, and that the oldest rocks in the ocean were never older than 200 million years! Compare that to the oldest rocks on land, with an age reaching 4.4 billion years. Geologically speaking, the ocean crust is young!

Armed with this new information, geologic theory cohered around the new theory of plate tectonics in the 1960s. Geologists hypothesized that Earth is composed of the thin, rigid lithosphere, broken into at least twelve tectonic plates. These rigid plates are in constant motion as they slide on top of the more plastic asthenosphere. New oceanic crust is formed at mid-ocean ridges, and the plates move away from the ridges. Plates collide at deep trenches where one plate *subducts* under the other and dips down to eventually reach the mantle. Therefore, crust is constantly being created at ridges, and constantly being consumed at trenches. At other boundaries, plates slide past one another, causing earthquakes. The San Andreas Fault in California is one such plate boundary.

While crust is destroyed at *subduction zones*, it is also being created. At depths of about sixty miles, the subducting plate melts, and magma works its way up through the crust to form volcanoes. Through time, an arc-shaped set of volcanic islands, like the Aleutian Islands in Alaska, emerges above the subduction zone. If the crust above the subducting crust is continental crust, it will be deformed into a linear mountain range with volcanoes, like the Cascades of the Western United States. If both plates are thick continental crust, the result might be the creation of a massive mountain range like the Himalayas, or mountain-building episodes (*orogenies*) like the one responsible for creating the Appalachian Mountains. *Plate boundaries* are where the action of plate tectonics is observed as earthquakes and volcanic activity. The Ring of Fire is a term used for the plate boundaries surrounding the Pacific Ocean, where many volcanic eruptions and earthquakes occur.

ROCKS AND MINERALS

Rocks, and the minerals they contain, are the fundamental units for geologic studies. The study of rocks has taught us much about the composition of the Earth, its history, and its structure. In nature, *minerals* exist in rocks, with most rocks being composed of at least a few different minerals. A mineral is a naturally occurring substance with a specific composition of atoms arranged in a *lattice*, a repeating three-dimensional structure.

Igneous rocks crystallize from underground molten rock (intrusive rocks from magma) or from volcanic eruptions (extrusive rocks from lava) at the Earth's surface. *Metamorphic rocks* form when heat and/or pressure on a rock causes some of the minerals to become unstable and new minerals to form. *Sedimentary rocks* are created by the accumulation of material on the land surface or ocean bottom resulting from the weathering of rocks and chemical precipitation from water.

Igneous and metamorphic rocks, often lumped into the term *crystalline rocks*, make up most of the Earth's crust by volume (95 percent), but only represent about 30 percent of the rocks near the land surface. Sedimentary rocks dominate (at about 70 percent) the volume of rocks on the land surface, but represent only 5 percent of the total volume of the Earth's crust.[4] The *rock cycle* explains this disparity. Buried *sediments* only become sedimentary rocks at the land surface.[5] Most igneous rocks crystallize at great depth. If buried

FIGURE 2.2. Earth's rock cycle. (Earle, *Physical Geography*, opentextbc.ca/geology/)

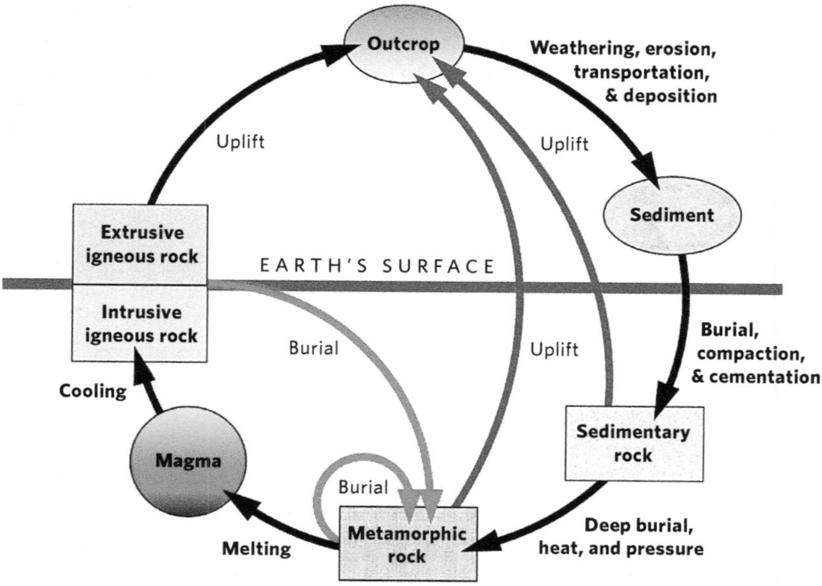

deeper, metamorphism alters them to form metamorphic rocks. If they are uplifted to the surface, *weathering* breaks them down and the weathered bits are deposited locally or transported far distances to accumulate as sediments that are then buried to become sedimentary rocks. Sedimentary rocks may be buried and metamorphosed into metamorphic rocks or uplifted to the land surface and weathered to become new sedimentary rocks. Metamorphic rocks may be buried to greater depths and melt to become *magma* that crystallizes into igneous rock or they may be uplifted and weathered to create new sedimentary rock.

Minerals are classified by their composition and crystal structure. Silicate minerals, containing silicon and oxygen, account for more than 90 percent of the rock-forming minerals in the Earth's crust. Examples include quartz, feldspar, mica (muscovite, biotite), hornblende, olivine, and pyroxene. Non-silicate minerals include carbonates, sulfides, chlorides, and oxides. Feldspar is the most common mineral in the Earth's crust and accounts for about half of its volume. At the land surface, quartz is the most common mineral; in the mantle, the most common is olivine.

Identifying minerals is the key to identifying rocks. Igneous rocks are classified based on their texture and mineral composition. *Extrusive* igneous rocks are typically composed of fine-grained minerals (not visible to the naked eye) while *intrusive* rocks, having taken longer to cool and having grown larger crystals, are composed of coarse-grained minerals. Compositional classification is done by estimating the mineral percentage in hand samples or in thin sections using a petrographic microscope.[6]

Igneous rocks in Delaware include gabbro and granitic pegmatite. Gabbro is a dark-colored igneous rock that contains feldspar and pyroxene. Examples are the Iron Hill Gabbro seen in Iron Hill and Rittenhouse parks in Newark, and the Bringhurst Gabbro in Bringhurst Woods Park. A granitic pegmatite is an igneous rock that has the mineralogy of a granite (quartz, feldspar) and abnormally large grains. Pegmatites occur in veins within the Wissahickon Formation and can be found in outcrops in White Clay Creek State Park. Some of the large crystals are in pebbles and cobbles in the bed of White Clay Creek.

Metamorphic rocks form through the recrystallization of minerals in their solid state caused by changes in pressure and temperature. Minerals are stable under specific environmental conditions. When pressure and temperature change, some of the minerals in a rock become unstable and atomic bonds are broken. New bonds form, creating different, more stable minerals. Diagnostic features of the original *parent rock* are either masked or destroyed, and there is a change in the overall texture of the rock. Metamorphism typically occurs when the temperature is greater than 390° Fahrenheit and/or pressure is greater than 14,500 to 43,500 pounds per square inch (at approximately a five-mile depth). The source of pressure is the weight of the overlying rocks; it increases at an approximate rate of 97 to 115 pounds per square foot every one hundred feet, depending on the density of the overlying rocks. Temperatures below the surface of the Earth increase with depth at a rate of about 8° to 16° Fahrenheit every one thousand feet. Radioactive decay of uranium, thorium, and potassium in the mantle and crust causes heat from the mantle and lower crust to flow upward via thermal conduction or convection (by upwelling fluids and/or magma). Metamorphism ceases when melting begins at temperatures of about 1290° to 1650° Fahrenheit, typically at depths of about seventeen to twenty-two miles. All silicate rocks are molten at about 2200° Fahrenheit (about thirty-one to forty-five miles depth).

Metamorphic rocks are classified by texture and/or composition. Is the rock *foliated* or *nonfoliated*? Foliation is a planar surface caused by oriented minerals and formed under *differential stress* (compression during metamorphism being different in different directions). These rocks have significant amounts of *platy* minerals (like mica), and are classified by foliation type, composition, and grain size (which can be coarse- or fine-grained). The types of foliation are *slaty cleavage, schistosity*, and *mineral banding*. Slaty cleavage is a pervasive, parallel foliation produced by fine-grained platy minerals. Rocks with slaty cleavage are slate and phyllite. Slate breaks easily along the foliation, and phyllite has a satiny luster.

Schistosity is a type of foliation in a coarse-grained rock resulting from the parallel arrangement of platy minerals (e.g., muscovite and biotite). *Schist* is a rock with shistocity. Schists typically contain quartz and feldspar, plus a variety of other minerals such as garnet, staurolite, kyanite, and sillimanite that are indicative of the pressure and temperature conditions of rock formation. Mineral banding is layering in a coarse-grained rock where bands of granular minerals (quartz and feldspar) alternate with bands of platy or elongate minerals (e.g., amphibole). A *gneiss* is a metamorphic rock with mineral banding. Nonfoliated rocks are composed of equant minerals typically classified by their composition. Examples are marble, amphibolite, and quartzite.

Metamorphic rocks in Delaware include gneiss, amphibolite, and marble. The gneisses are coarse-grained rocks with light layers containing quartz and feldspar and dark layers containing amphibole, pyroxene, mica, garnet, and sillimanite. Examples are the Brandywine Blue Gneiss seen in the old quarry in Alapocas Run State Park and the Windy Hills Gneiss seen in White Clay Creek where it flows under Kirkwood Highway. Amphibolite is a dark color and is made of amphibole and feldspar. It occurs as fragments in the Wissahickon Formation. Marble is a massive, coarse-grained, white rock composed mostly of calcite and/or dolomite. An example in Delaware is the Cockeysville Marble. There are no accessible outcrops of the Cockeysville Marble in Delaware but a large sample is on display outside the Delaware Geological Survey building in Newark.

Sedimentary rocks cover a large portion of the Earth's surface and ocean bottoms. They commonly exhibit *layering*. Sequences of layers tell us a lot about the environmental conditions, climate, and sea level when deposition occurred. They often contain indicators of the environment in which the rocks were deposited, such as along a river, a delta, or along the coast of an ocean. Some sedimentary rocks also contain animal and plant fossils that can be used to determine the ages of the rocks as well as their depositional environment.

Sedimentary rocks are typically classified as *clastic* or *chemical* rocks. Clastic rocks are dominated by *clasts* (clay-, silt-, sand-, or larger-sized sediment particles) that are single minerals or rock fragments. A clast is a fragment of rock or mineral that typically ranges in size from less than a micron up to one foot and larger. Clastic sedimentary rocks are the result of the weathering and erosion of other rocks; transport by gravity, water, or wind; or deposition, burial, and lithification. *Lithification* is the process or processes by which loose sediment becomes coherent rock by *compaction* and/or *cementation*. Most of the sedimentary rocks penetrated by drilling in the Delaware Coastal Plain are somewhat compacted, but rarely cemented. For our purposes and for most geologic applications, the classification of either sediment or rock is the same and based on texture (grain-size) and composition.

Variations in grain size are a primary factor in interpreting how grains are transported and where they are deposited. Composition is key to determining the source of the grains. Most geologists use a grain-size scale called the Udden-Wentworth scale. It differs from the grain-size scale widely used in engineering. Grain sizes from small to large are (in inches): *clay* (< 0.00016); *silt* (< 0.002); *sand* (< 0.079); *granule* (< 0.16); *pebble* (< 0.24); *cobble* (< 10.1); and *boulder* (> 10.1). Most silt and clay grains are composed of clay minerals. In many rocks, sand-sized grains are predominantly quartz because quartz resists weathering more than other common minerals do. Most grains larger than sand are rock fragments.

Chemical sedimentary rocks are dominated by components that have been transported as ions in solution. There is some overlap between clastic and chemical rocks because almost all clastic sedimentary rocks contain cement formed from dissolved ions, and many chemical sedimentary rocks include some clasts (e.g., fossil shells). Chemical sedimentary rocks form from ions in water, converted into minerals by biological and/or chemical processes. The most common chemical rock, limestone, typically forms in shallow tropical environments, where there is a lot of biological activity. Limestone is almost entirely composed of fragments of marine organisms that make calcite for their shells and other hard parts. Animals like corals, urchins, sponges, mollusks, and crustaceans die and their shells are buried in place or transported like other clasts. Dolomite, chert, halite, and gypsum are other types of chemical sedimentary rocks.

Most of the sedimentary rocks older than the Pleistocene age in Delaware are not readily seen at the land surface except in excavations for construction projects. Some exposures are found in natural bluffs along the coastline and in creek banks. There used to be large exposures along the C&D Canal but they have been graded for safety purposes. After large coastal storms, outcrops of Holocene age muds and sands can be seen on Atlantic Ocean and Delaware Bay beaches.

For mapping purposes, rocks are classified into *formations* and *groups*. A formation is a distinct rock that is mappable (identifiable in relation to surrounding rock layers) and large enough to be shown on a geologic map. Formations are combined into groups that facilitate correlations over broader regions and have areas large enough to show on a regional map. Formations and groups typically have two-part names. The first part is named for a site where the rocks are well exposed and described (type locality). The second part is the *lithology*, a visual description of the rock type in the field that is plainly visible in an outcrop or with low magnification. If a formation contains a variety of rock types (e.g., alternating layers of sandstone and shale), then the second part of the name can be simply "Formation" or "Group." An example of a formation name is the Vincentown Formation that crops out in a creek bank in Vincentown, New Jersey.

GEOLOGIC HISTORY OF DELAWARE

If you drive south in Delaware from the Pennsylvania border, you notice a change in the landscape from rolling uplands with occasional outcroppings of rock to gently sloping low-lying lands with sandy soils. The change occurs at the approximate location of Kirkwood Highway (Route 2) between Newark and Wilmington. It may be subtle, because urbanization has subdued the landscape, but think of the hilly streets in Wilmington and the flatter land south of New Castle. Here, you have crossed the boundary between two physiographic provinces, each with a characteristic landscape and geology.[7] You have just crossed from the *Piedmont Province* in the Appalachians Highland Physiographic Region into the *Coastal Plain Province* in the Atlantic Physiographic Region. The boundary is also known as the *Fall Line* or *Fall Zone* because it was the farthest upriver that early settlers could navigate with boats before waterfalls (or rapids) impeded their progress. Wilmington, Philadelphia, Baltimore, and Washington, D.C. all sit astride the Fall Line.

The Piedmont Province extends from Alabama to the Hudson River in New York. The Piedmont (the word means "foot of the mountains") is often referred to as a "rolling upland," and we see this landscape in the Brandywine, White Clay, and Red Clay Creek watersheds in Northern New Castle County. The topographic relief is about ninety meters with elevations ranging from approximately thirty to 120 meters.[8] The rocks in the Piedmont Province formed in the Paleozoic Era during the collision of two tectonic plates.[9] These rocks are not exposed south of the Fall Line but extend underneath the Coastal Plain, where they are referred to as *basement rocks*.

The Coastal Plain Province is gently sloping land extending from Cape Cod to the Yucatan Peninsula in Mexico. It slopes gently seaward, with many beaches, large embayments, drowned river valleys, and tidal wetlands on its margins. It has a relief of about 295 feet with elevations ranging from zero to 295 feet. The Coastal Plain is made up of sediments that are not quite hard enough for what most would call a rock. They were deposited during the Mesozoic and Cenozoic Eras following the breakup of the supercontinent Pangaea and the opening of the Atlantic Ocean.

If you continue driving southeast through Delaware and hop on a boat at Indian River Inlet to go fishing at the offshore Baltimore and Wilmington canyons, you are boating over the *continental shelf* that extends about eighty miles offshore. Geologically, the continental shelf is really just a continuation of the Coastal Plain. It just happens to be underwater at this point in geologic time. About twenty thousand years ago in the last ice age, sea level was about four hundred feet lower than it is today and the Delaware shoreline was eighty miles farther offshore, at the edge of the continental shelf. At that time, the head of Wilmington Canyon was out of the water and the Delaware River flowed across the continental shelf and into the canyon.

The rolling uplands of Northern New Castle County owe their character to the rocks of the Piedmont Province. Sandy Schenck and Peg Plank from the Delaware Geological Survey traveled the roads of the Piedmont for a number of years to unravel the complex history of the Piedmont rocks. They coauthored several reports and maps on Piedmont geology, and their work forms the basis for much of the following discussion.[10]

The Piedmont Province is part of the Appalachian Mountain system that extends over 1,860 miles from Alabama to Newfoundland, Canada. The mountains came to be in three mountain-building episodes, or *orogenies*, in the Paleozoic Era: the Taconic, Acadian, and Alleghanian orogenies.[11] The results of these orogenies are the rocks we see in the Piedmont, Blue Ridge, Valley and Ridge, and Appalachian Plateau physiographic provinces (from east to west). The Piedmont Province is the only one we experience here in Delaware, but drive about 250 miles west to Morgantown, Virginia, and you will have visited all four provinces. The assembly of the Appalachians spanned about 290 million years in the Paleozoic Era. Metamorphic rocks of the Piedmont were created deep down in the roots of the mountains and were uplifted and juxtaposed with rocks from a Precambrian orogeny.

About 1.2 billion years ago, in the Proterozoic Eon, the proto-North American continent collided with Amazonia (South America), creating the supercontinent *Rodinia* and uplifting an enormous mountain chain. This orogeny created the *Baltimore Gneiss*, the oldest rocks in Delaware. Rodinia wandered around the Southern Hemisphere for six hundred million years before it broke into multiple plates near the end of the Proterozoic. This begat the *Laurentia* (North America) and *Baltica* (Eurasia) continents and the newly formed *Iapetus Ocean* in between them. Geologically speaking, the Iapetus was short-lived. By five hundred million years ago, in the Late Cambrian Period, Laurentia was on a collision course with Baltica in the narrowing Iapetus Ocean.

TABLE 2.1. GEOCHRONOLOGY OF THE DELAWARE PIEDMONT

Era / Period or Epoch	Geologic Units		Events
HADEAN EON	No rocks in Delaware		Earth forms.
ARCHEAN EON	No rocks in Delaware		Earth's crust forms. Oldest rocks on Earth (4.4 billion years before present). Life (bacteria) begins at 3.5 billion years before present.
PROTEROZOIC EON	Baltimore Gneiss		Simple and complex single-celled organisms. *Closing of sea between Laurentia and Amazonia continents (Ottowan Orogeny) forms supercontinent Rodinia. Rifting of Rodinia later in eon.*
PALEOZOIC — Cambrian	WISSAHICKON FORMATION: Amphibolites Metapyroxenite & metagabbro suite Serpentinite		Explosion of life. First fish and chordates (animals with backbones). *Carbonate bank and volcanic arc in Iapetus Ocean east of proto-America. Subduction begins to east of proto-America.*
PALEOZOIC — Ordovician	GLENARM GROUP: Cockeysville Marble Setters Formation	WILMINGTON COMPLEX: Brandywine Blue Gneiss, Rockford Park Gneiss, Mill Creek Metagabbro, Montchanin Metagabbro, Barley Mill Gneiss, Christianstead Gneiss, Faulkland Gneiss, Windy Hills Gneiss	Major diversification of animal life. First land animals and fish. *Taconic Orogeny: proto-America collides with island arc, causes metamorphism and deformation.*
PALEOZOIC — Silurian	Pegmatite	WILMINGTON COMPLEX: Iron Hill Gabbro, Bringhurst Gabbro, Arden Plutonic Supersuite, Biotite Tonalite	First land plants.
PALEOZOIC — Devonian	No rocks in Delaware		Fish diversify. First amphibians and trees. *Acadian Orogeny.*
PALEOZOIC — Mississippian			Great coal forests.
PALEOZOIC — Pennsylvanian			Great coal forests. First reptiles. *Alleghanian Orogeny.*
PALEOZOIC — Permian			Greatest extinction. Extinction of trilobites. Reptiles diversify. *Final assembly of supercontinent Pangaea.*

Laurentia was moving east, with its crust subducting under Baltica. The subduction produced magma that fueled the creation of an extensive arc-shaped chain of volcanoes eastward of the subduction zone in the late Proterozoic and Cambrian. Sediments shed from the volcanoes were deposited in the Iapetus Ocean. The continents were in the southern equatorial region and, with the explosion of life in the Cambrian Period, a carbonate bank developed in the warm waters east of Laurentia.

At the same time, quartz-rich sediments shed by Laurentia were deposited near the Laurentia shoreline. As subduction continued, the sediments were scraped off the Laurentia plate to form a thick pile of deformed and metamorphosed rocks. These rocks are the Wissahickon Formation of Delaware, sediments deposited in the deep ocean basin between the Laurentia continental shelf and the volcanic arc. Amphibolites in the Wissahickon Formation suggest that ash falls or basalt flows from the volcanos or slivers of oceanic crust were mixed in with the sediments. As subduction continued, Laurentia got closer to the subduction zone. In the Ordovician and Silurian, from 470–430 million years ago, the island arc collided with Laurentia during the *Taconic Orogeny* and the volcanic arc welded onto Laurentia.

This orogeny created all of the other rocks in the Delaware Piedmont. The amphibolites and "blue rocks" of the *Wilmington Complex* are the metamorphosed volcanic rocks that were deeply buried and heated to temperatures in excess of 1500° Fahrenheit. These rocks include the *Brandywine Blue Gneiss (Blue Rock)*, *Windy Hills Gneiss*, and *Rockford Park Gneiss*. The crystalline Cockeysville Marble is a metamorphosed shallow water carbonate (limestone) composed primarily of shells, skeletal remains, and precipitates. The impure quartzites of the *Setters Formation* are the nearshore, muddy beach sand.

Now there was only the open Iapetus Ocean, between Laurentia—with its new Taconic arc growth—and the Baltica continent. Beyond Baltica, across the *Rheic Sea*, loomed the massive *Gondwana* continent, aiming its western side (Africa) toward Laurentia. In the Devonian Period, about four hundred million years ago, Baltica itself collided with Laurentia in the *Acadian Orogeny* with a deformation and metamorphism event extending into the Early Mississippian Period. It resulted in Baltica welding onto the Taconic volcanic arc and Laurentia. The large Gondwana continent was coming from the east and subducting under Laurentia. In the Pennsylvanian and continuing into the Permian Period, Gondwana hit Laurentia in the *Alleghanian Orogeny*. At the end of this orogeny, the formation of the supercontinent Pangaea was complete. The Appalachians were created, and left to weather and erode over the next 260 million years. The eroded remnants washed into the creeks and took a journey, first to the Appalachian Basin toward the west, then to the Coastal Plain and the continental shelf, where they continue to follow the rock cycle to become sedimentary rocks. The process continues today in the Delaware Piedmont, the rocks fractured and covered with moss and lichen slowly weathering and crumbling.

Significant time went by between the Silurian and Triassic Periods, before the creation of more rocks in Delaware. These time intervals, when no sediments were preserved, are called *unconformities*. The primary process happening during these times was erosion of the Appalachian Mountains, including the Delaware Piedmont. The Piedmont rocks we see today were once buried to a depth of about six to twelve miles! Erosion slowly wore down the rocks overlying them and they rose to the surface to compensate for the loss of that huge amount of overlying weight. Eroded sediments from the Appalachians were being transported and deposited in the Appalachian Basin west of the mountain chain rather than in the Coastal Plain. The supercontinent Pangaea began to rift apart in the Late Triassic Period, about 220 million years ago. By the start of the Cretaceous Period, 145 million years ago, North America was an independent continent with a spreading center and widening Atlantic Ocean to the east. This period—beginning the deposition of sediments in the Atlantic Coastal Plain and continental shelf from the Cretaceous to the present day—was the second geologic phase in the evolution of Delaware.

Coastal Plain sediments are feather-thin at the Fall Line in New Castle County and slope gently and thicken toward the southeast, reaching a thickness of over 1.5 miles in Sussex County and up to ten miles on the continental shelf! The sediments are mostly silt, sand, and gravel eroded from the Appalachian Mountains. Kelvin Ramsey of the Delaware Geological Survey, in a recent collaboration with Jaime Tomlinson, is methodically mapping the surficial deposits in the Coastal Plain. Their work guided the following discussion.[12]

The Triassic rifting event split Pangaea into the North American and African plates. The Earth's crust was under tension, and a huge block of land dropped down along a series of faults. Some of these rifts shut before the Atlantic Ocean opened, and small basins called *grabens* were formed on the downthrown side of normal faults. Sediments filled the grabens during the Triassic and Jurassic Periods. Triassic *sedimentary basins* formed by this process exist from Georgia to Newfoundland, Canada. Two close by are the Newark Basin in New Jersey and the Gettysburg Basin in Pennsylvania. No rocks of this age are found in Delaware, but geophysical evidence suggests that a graben lies beneath the large pile of Coastal Plain sediments.

Pangaea finally broke apart in the Jurassic Period to create the Atlantic Ocean. North America and Africa spread apart, and new oceanic crust formed at the Mid-Atlantic Ridge. The new crust cooled and *subsided* as heavy sediments piled on top. Compaction increased the space to accommodate even more sediment. A thick package of *sedimentary strata* was deposited from the Early Cretaceous to Late Miocene Periods. Unfortunately, the units are not well exposed at the land surface in Delaware, but their formation names

TABLE 2.2. GEOCHRONOLOGY OF THE ATLANTIC COASTAL PLAIN

Era	Period or Epoch	Geologic Units	Events
MESOZOIC	Triassic	Rift-basin rocks (inferred)	First dinosaurs. *Breakup of supercontinent Pangaea by rifting.*
MESOZOIC	Jurassic	Post-rift unconformity sediments (inferred)	Dinosaurs diversify. First birds and mammals. *Atlantic Ocean begins to form.*
MESOZOIC	Cretaceous	Navesink Formation Mt. Laurel Formation Marshalltown Formation Englishtown Formation Merchantville Formation Magothy Formation Potomac Formation	Extinction of dinosaurs. First primates and flowering plants. *Sediment deposition begins on Coastal Plain.*
CENOZOIC	Paleocoene	Vincentown Formation Hornerstown Formation	Age of Mammals begins.
CENOZOIC	Eocene	Piney Point Formation Shark River Formation Manasquan Formation	Highest mean temperature in Cenozoic Era, with temperatures about 86°F (Early Eocene).
CENOZOIC	Oligocene	Glauconitic unit	First horses and many grasses.
CENOZOIC	Miocene	Bryn Mawr Formation Bethany Formation Cat Hill Formation St. Marys Formation Choptank Formation Calvert Formation	First grasslands and kelp forests.
CENOZOIC	Pliocene	Beaverdam Formation	Spread of grasslands. Rise of long-legged grazers.
CENOZOIC	Pleistocene	Delaware Bay, Assawoman Bay, and Nanticoke River Groups Old College Formation Columbia Formation	Humans arrive. *Sea level rises with glacial advance, falls with glacial retreat. Last glacial maximum at 20,000 years before present with sea level 140 meters below present level. Delaware's beaches and salt marshes start to form.*
CENOZOIC	Holocene	Cypress Swamp Formation (Holocene and Pleistocene)	Humans become agents of geologic change.

tell us that the locations of the type-sections are in Maryland and New Jersey. What we know of these deposits is mostly from the study of geologic samples and geophysical logs collected during drilling for research and the construction of water supply wells, and from temporary excavations for construction.

From the Early to Late Cretaceous Period, anastomosing cross-channel rivers flowed southeastward across Delaware in a tropical to subtropical climate. They deposited the sands and muds of the *Potomac Formation*. The sea transgressed onto Delaware in the Late Cretaceous Period. Marine sediments were deposited as the sea level rose and fell multiple times, creating the Cretaceous *Magothy Formation* and ending with the Miocene *Bethany Formation*. Except for the Bryn Mawr Formation, the sediments all slope gently and thicken toward the southeast. The Bryn Mawr Formation is sand and gravel isolated within the Piedmont that were deposited by a paleo-Delaware River.

The geometry produced by deposition is a wedge of sediment dipping and thickening to the southeast. After deposition of the Late Miocene sediments, there was a period of erosion marked by an angular unconformity, where the beds in the younger units cut across those of the older units at an angle. The unconformity cuts across all of the older Coastal Plain units in all three Delaware counties, forming an irregular topography as the base for more sediments. In the Pliocene Epoch, the paleo-Susquehanna and paleo-Delaware rivers flowed across Delaware depositing sediments of the Beaverdam Formation in Kent and Sussex counties. Some strata in the southeastward-dipping sediments are permeable and one hundred percent of the water supply south of the Chesapeake and Delaware Canal comes from *confined aquifers* in these sediments. New Castle County uses Cretaceous and Paleocene Period aquifers and Kent County uses the Piney Point aquifer. The Cheswold, Federalsburg, Frederica, and Milford aquifers supply water in Kent and Sussex counties as do the Maonkin and Pocomoke aquifers in Sussex County.

In the Pleistocene Epoch, ice caps grew in the Northern Hemisphere and came and went periodically. The ice came as far south as Northern New Jersey and Pennsylvania but never into Delaware. However, Delaware did experience a colder climate then. Two of the major advances of ice were the *Illinoian Glaciation*, reaching its maximum southern advance 140 thousand to 160 thousand years ago, and the *Wisconsinan Glaciation*, the latest one, which lasted from seventy-five thousand to eleven thousand years ago, reaching its maximum about twenty-five thousand to twenty-one thousand years ago. As glaciers advanced and retreated, sea levels lowered and rose.

The Columbia Formation of New Castle and Kent counties is *glacial outwash*, the sand, gravel, pebbles, and boulders deposited by high-energy rivers. In addition to the glacial outwash, there are several other kinds of evidence of the nearby Pleistocene glaciers and a colder climate in Delaware. Fossil pollen from trees living in colder climates are in the sediments. There are *periglacial* features like a boulder field in Rittenhouse Park, a small version of the largest one in the Appalachians at Hickory Run State Park in Pennsylvania, *polygonal ground*, as seen in aerial photographs from the 1930s, and distinctive round surface depressions called *Carolina Bays*, sometimes associated with old sand dunes. The Carolina Bays and dunes overlie the Beaverdam Formation and Pleistocene units. In the late Pleistocene age, during periods of high sea level, fluvial and estuarine sediments were deposited: the Delaware Bay and Nanticoke River groups. These deposits cut into the Beaverdam and Columbia formations with an unconformity at their lower boundary. The Columbia aquifer, a shallow, *unconfined aquifer* widely used for irrigation, is within the Columbia and Beaverdam formations, as well as the Delaware Bay, Assawoman Bay, and Nanticoke Rover groups.

From the last glacial maximum in the Pleistocene about twenty-thousand years ago to the present day, we are undergoing a *marine transgression* as sea level rises. Over time, the sea is moving over the shoreline onto the land. Sea level was about four hundred feet lower during the glacial maximum. As the sea level rises, younger sediments move toward the upland and up in the layers of rock. Based on years of working the coastline of Delaware as a coastal geologist at the University of Delaware, Dr. J. Chris Kraft coined the phrase "onward and upward through time and space."[13] Recent beach sands have moved landward over older marsh muds that formed on the landward side of the beach and are now bayward of the beach. Transgression is evident by the "*washovers*" of beach sands over the modern saltmarsh. These features are evident along many areas of the Delaware Bay coastline. They can be observed at South Bowers in Kent County. Holocene Era marsh muds that are now ten feet below present sea level and 0.9 miles offshore of South Bowers were dated as being 2,740 years old.

ROCKS IN THE FIELD AND ON DISPLAY

If you would like to learn more and get hands-on experience with Delaware rocks, check out the Piedmont Field Trips—GeoAdventures and Selected Outcrops of the Delaware Piedmont sections of the Delaware Geological Survey website (www.dgs.udel.edu). Interested in rocks and gardens? Visit the rock outcrop garden at the Mt. Cuba Center near Wilmington. The rock in the garden is folded gneiss of the Wissahickon Formation.

Coastal environments are where dynamic sedimentary processes happen in real time. Next time you are standing on a beach, look out at the waves. Where do they come from? Many people say the tides create them but that is a longer time-scale process controlled by the gravitational attraction between the Earth, the Moon, and the Sun. In Delaware, the time between consecutive high tides is 12.5 hours. Waves reach the beach about every three to twelve seconds. Their force is the wind that blows across open water in an area called the *fetch*. The friction of that air on the water creates the waves. The fetch may be local or it may be the faraway fetch creating the long-period swell of the surfers. Are the waves breaking at an angle to the beach? That angle results in the *longshore currents* you feel pulling you to the side while you are playing in the ocean. The waves are hitting the beach first on the upstream side and later on the downstream side, forcing the water to flow downstream. This is caused by the direction of the wind or by the refraction of the waves around an object in the water.

When you come out of the water from a swim, you may have sand in your bathing suit. This is from the sand particles being transported in the water as *suspended load*. This energetic environment is one of the rare locations where sand is transported in suspension. When a wave comes in, watch the swash run up and down the beach. Look closely. See the sand grains rolling along the beach? This is a type of sediment transport called *bedload transport*, where the particles bounce along on the bed instead of being suspended in the water. While you are examining the sand, take a look at the *sand grains*. The light grains are most likely quartz. Recall that it is a relatively hard mineral. This high-energy environment breaks down other common rock-forming minerals. You may see some small, dark black, purple, or reddish grains. Sometimes you will see these as "black sand" on specific parts of the beach or in specific layers when you dig a hole in the sand. These are *heavy minerals* like garnet, magnetite, and ilmenite. Water flowing with a given amount of energy will be able to transport relatively large quartz grains but only smaller grains of the heavy minerals. If you visit the beach during or after a storm, you may see *scarps*, small cliffs on the beach or in the dune. The storm eroded the beach and dune and the sand was transported to offshore sandbars. It is common for sand to be stored in a geomorphic feature called an *offshore bar* in the winter and then slowly moved onshore by wave action. The bar welds onto the beach in summer.

If you happen to be on a beach at a time when the bar is almost welded to it, you may see a *ridge* and *runnel* where the ridge is the bar and the runnel is the low spot between the bar and the beach. Water flows into the runnel when waves wash over the bar and then flows to a location where a break in the bar allows it to go back into the ocean. Look closer at the runnel. You will likely see ripples in the sand. Geologists often see ripples in sediments and sedimentary rocks, both in outcrops and in cores collected by drilling into the earth. Ripples and other sedimentary structures help geologists interpret the depositional environment that forms a sedimentary rock. More information on the geology of coastal environments is available on the Santa Aguila Foundation's Coastal Care website, at coastalcare.org/educate/beach-basics.

LEARNING RESOURCES

Rocks, minerals, and fossils from Delaware are on display at the Delaware Geological Survey, and an excellent collection of minerals is displayed at the Mineralogical Museum, both on the University of Delaware campus in Newark (see more information at www.dgs.udel.edu and library.udel.edu/special/collections/mineralogical-museum).

The Iron Hill Science Center in Iron Hill Park near Newark also has a mineral display. While there, you can visit the information kiosks at the top of the hill to learn about the history of iron mining and the historic Native American jasper (type of quartz) quarry on Iron Hill (see http://ironhillsciencecenter.org and www.nccde.org/Facilities/Facility/Details/Iron-Hill-Park-54).

While in Newark, you can see some rocks that come from outside of Delaware at these places:

- Green serpentinite: building, corner of Main and Academy Streets
- Sandstone: University of Delaware Trabant Center, outer wall of Daugherty Hall, facing West Main Street
- Limestone: base of Recitation Hall, University of Delaware, corner of Main Street and College Avenue
- Labradorite, an iridescent feldspar: Planet Mercury plaque, front of Recitation Hall
- Granite: statue at University of Delaware Mentor Circle at South College Avenue and Kent Way

If you really like minerals and fossils and want to see a wide variety of them, try attending the annual Gem and Mineral Show put on by the Delaware Mineralogical Society (DMS). Want to collect minerals and fossils? DMS arranges five or more field trips every spring and fall (www.delminsociety.net). Other fossil and mineral collections and geology exhibits are a bit farther afield but well worth the effort. The Museum of Natural History in New York City (www.amnh.org) and the Smithsonian National Museum of Natural History in Washington, D.C. (naturalhistory.si.edu) both have stunning collections. If you visit D.C., you can also view building stones using the "Stones of our Capitol" guide (pubs.usgs.gov/gip/stones). Up the road in Baltimore is another guided walk of building stones (www.mgs.md.gov/geology/geology_tour/walking_tour_baltimore.html).

Here is a list of other useful resources for the budding geologist:

- Online textbooks *Geology* (www.oercommons.org/courses/geology-4/view) provided by Lumen Learning, and Steven Earle's *Physical Geology* (Victoria, BC: University of British Columbia, 2015, opentextbc.ca/geology).
- The *Roadside Geology* series from Mountain Press is a way to enjoy geology during your travels (mountain-press.com/collections/roadside-geology).
- *Basin and Range* by John McPhee (New York: Macmillan, 1981). McPhee authored four books about his travels across the United States with field geologists. A compilation of these, *Annals of the Former World*, won him the Pulitzer Prize for Nonfiction in 1999. *Basin and Range* is the first in the set and focuses on plate tectonics.
- *Krakatoa: The Day the World Exploded: August 27, 1883* by Simon Winchester (London: Viking, 2003), the story of the catastrophic eruption off the coast of Java, Indonesia.
- Delaware Geological Survey (www.dgs.udel.edu) publications: Delaware Piedmont Geology—for nongeologists (SP20), New Castle and Kent county geologic maps (GM13, GM14), Piedmont geologic map and report (RI59), generalized state geologic map (SP16), geomorphic features (SP24), Digital Elevation Model (SP28). See also Delaware Geology and Story Maps sections of the DGS website.
- *Discover Geology* by the British Geological Survey provides information on a range of topics with a good set of pages on geological time that have a focus on fossils (www.bgs.ac.uk/discoveringGeology).
- Paleomap Project videos, plate tectonics videos with continents moving on a globe, such as "Geologic time" (www.youtube.com/user/cscotese).
- The Society for Sedimentary Geology's stratigraphy website SEPM Strata, for understanding sedimentary geology (www.sepmstrata.org).

- The National Park Service's Geology website, which covers many excellent topics (www.nps.gov/subjects/geology).
- The United States Geological Survey (USGS) Geologic Map of North America (www.usgs.gov/media/images/geologic-map-north-america). The USGS National Map, including geologic, topographic, and land cover layers (viewer.nationalmap.gov/advanced-viewer).

NOTES

1. Kirk Johnson and Ray Troll, *Cruisin' The Fossil Freeway* (Golden, CO: Fulcrum Publishing, 2007), 208.
2. Geological Society of America, GSA Geologic Time Scale (2018), www.geosociety.org/GSA/Education_Careers/Geologic_Time_Scale/GSA/timescale/home.aspx.
3. Chris Johnson, Matt Affolter, Paul Inkenbrandt, and Cam Mosher, *An Introduction to Geology* (Salt Lake City, UT: Salt Lake Community College, 2017), https://www.oercommons.org/courses/an-introduction-to-geology-free-textbook-for-college-level-introductory-geology-courses/.
4. J. Douglas Walker, and Harvey A. Cohen (compilers), *The Geoscientist Handbook, AGI Data Sheets*, 4th ed. (Alexandria, VA: American Geological Institute, 2006), 302.
5. Steven Earle, *Physical Geology* (Victoria: University of British Columbia, 2015), available through BCcampus, https://opentextbc.ca/geology/.
6. Walker and Cohen, *The Geoscientist Handbook*, 302.
7. Nevin M. Fenneman, and Douglas W. Johnson, *Physical Divisions of the United States* (Denver: United States Geological Survey, 1946), 1:7M.
8. Lillian T. Wang, *Digital Elevation Model of Delaware*, Special Publication 28 (Newark: Delaware Geological Survey, 2017).
9. Delaware Geological Survey, *Generalized Geologic Map of Delaware*, Special Publication 16 (Newark: Delaware Geological Survey, 2018).
10. Margaret O. Plank and William S. Schenck, *Delaware Piedmont Geology*, Special Publication 20, 64 (Newark: Delaware Geological Survey, 1998), and Margaret O. Plank, William S. Schenck, and LeeAnn Srogi, *Bedrock Geology of the Piedmont of Delaware and Adjacent Pennsylvania*, Report of Investigations 59, 52 (Newark: Delaware Geological Survey, 2000).
11. Robert D. Hatcher, Jr., "The Appalachian Orogen; A Brief Summary," in *From Rodinia to Pangea: the Lithotectonic Record of the Appalachian Region*, eds. Richard P. Tollo, et al., Geological Society of American Volume 206 (Boulder, CO: Geological Society of America, 2010), 1–19.
12. Kelvin W. Ramsey, *Geologic Map of New Castle County*, Geologic Map 13 (Newark: Delaware Geological Survey, 2005), 1:100,000; Ramsey, *Geologic Map of Kent County*, Geologic Map 14 (Newark: Delaware Geological Survey, 2007), 1:100,000; Ramsey, *Stratigraphy, Correlation, and Depositional Environments of the Middle to Late Pleistocene Interglacial Deposits of Southern Delaware*, Report of Investigation 76 (Newark: Delaware Geological Survey, 2010), 43; and Ramsey and Jaime L. Tomlinson, *Geologic Map of the Millington, Clayton and Smyrna Quadrangles, Delaware*, Geologic Map 24 (Newark: Delaware Geological Survey, 2018), 1:24,000.
13. John C. Kraft, "Sedimentary Facies Patterns and Geologic History of a Holocene Marine Transgression," *Geological Society of America Bulletin*, vol. 82, no. 8 (1971): 2131–58; and John C. Kraft, Elizabeth A. Allen, Daniel F. Belknap, Chacko J. John, and Evelyn M. Maurmeyer, *Delaware's Changing Shoreline*, Technical Report 1 (Dover: Delaware Coastal Zone Management Program, 1975), 319.

CHAPTER THREE

Watershed Ecology

GERALD McADAMS KAUFFMAN

THE WATER STATE

Delaware is a water state. Sitting on the Delmarva Peninsula and surrounded on three sides by water, it is one of just three *peninsular* states, and with the 1829 cutting of the C&D Canal, many consider it to be technically an *island*. At a mean elevation of only sixty feet above sea level, the First State is also the lowest state in the United States, with a beautiful and bounteous coastline along the Atlantic Ocean. But this profile also leaves the state vulnerable to worsening coastal storms and accelerating sea level rise—perhaps more so than most other places. It is fortuitously situated by geography and hydrology between two great estuary systems in America, the Chesapeake and the Delaware, that support abundant ecology and economy. In 2010, more than three hundred million gallons per day of drinking water and industrial water supplies were drawn from the rivers, streams, and aquifers in Delaware's watersheds to sustain the state's domestic, commercial, and industrial economy. But more than 90 percent of Delaware water is also so polluted it does not pass federal standards, largely due to a high population density in the metropolitan corridor to the north and the substantial agricultural economy to the south. While only the second smallest state in the Union, almost a million people in Delaware draw drinking water from just four small streams that originate upstream in the Appalachian Piedmont of Pennsylvania, and from Atlantic Coastal Plain aquifers that reach a mile down to bedrock. Delaware is diminutive, but its waters run deep.

The state owes its history and formative years to the waters that surround it. About fifteen thousand years ago, the North American glaciers melted and the sea rose to form more or less the watershed geography of present-day Delaware. Before the last ice age, ocean waters covered most of what is now Delaware. Over time, as the polar ice caps grew and continental glaciers drifted southward, sea level dropped significantly to a point about four hundred feet lower than present-day sea level. Since then, with increasing global temperatures, the polar ice caps have decreased in size and the continental glaciers have retreated, causing sea level to rise to its current position.

Although Delaware is located south of the maximum extent of the last continental glaciers, it is believed that the great weight of the glaciers depressed the land they overrode, causing the formation of a marginal bulge in the area of present-day Delaware. As the glaciers retreated, the Earth's surface rebounded upward in areas to the north, causing the marginal bulge previously formed in the area of Delaware to subside and leading to a relative rise in sea level. The rising sea flooded the ancestral valleys of the Delaware and Susquehanna Rivers, and created the two great estuaries that bound either side of the state. While the terminal *moraine* (or southerly extent of the ice sheet) from the North American glacier lies a hundred miles north of Delaware, glacial outwash formed the incised valleys of the Appalachian Piedmont (Italian for "foot of the mountains"), and of creeks like the Brandywine to the north; and the sand and gravel deposits that are sole source aquifers in the Atlantic Coastal Plain to the south. The seas are still rising: Delaware Bay rose a foot over the last century.

The current marine transgression began approximately fourteen thousand years ago, when the polar ice caps began melting. The Delaware coastline at that time was approximately eighty to one hundred miles east of its current location. A rapid rise in sea level—about three inches per year—lasted until about seven thousand years ago, when sea level was about thirty-three feet below its present level. Since then sea level rose at a slower rate, until about three thousand years ago when it reached its present level.

As the ocean advanced across the continental shelf, it flooded ancient river valleys and moved large masses of Pleistocene age sediments in a landward direction, overtopping previous lagoons and marshes. Over time, Delaware's coastline, including both the Delaware Bay and the Atlantic Ocean coastlines, began to evolve to its present-day configuration. The present coastline is moving landward and upward in response to longshore transport of sediments and storms. As sea level rises, waves attack the beach at higher elevations, which concentrates erosion on headland areas and works to straighten the coastline.

The indigenous people of Delaware, the Lenni Lenape, have lived here for the last three millennia, settling in villages like: Queonemysing, along the Brandywine (which they called the "Wawaset") where Pennsylvania and Delaware now join together; Opasiskunk, at the confluence of the east and west branches of White Clay Creek; and on the Leipsic River along Delaware Bay (the ancestral home of the Delaware Lenni Lenape). The Europeans arrived when Englishman Henry Hudson stranded the *Half Moon* on an oyster bar at the mouth of Delaware Bay in 1608, then sailed north to explore the river that bears his name.[1] In 1631, Dutch whalers landed at Cape Henlopen, where the ocean meets the bay, and founded Zwaanendael (later named Lewes) as the first European settlement in Delaware. In 1638, Swedes on the *Kalmar Nyckel* landed on the "Rocks" at the junction of the Brandywine (Fishkill) and

Christina (Christinakill) Rivers and named Fort Christina after their teenage queen, and the town we now know as Wilmington became the oldest permanent European settlement in the Delaware Valley.

The Dutch influence can be found in the dykes along the Delaware River at New Castle and along rivers such as the Broadkill and Murderkill along Delaware Bay. In 1682, to settle a conflict between the Philadelphia Quakers and Baltimore Catholics, William Penn drew an arc boundary twelve miles in radius from the town of New Castle that bisected the watersheds of the Brandywine, Red Clay, and White Clay Creeks into two separate jurisdictions: Delaware and Pennsylvania. In 1763, Charles Mason and Jeremiah Dixon surveyed a north-south line midway between the Atlantic and Chesapeake to join the arc boundary, creating (at least until the 1920s) a stateless "wedge," and this boundary landlocked Delaware from the shores of the Chesapeake.

In August 1777, the British landed at the head of the Chesapeake near Head-of-Elk, Maryland, then invaded Delaware and fought the Americans at the Battle of Cooch's Bridge along the Christina River near Newark on the way to the September 11, 1777, Battle of the Brandywine, the largest battle of the Revolution.[2] In 1802, the DuPonts searched up and down the Eastern seaboard and selected the Brandywine at Wilmington—as the river descends from the heights of the Piedmont to sea level with a higher fall than the Niagara—to site their gunpowder mills. During the 1850s, Harriet Tubman (the "Moses of the New World") led the Underground Railroad from the Chesapeake northward to cross the Christina and Brandywine rivers at Wilmington, where slaves would finally be free.

It seems that every stream in Delaware has a story. "Christina" and "Christiana" both derive from the Swedish queen and provide the names of the Christina River and school district, and the Christiana mall, as well as the Colonial-era town of Christiana. Such was the confusion between the two names that the Delaware General Assembly passed a 1930s law against referring to the river as Christiana in public. Murderkill is the "mother river," from the Dutch *Moeder*. The Nanticoke (named for the Tidewater People tribe) is the largest river in Delaware and one of the largest tributaries to the Eastern Shore of the Chesapeake Bay. Naamans Creek is named after a Minqa chief who traded with the Swedes. One of the oldest water debates in Delaware is whether the Brandywine is a river (wide enough that one must swim across) or a creek (small enough to wade across).

DELAWARE HYDROLOGY

Hydrology is the study of water as it is circulates between the earth and the atmosphere.[3] The *hydrologic cycle* (or "water budget") tracks how the precipitation that falls on the earth and collects in a watershed (runoff), permeates into the ground (infiltration), disperses back into the atmosphere (evaporation), or is absorbed by plants (transpiration). Key factors that affect

FIGURE 3.1.
Nested watersheds in the Delaware River Basin. (Courtesy of the University of Delaware Water Resources Center.)

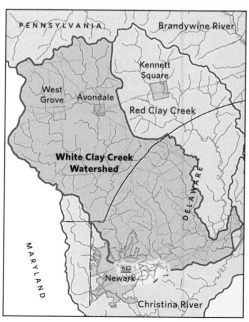

Delaware hydrology include the characteristics and size of a watershed, the volume and frequency of precipitation, and the rate of evaporation (evapotranspiration).

The *watershed* is the fundamental hydrologic unit for managing water resources, and is defined as the geographical area that provides the water flow to a particular waterway such as a lake, stream, river, or ocean. Watersheds can be nested from small *catchments* to large watersheds (Figure 3.1 and Table 3.1). In Delaware, a half-square-mile urban drainage in Newark is nested within the seven-square-mile Cool Run *subwatershed*, which in turn is nested within the ninety-eight-square-mile White Clay Creek watershed. The White Clay Creek watershed is one of the four major streams in the 565-square-mile Christina River *subbasin*, which is part of the thirteen thousand-square-mile Delaware River *Basin*. Basin, watershed, and catchment are often used interchangeably.

TABLE 3.1. WATERSHED HIERARCHY

Unit	Area (square miles)	Example
Catchment	0.5 to 1.0	Urban drainage (0.5 sq. mi.)
Subwatershed	1.0 to 10	Cool Run (7 sq. mi.)
Watershed	10 to 100	White Clay Creek (98 sq. mi.)
Subbasin	100 to 1000	Christina Basin (565 sq. mi.)
Basin	over 1000	Delaware River (13,000 sq. mi.)

The United States Geological Survey has delineated twenty-one major watersheds (river basins) in the United States (Figure 3.2). Watershed boundaries rarely coincide with political boundaries. Because physical watersheds do not follow state boundaries, one must have knowledge of the scientific and the socio-political aspects of water resources management in order to execute that management effectively. The interstate and interdepartmental management of our nation's river and stream systems adds to the challenge and complexity of watershed management.

A watershed can be delineated on a topographic map according to the following three-step process (see Figure 3.3):

1. Identify the point of interest (P.I.) at the outlet of the watershed in question.
2. Highlight the streams on the topographic map.
3. Starting at the P.I., delineate the watershed with the boundary crossing perpendicular to the contour lines. Look for closed contour lines at the top of the ridgelines, which usually indicate the watershed boundary. Once the boundaries of the watershed are delineated, calculate the drainage area in acres or square miles.

FIGURE 3.2. Major river basins in the United States. (Courtesy of the U.S. Geological Survey.)

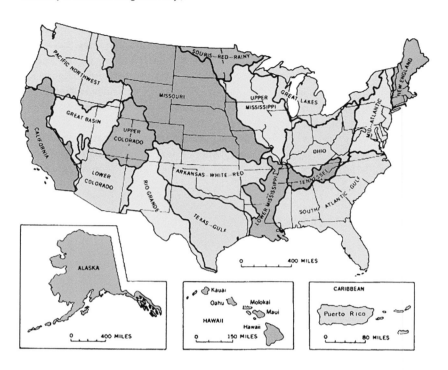

FIGURE 3.3. Wilson Run watershed tributary to the Brandywine River at Winterthur, Delaware. (Courtesy of the University of Delaware Water Resources Center.)

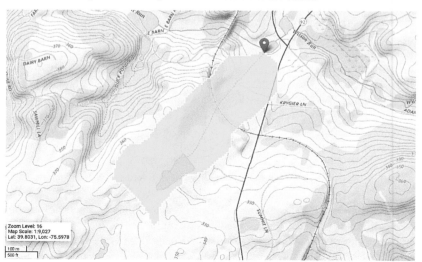

FIGURE 3.4. Mean annual precipitation in the continental United States in inches. (Courtesy of the National Weather Service.)

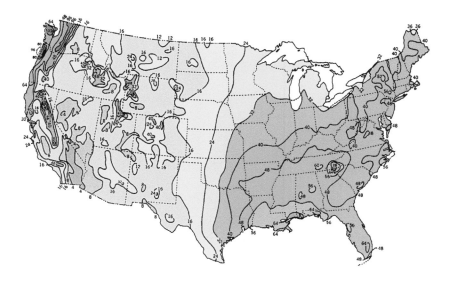

In Delaware, annual *precipitation* ranges from forty to forty-eight inches per year in a humid continental climate moderated by the influence of the Atlantic Ocean. Figure 3.4 depicts the mean annual precipitation in the United States for 1971–2000. Annual precipitation is highest (over sixty inches per year) along the Gulf Coast and Southeastern United States, where it is fueled by tropical moisture, and in the Northwest, where it is powered by the moist Pacific marine currents. Generally, as one proceeds inland, the climate becomes drier, leading to the semiarid Great Plains (receiving ten to twenty inches of precipitation per year), and dryer still in the arid climates of the desert Southwest (four to ten inches per year). The increase in precipitation with elevation in mountainous areas is due to the lifting or *orographic* effect (rising air cools, causing moisture to condense).

In Delaware, the precipitation depth for a *100-year storm* ranges from 8.0 inches in Northern Delaware to 9.2 inches in Southern Delaware in a twenty-four-hour event (Table 3.2). A one hundred-year storm is often defined as a storm likely to occur just once every one hundred years, but it is more appropriate to describe such a storm by its probability, that is, this type of storm has a 1 percent chance of occurring within any given year. The frequency, T, of a storm is the inverse of the probability, p. A one hundred-year storm has a probability of $1/T$ or $1/100$, which is a 0.01 or a 1 percent chance of occurring in any given single year. While there is a low probability for a one hundred-year storm to occur twice within a short period, such frequent occurrences are not uncommon.

TABLE 3.2. PRECIPITATION DEPTHS IN DELAWARE FOR NRCS TYPE II, 24-HOUR DURATION EVENT (National Weather Service)

Storm Event	New Castle County (in.)	Kent County (in.)	Sussex County (in.)
1-year	2.7	2.7	2.8
2-year	3.2	3.3	3.4
5-year	4.1	4.3	4.4
10-year	4.8	5.2	5.3
25-year	6.0	6.5	6.7
50-year	6.9	7.6	7.9
100-year	8.0	8.9	9.2
500-year	10.9	12.6	13.0

In Delaware, mean *evapotranspiration* ranges from thirty-five to forty inches per year. Evapotranspiration is defined as all water vapor emitted into the atmosphere from plants and other sources. It increases with warmth and plant foliage during the spring, summer, and early fall months and decreases during the cooler months after the leaves fall off of the trees and plants.

DELAWARE WATERSHEDS

Delaware is split by a barely perceptible subcontinental divide where two-thirds of Delaware water flows east to the Delaware Estuary and Atlantic Ocean and one-third flows west to the Chesapeake Bay. The state occupies four major basins—the Piedmont, Delaware Bay and Estuary, the Inland Bays, and Chesapeake Bay (see Figure 3.5) and includes:

- 25 miles of ocean coastline
- 841 square miles of bay
- 2,509 miles of rivers and streams
- 2,934 acres of lakes/ponds
- 46 watersheds

The four major basins in Delaware flow from the Piedmont and Coastal Plain physiographic provinces to the tidal rivers and bays:

- The Piedmont Basin—Includes the Brandywine, Red Clay, White Clay, and Christina Creeks that flow to the Delaware River, and comprises 605 square miles, 80 percent of which lies in Pennsylvania and a small sliver of Maryland.
- The Delaware Estuary Basin—Includes the Appoquinimink, St. Jones, and Broadkill Rivers in Eastern New Castle, Kent, and Sussex Counties, wherein 814 square miles drains to Delaware Bay.

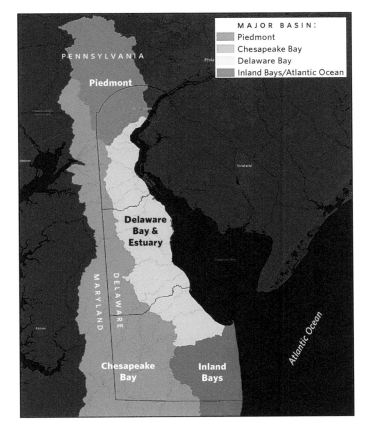

FIGURE 3.5. Major basins and watersheds in Delaware. (Courtesy of the University of Delaware Water Resources Center.)

- The Inland Bays/Atlantic Ocean Basin—Includes Rehoboth, Indian River, and Little Assawoman Bays watersheds that cover 313 square miles in Eastern Sussex County.
- The Chesapeake Basin—Includes Eastern Shore rivers such as the Sassafras, Chester, and Nanticoke that flow west to the nation's largest estuary, the Chesapeake Bay, and encompasses 769 square miles in Western New Castle, Kent, and Sussex Counties.

A great many of Delaware's rivers are troubled by *impaired waters*: they are so polluted they do not meet federal water quality standards. The state's list of impaired waters, filed with the Environmental Protection Agency, includes 377 bodies of water that suffer from eleven different impairments, the most common of which are pathogens and nutrients. Pollutants in Delaware waters are often chemicals, such as nitrogen and phosphorus from fertilizer runoff. They can come from specific "point" sources, such as sewage treatment plants, or from "nonpoint" sources, like runoff from lawns, farms, parking lots, and golf courses. Most nonpoint impairments are notoriously difficult to control. Along Delaware rivers and streams,

- 86 percent are impaired for swimming due to high bacteria
- 97 percent do not meet fish and wildlife water quality standards
- 44 percent of ponds and lakes do not meet swimming uses, and
- > 100 miles of waters have fish-consumption advisories from high levels of polychlorinated biphenyls (PCBs), metals, and pesticides

As of 2007, 39 percent of Delaware's land was in agriculture, 18 percent was forest, 17 percent was saltwater/freshwater wetland, 15 percent was urban, 8 percent was marine, and 3 percent was open freshwater (Table 3.3).

TABLE 3.3. LAND USE IN DELAWARE WATERSHEDS
(National Oceanic and Atmospheric Administration Coastal Services Center, 2007)

Land use	Area (acres)	Area (percent)
Freshwater wetlands	178,632	11.8%
Marine	124,879	8.2%
Farmland	590,150	38.9%
Forest	265,476	17.5%
Saltwater wetland	71,001	4.7%
Barren land	6,459	0.4%
Urban	229,827	15.2%
Beach/Dune	588	0.0%
Open freshwater	48,253	3.2%
Total	1,515,263	100.0%

According to the U.S. Census Bureau, from 2000 to 2010, the Delaware population grew by 14.6 percent, then from 897,934 people in 2010 to 967,000 in 2018. Approximately 60 percent of the population reside in New Castle County, 20 percent live in Kent County, and 20 percent make Sussex County their home. About 400,000 people are employed in Delaware: 70 percent in New Castle County, 15 percent in Kent County, and 15 percent in Sussex County (Table 3.4). By 2030, the population in Delaware is projected to grow by 156,697 (18 percent) to over a million people.

Delaware's waters and watersheds are rich in natural resources and habitat, as measured by the economic value of ecosystem goods and services. Ecosystem goods are benefits provided by the sale of watershed products, such as drinking water and fish. Ecosystem services are economic benefits provided to human society by nature, such as water filtration, flood reduction, and carbon storage. A study by the University of Delaware indicates

TABLE 3.4. LAND AREA, POPULATION, AND EMPLOYMENT
(U.S. Census Bureau, 2010; U.S. Bureau of Labor Statistics, 2011)

County/State	Area (sq. mi.)	Population	Employment
New Castle	426	538,479	267,683
Kent	590	162,310	60,964
Sussex	940	197,145	67,447
Delaware	1,956	897,934	396,094

TABLE 3.5. ECOSYSTEM SERVICES VALUE OF DELAWARE WATERSHEDS IN 2010 DOLLARS

Ecosystem	Area (acres)	Value ($/acre/year)	Value ($/year)
Freshwater wetlands	178,632	$13,621	$2,433,081,000
Marine	124,879	10,006	1,249,541,955
Farmland	590,150	2,949	1,740,640,688
Forest land	265,476	1,978	525,143,567
Saltwater wetlands	71,001	7,235	513,691,702
Barren land	6,459	0	0
Urban	229,827	342	78,511,742
Beach/dune	588	48,644	28,579,665
Open water	48,253	1,946	93,891,133
Total	1,515,265		$6,663,081,452

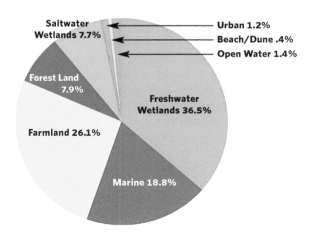

FIGURE 3.6.
Relative ecosystem services value within Delaware watersheds in 2010.
(Adapted from Narvaez and Kauffman, "Economic Benefits and Jobs Provided by Delaware Watersheds.")

that the value of natural goods and services from ecosystems in Delaware watersheds is $6.7 billion (in 2010 dollars) with a net present value (NPV) of $216.6 billion, using a discount rate of 3 percent over one hundred years (Table 3.5).[4]

Freshwater wetlands, farms, marine habitats, forests, and saltwater wetlands provide the highest total ecosystems goods and services values (Figure 3.6). The Delaware Estuary provides the highest value (at $2.4 billion) of annual ecosystem services, and the Chesapeake Bay and Inland Bays follow close behind at $2.0 billion each. Delaware watersheds with the highest value of annual ecosystem services per acre include the Inland Bays ($6,147 per acre), Chesapeake Bay ($4,562 per acre), and the Delaware Estuary ($3,878 per acre) watersheds, as over 75 percent of these systems are covered by forests, marine, and wetlands habitats.

PIEDMONT

The Piedmont Basin flows to the Delaware River, and while 80 percent of the watershed drains the headwaters in Pennsylvania (and a small sliver of Maryland), the lower 20 percent drains Northern Delaware[5] (Figure 3.7). The Piedmont Basin contributes freshwater from surface water (streams) and groundwater (bedrock wells) and includes hilly, rocky watersheds such as:

- Brandywine Creek
- Red Clay Creek
- White Clay Creek
- Christina River
- Naamans Creek
- Shellpot Creek

The geologically unique Fall Line located in the Piedmont Basin runs along a line between Newark and Wilmington and separates the hilly, rocky Piedmont Plateau from the flat, sandy Coastal Plain provinces. This transition between the two geological provinces supports diverse plant (flora) and animal (fauna) habitats.

While 90 percent of the watershed lies upstream in Pennsylvania, the Brandywine River is the largest drinking water supply in Delaware and serves the state's largest city, Wilmington. The city of Newark and SUEZ Delaware water utilities draw drinking water for 200,000 people (or one-fifth of the state's population) from the Red Clay, White Clay, and Christina Creeks. In contrast to Southern Delaware, where groundwater is the sole source of drinking water, the Piedmont Basin includes the only four streams in Delaware that provide drinking water to two-thirds of the state's population.

The tidal Christina River supports spawning grounds for a striped bass fishery and the nontidal Brandywine Creek provides good smallmouth bass fishing. State surveys in 2010 indicate that abundant anadromous (fish that live in the ocean and spawn in freshwater streams) fisheries exist in the

FIGURE 3.7. Piedmont watersheds in Delaware.
(Courtesy of the University of Delaware Water Resources Center.)

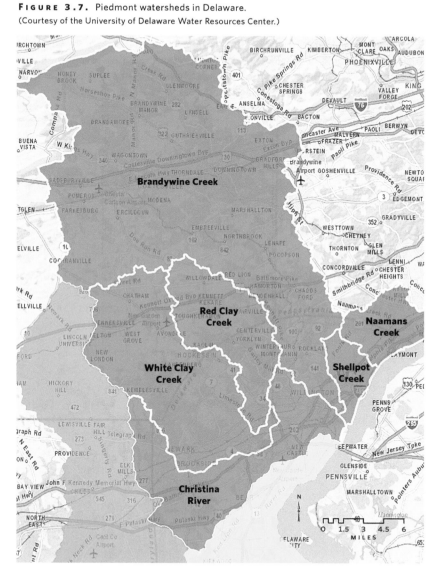

White Clay Creek, including for American shad, hickory shad, white perch, striped bass, alewives, and blueback herring. The Brandywine Shad 2020 initiative is removing dams from the Colonial era to allow anadromous fish such as the American shad and striped bass to swim up the Brandywine and White Clay Creeks to their ancestral spawning grounds for the first time in centuries.

Parks, recreational facilities, and open space are an important part of the social, cultural, and physical fabric of the community and are abundant in the Piedmont Basin. The Brandywine Creek hosts many canoe and kayak enthusiasts at public boat landings and commercial liveries. The White Clay Creek State Park, Brandywine Creek State Park, and numerous municipal and state parks provide hiking and biking trails for the community. Freshwater trout fishing is available at six designated trout streams within the basin and nineteen miles of streams are stocked annually with trout—the only trout-stocked streams in Delaware. One of these trout streams, White Clay Creek, has a fly fishing-only section, providing anglers an unparalleled experience within the state. Another unique fishing opportunity in the Piedmont Basin is on the Brandywine Creek, which provides the only sustainable smallmouth bass fishery within the state. Public access for boaters is limited, with only three public boat-launching facilities, although recreational opportunities for non-motorized boats—canoes and kayaks—are quite popular.

The Piedmont supports the highest acreage of mature, deciduous woodlands in the state, including oaks, beech, tulip poplar, hickories, and sweet birch on the steep slopes and dry ridge tops, and box elder, sycamore, sweet gum, slippery elm, red maple, tulip poplar, and sometimes river birch and black willow along narrow streamside forests.

DELAWARE ESTUARY

The Delaware Estuary Basin drains 814 square miles in Eastern New Castle, Kent, and Sussex Counties and includes the

- Delaware River
- Army Creek
- Red Lion Creek
- Dragon Run Creek
- Chesapeake & Delaware Canal
- Appoquinimink River
- Blackbird Creek
- Smyrna River
- Leipsic River
- Little Creek
- St. Jones River
- Murderkill River
- Mispillion River
- Cedar Creek
- Broadkill River

A significant farm economy is supported by the fertile soils, plentiful irrigation supplies, and productive weather on the gentle Coastal Plain along the Delaware Bay.

FIGURE 3.8. Delaware Estuary Basin in Delaware.
(Courtesy of the University of Delaware Water Resources Center.)

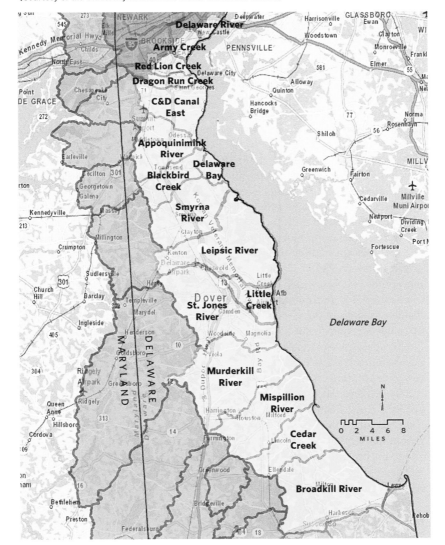

The Delaware Estuary Basin supports diverse habitats in the large expanse of freshwater and saltwater wetlands that rim the bay. The federally endangered Delmarva fox squirrel was extirpated from its forest habitat but is now reemerging in Eastern Sussex County. The Delaware Natural Heritage Program (DNHP) in the Division of Fish and Wildlife inventories natural communities, as well as plants, birds, insects, mussels, reptiles, and amphibians that are rare in the state and globally.

The Division of Fish and Wildlife classifies fifty-eight game species in the Delaware Estuary Basin, including forty-four birds, eleven mammals, two reptiles, and one amphibian. The U.S. Fish and Wildlife Service protects habitats that serve as wintering areas or migration corridors for migratory birds at Bombay Hook and Prime Hook National Wildlife Refuges along Delaware Bay. The basin provides for migratory species protected by the North American Waterfowl Conservation Act (NAWCA) such as black duck and mallard.

About 45 percent of the Delaware Estuary Basin is covered by agriculture, which is Delaware's most valuable industry, with sales of $1.1 billion annually. Of this amount, poultry comprises 90 percent of agricultural sales, especially in Sussex County, which is the number one poultry county in the U.S. with over three hundred million broilers/roasters raised annually. The dairy industry is the second largest animal-based waste generator in Delaware, and includes approximately one hundred registered dairy farms in the state.

The Delaware Estuary Basin lies entirely in the Atlantic Coastal Plain physiographic province, which consists of a series of southeastward-thickening, unconsolidated sediments deposited over "basement" rocks of the Appalachian Piedmont: the coastal lowland belt contains extensive tidal marshes up to five feet above mean sea level that are separated by narrow necks up to twenty feet above mean sea level. The inland plain consists of flat highlands between twenty-five and seventy-five feet above mean sea level, cut by narrow, steep stream valleys.

The soil and landscapes of the Delaware Bay and Estuary Basin are the product of soil-forming factors operating on surficial materials over geologic time. Factors influencing soil formation include parent material (geology), topography (relief), climate (temperature, moisture, and wind), vegetation, and living organisms (including human beings). The degree and intensity of soil development is dependent on the climate and on the biological agents (plants, living organisms) in it.

Topography in the northern part of the basin is dominantly undulating and rolling with moderate dissection. In the southern portion of the drainage basin, flatter landscapes dominate, but constitute less than one percent of the basin's topography. This relief affects landscape distribution of soils, landscape distribution of moisture, erosion and alleviation patterns, and degree of soil development. The presence and height of the seasonal high water table also correlates to the surface of the landscape.

The natural resources of the Delaware Estuary Basin provide opportunities for recreational fishing, camping, hunting, and hiking in the forests, marshes, and open waters that are popular destinations for residents and visitors alike. Over seventy-four thousand acres of land are protected by state, federal, or private conservation ownership, including four state parks, two national wildlife refuges, and twelve state wildlife areas.

In the Delaware Estuary, popular recreational fisheries include: striped bass, bluefish, carp, catfish, drum fish, summer flounder, white perch, and yellow perch. Bluefish and summer flounder are also sought-after species in the estuary.

INLAND BAYS/ATLANTIC OCEAN

The Inland Bays/Atlantic Ocean Basin covers 313 square miles of Eastern Sussex County in the Atlantic Coastal Plain physiographic province (Figure 3.9). Three "inland bays" flow east to the Atlantic Ocean at the Indian River Inlet and include Rehoboth Bay, Indian River Bay, and Little Assawoman Bay.

Distinctive physiographic characteristics include the flat topography and manmade drainage ditches that drain soils and farms with high water tables south of Millsboro and Indian River Bay.

The Inland Bays Watershed supports Delaware's $2 billion ocean and coastal tourism economy and contains rapidly growing industries of poultry farming and residential development, contributing to dramatic changes in land use and degradation of the basin's land and water resources. More than eleven miles of Delaware's ocean coastline are developed with homes and businesses, many situated on the barrier island—a thin strip of land separating the ocean from the Inland Bays. Sea level rise, storms, and natural coastal transport processes cause the barrier island system to migrate westward in a landward direction.

The Inland Bays/Atlantic Ocean Basin extends from Lewes twenty-four miles south along the Atlantic Ocean to Maryland, and includes the coastal towns of Rehoboth Beach, Dewey Beach, Bethany Beach, South Bethany Beach, and Fenwick Island, connected by State Route 1. The land surface of the basin rises from sea level in the east to an average elevation of twenty-five to thirty feet in the west.

At the Maryland line, the basin extends west sixteen miles to Cypress Swamp and then over nineteen miles to Georgetown, the county seat of Sussex County. U.S. Route 113 connects Selbyville, Frankford, Dagsboro, Millsboro, and Georgetown. The northern border of the Inland Bays/Atlantic Ocean Basin parallels State Route 9 from Georgetown to Lewes and Cape Henlopen State Park.

Waters draining into the Inland Bays are high in nitrogen and phosphorus, nutrients that emanate from agricultural runoff and sewage overflows. The high concentration of nutrients has led to dramatic water quality impairment,

FIGURE 3.9. Inland Bays/Atlantic Ocean Basin in Delaware.
(Courtesy of the University of Delaware Water Resources Center.)

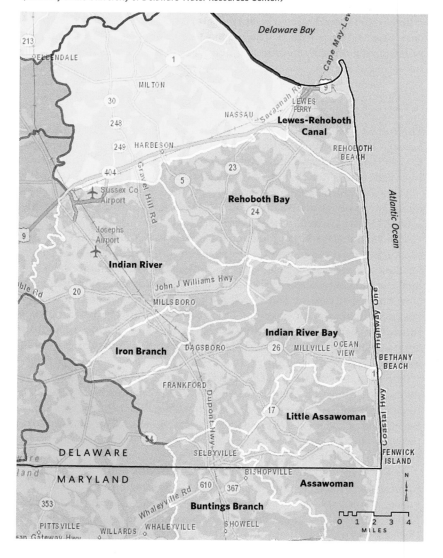

including algae blooms, dissolved oxygen depletion, increased turbidity, and fish kills.

The nutrients entering the state's watersheds come from agricultural runoff, urban runoff, and municipal and industrial point source discharges. Poultry is the primary agricultural product in Delaware, and within Sussex Country, the Indian River, Nanticoke River, and Broad Creek watersheds are the main poultry producers. The Broad Creek watershed has the highest density of poultry per acre of any watershed in the state, causing the Nanticoke River and Broad Creek to have an abundance of manure.

Nitrogen and phosphorus originating from agricultural activities have long been identified as key factors in nonpoint source pollution in the Inland Bays/Atlantic Ocean Basin. There are approximately seventy-two thousand acres of agricultural land in the basin, representing more than forty percent of the total land area. The majority of croplands are devoted to growing corn, soybeans, and sorghum, which go to feed the basin's thriving poultry industry. Unfortunately, agricultural lands are highly susceptible to "nutrient loss" (that is, nitrogen and phosphorus running off the fields where they are wanted, and into streams and bays where they are not wanted).

Water quality samples collected from several agricultural drainage waters within the basin have shown elevated levels of both nitrogen and phosphorus that exceed water quality standards for streams in the basin.

Recreational saltwater fishing within the Inland Bays and nearshore Atlantic Ocean is extremely popular. Key game species in the Inland Bays include summer flounder, sea trout, bluefish, tautog, white perch, rockfish, and winter flounder. In addition to these species, smooth dogfish, sandbar sharks, and kingfish can be caught in the surf. The fishery in the ocean has an added variety of pelagic fish such as white marlin, yellowfin tuna, mako shark, cod, and ling.

The Inland Bays/Atlantic Ocean Basin includes four of Delaware's fourteen state parks: Delaware Seashore, Fenwick Island, Holts Landing, and Cape Henlopen. They include 8,600 acres or 43 percent of the acreage in Delaware's state park system and preserve fourteen miles of Delaware's twenty-four miles of Atlantic Coast beaches. The Division of Fish and Wildlife manages more than 3,200 acres of land and water in the basin that draw visitors for fishing, hunting, and other forms of recreation, such as Assawoman Wildlife Area and fishing sites such as Love Creek, Ingram Pond, Massey's Landing, Pepper Creek, Millsboro Pond, and Rosedale Beach.

Hiking, bicycling, and camping continue to grow in popularity. Trails and pathways are a growing part of Delaware's recreation infrastructure. Over the past ten years, the public has recognized a need to develop and expand trails and pathways to use for fitness, recreation, nature exploration, and transportation. More than twenty private campgrounds and two public campgrounds at Delaware Seashore and Cape Henlopen State Parks provide more than five thousand campsites.

Forests in the Inland Bays/Atlantic Ocean Basin include loblolly pine (*Pinus taeda*), Virginia pine (*Pinus virginiana*), white oak (*Quercus alba*), southern red oak (*Quercus falcata*), scarlet oak (*Quercus coccinea*), willow oak (*Quercus phellos*), tulip tree (*Liriodendron tulipifera*), red maple (*Acer rubrum*), sweet gum (*Liquidamber styraciflua*), black gum (*Nyssa sylvatica*), sassafras (*Sassifras albidum*), dogwood (*Cornus florida*), and others. Of the sixty thousand acres of forested lands in the basin, thirty-five hundred are owned by private landowners in the forest industry for timber production, wildlife habitat, and recreation.

CHESAPEAKE BAY BASIN

The Chesapeake Bay Basin drains to the nation's largest estuary, the Chesapeake Bay (Figure 3.10). As an estuary, the Chesapeake Bay contains a mixture of freshwater and saltwater, creating an ideal habitat for a diverse array of plants and animals. The basin drains 769 square miles (about one percent of the sixty-four thousand square miles in the Chesapeake Bay watershed) in Western New Castle, Kent, and Sussex Counties. It encompasses sixteen watersheds, including Bohemia Creek, Broad Creek, the C&D Canal, Deep Creek, Elk Creek, Gravelly Branch, Gum Branch, Marshyhope Creek, the Nanticoke River, Perch Creek, the Pocomoke River, the Sassafras River, and the Wicomico River.

The Chesapeake Bay's welfare is heavily reliant on the land use of the basin, since Delaware's portion of it contains headwater areas, the area where a waterway originates. The streams and rivers that drain into the Chesapeake Bay support many species of fish harvested for both food and profit. Substantial commercial fishing efforts take place in the Nanticoke River, with American shad, blueback herring, alewife, white catfish, channel catfish, striped bass, and white perch representing the highest percentage of the catch. Many of Delaware's residents and visitors depend on water for their recreational enjoyment. Fishing, swimming, and boating are popular activities throughout Delaware. Delaware's portion of the Chesapeake Basin includes a dozen publicly owned ponds and lakes, comprising nearly seven hundred acres that serve recreational needs. The health of Delaware's waters affect the recreation potential of the included ponds and streams. Delaware's wildlife represents a vital recreational resource base as well. Both hunting and birding depend on the health of the state's natural resources.

In addition to the nutrients running off farm fields, septic systems also contribute high amounts of nutrients to the Chesapeake Bay. The Chesapeake Basin has one of the highest percentages (95 percent) of land area covered by septic systems. From research done on the Nanticoke watershed, a notable amount of nitrogen loading was found to be originating from septic systems. As the soil type and water table depths in the Nanticoke watershed

FIGURE 3.10. Chesapeake Bay Basin in Delaware.
(Courtesy of the University of Delaware Water Resources Center.)

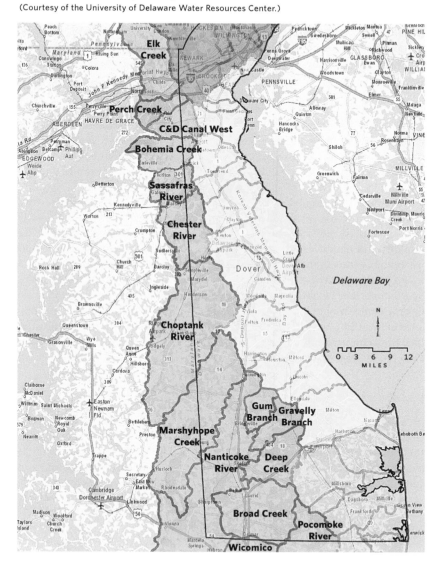

are similar to those in the rest of the Chesapeake Basin, similar nutrient loads can be predicted throughout the basin. Raw or inadequately treated sewage is also a severe contributor to the declining health of the Chesapeake Bay. This sewage contains pathogens, which are disease-causing bacteria and viruses. The potential daily pathogen output from one person's untreated sewage can equal that of treated sewage from hundreds to even thousands of people, depending on the level of treatment. Wildlife and animal operations can also be significant sources of pathogens. High levels of bacteria can cause serious illness to swimmers, people that eat shellfish, and others who come in contact with contaminated waters.

The Environmental Protection Agency established a Total Maximum Daily Load (TMDL) for nitrogen, phosphorus, and sediment for the entire Chesapeake Bay Watershed. This TMDL requires reductions of approximately 24 percent for nitrogen and 20 percent for phosphorus between 2009 and 2025 in all of the Chesapeake watersheds within Delaware. Sediment loads from Delaware's portion of the Chesapeake must remain at 2009 levels under this TMDL.

Delaware's Chesapeake Basin contains a variety of habitat types that provide homes to diverse flora and fauna. The northernmost forests usually contain a mixture of hardwoods, dominated mostly by oaks, beeches, tulip poplars, and hickories on the drier sites. The predominant tree species in wetland habitats include box elder, sycamore, sweet gum, slippery elm, red maple, tulip poplar, ash, pin oak, and sometimes river birch and black willow. The oldest trees in Delaware reside in this basin, reaching ages estimated to be around five hundred years old! Moving south along the Chesapeake Basin, a transition in forest species begins. Entering Sussex County, where the deciduous hardwood-dominated forest gives way to an evergreen forest, the Oak-Pine Forest Region.

RAIN GARDENS

An ecologically effective way to restore watersheds is to plant a *rain garden* (Figure 3.11). A rain garden is a shallow landscaped depression that captures the rain as it runs off impervious surfaces.[6] This allows the collected water to evaporate into the air, soak into the ground, or be absorbed by plants and turned into oxygen. Rain gardens provide the following benefits:

- Reduce the opportunity for flooding
- Create habitat for wildlife
- Protect rivers and streams
- Conserve drinking water
- Promote infiltration of freshwater
- Enhance aesthetics and improve property value
- Reduce lawn maintenance using fertilizers

FIGURE 3.11. University of Delaware rain garden.
(Courtesy of the University of Delaware Water Resources Center.)

Rain garden design typically follows these six steps:

1. Determine rain garden site at the low point of the property near the stormwater drain.
2. Determine the amount of rain in cubic feet draining to the garden by calculating the area of the property in square feet multiplied by the two-year storm depth in Delaware (3.2 inches).
3. Determine the desired size of the rain garden in square feet and mark off.
4. Determine the infiltration rate of rain (should be at least one half-inch per hour) as it soaks into the garden by digging a small hole and recording the time it takes for water to drain one foot in depth.
5. Determine the depth of the rain garden, one to two feet maximum, that allows the rain to drain into the ground within four days or less.
6. Select native plants from the following list, plant them, then mulch in accordance with the plan.
    - Virginia sweetspire
    - Buttonbush
    - Marsh Marigold
    - Blackhaw
    - Swamp Milkweed
    - Arrowwood
    - Possomhaw
    - Spicebush
    - Swamp Azalea
    - Bald Cypress
    - Winterberry Holly

NOTES

1. Gerald J. Kauffman and Martin W. Wollaston, *Along the Fall Line: Histories of Newark, 1758–2008* (Newark, DE: Wallflowers Press, 2007), 137.
2. Gerald J. Kauffman, "The Delaware River Revival: Four Centuries of Historic Water Quality Change from Henry Hudson to Benjamin Franklin to JFK," *Pennsylvania History, A Journal of Mid-Atlantic Studies* 77(4): 432–65. A detailed description and diagram of "The Water Cycle" is available via the U.S. Geological Survey at https://www.usgs.gov/media/images/water-cycle-natural-water-cycle.
3. Jonathon Brant and Gerald J. Kauffman, *Water Resources and Environmental Engineering Depth Reference Manual* (Belmont, CA: Professional Publications, Inc., 2011).
4. Martha C. Narvaez and Gerald J. Kauffman, "Economic Benefits and Jobs Provided by Delaware Watersheds." Published by Institute for Public Administration, School of Public Policy and Administration, University of Delaware. (Newark: University of Delaware, 2012).
5. DNREC Division of Watershed Stewardship and University of Delaware Water Resources Center, "Delaware Watersheds," www.delawarewatersheds.org, accessed May 2019.
6. Elaine Grehl and Gerald Kauffman, "The University of Delaware Rain Garden: Environmental Mitigation of a Building Footprint," *Journal of Green Building* 2(1): 53–67.

CHAPTER FOUR

Botany and Soil

SUSAN BARTON AND JOCELYN WARDRUP

Plants supply the very air we breathe. They provide a great deal of the natural beauty we see every day. They are also the only organisms on the planet that can make their own food. And since plants exist at the very base of the global food web, meaning that every other organism in the world either eats plants, or eats things that eat plants, we ought to know something about these amazing organisms. We also ought to know something about the three things that plants need most: sunlight, water, and (most importantly, for this chapter) soil. Learning about botany and soil will significantly deepen our appreciation for the Earth as a whole, and take us a big step forward in our ability to care for the Earth as well.

BOTANY

First, some terminology. There are five living kingdoms in the world. *Animalia* includes sponges, worms, insects, fish, amphibians, reptiles, birds, and mammals. *Monera* cannot be seen with the naked eye and includes bacteria, blue-green algae, and spirochetes. *Protista* includes protozoans and various types of algae. *Fungi* includes the funguses, such as molds, mushrooms, yeasts, mildews, and smuts. Finally, *plantae* includes mosses, ferns, gymnosperms, and woody and non-woody flowering plants.

Just as animals are characterized as the *fauna* living in a certain area, plants are categorized (and described) in books or online references called *floras*. We might talk about the *Flora of Delaware* or the *Flora of the Coastal Plain*. A flora might be used to confirm that a plant found in one's backyard is present throughout that region. It could also tell you if that plant could thrive in another area. If you were moving to a new part of the country, you could use a flora to learn which plants thrive in your new location. If you were backpacking, you could bring a pocket flora to help you identify plants you encounter, or to help you determine which plants are safe to eat, and which plants could be used for medicine if you become sick. Floras simply help you discern one plant from another, and are a great tool for the naturalist's backpack.

Let's start with the basics. The structure of a plant consists of the following five parts: leaves, stems, roots, flowers, and seeds. Although these parts differ visually from one plant to another, their functions are basically the same.

LEAVES

Leaves are the most familiar part of a plant (and the most commonly used in plant identification) and vary in their structure, color, vein distribution, arrangement, and many other characteristics. They perform three primary functions: photosynthesis, respiration, and transpiration.

Most of us remember the high school biology lesson that *photosynthesis* (derived from the Greek *photos*: light and *synthesis*: putting together) is the process that plants use to make food. We may even remember the equation we learned about this process: CO_2 + water + sunlight = O_2 + sugar. But there is a great deal more to understand. During photosynthesis, plants use the green chlorophyll present in their organelles called *chloroplasts* to absorb light, and their leaves pull carbon dioxide (CO_2) from the air, energy from the sun, and water from the soil, which combine with the energy from the chloroplasts into an organic chemical compound called glucose, or simple sugar. Plants use this compound for all growth processes. During daylight hours, photosynthesis also produces excess oxygen, which plants release into the atmosphere, and which humans and animals alike require to breathe. Understanding this process helps naturalists understand why you shouldn't prune trees when you transplant them. Plants photosynthesize in their green leaf tissue and that process is what makes sugars to help regrow their damaged root systems. The faster new roots are grown, the more quickly a plant will become acclimated and start growing in its new location.

During *respiration*, the photosynthesis process reverses: plants take in oxygen and release CO_2. During respiration, which happens in both the light and the dark, the bound energy (processed in the plant's mitochondria) is released and stored as ATP (adenosine triphosphate). As ATP, it is further transported to systems throughout the plant where energy is needed. During the day, photosynthesis exceeds respiration, and there is a net increase in CO_2 absorbed and O_2 released. At night, when photosynthesis has ceased, only CO_2 is released, but the quantity emitted never exceeds the daily intake.

Understanding the concept of "carbon balance" has become more critical in this time of climate change and excess global carbon production. It is now widely recognized that one way to efficiently extract atmospheric carbon, which causes the greenhouse effect (and leads to global warming) is to plant many more (some estimate 1.5 trillion) trees around the world, as they are highly effective at sequestering carbon. As trees grow, they continue to absorb and store CO_2; this carbon remains bound until a tree dies and decomposes into the soil.

Transpiration, the loss of water vapor from plant leaves, is responsible for the movement of much of the water used by a plant. Water is taken up through root hairs and moves up through *xylem* vessels to the leaves, where is it escapes through *stoma* on the undersides of the leaves. *Guard cells* open on either side of a stomata, creating a pore that allows water vapor to escape. In addition to pulling water up through the plant, this process also has a measurable cooling effect on both the plant and its immediate surroundings. This is one reason forests are cooler than lawns, which themselves are cooler than unplanted landscapes. It is also another reason, beyond providing shade, that trees can help keep cities cool. However, if water loss is more rapid than absorption, then a water-deficit *wilting* occurs. Eventually the plant will die if the situation is not remedied through rainfall or watering.

Leaf surfaces are covered with *cuticles*, which protect the leaves. Leaf surfaces may be smooth or covered in fine hairs. A hairy leaf surface will reduce water loss, improving drought tolerance, and leaves with thick, waxy cuticles are also less likely to lose water and are more drought tolerant.

All leaves senesce at some point, but deciduous plants lose their leaves once per year in the fall. Evergreens keep their leaves during the winter, but individual leaves fall off periodically and are replaced by new leaves. White pines, for example, keep their needles all winter but usually lose about one-third of their leaves in the spring, causing unnecessary alarm to people who see lots of yellow needles on branches close to the tree trunk and assume, incorrectly, that something is wrong with their tree.

STEMS

The primary functions of the *stem* are support, conduction, and storage. Stems hold leaves up to the light, so they can photosynthesize and manufacture food. Green stems contain chlorophyll and can also participate in photosynthesis and respiration, although to a lesser extent than leaves do. Some plants have a single stem, often called a *trunk*, from which branching occurs. Others have many stems arising from the *root crown*. In older stems, the interior, dense heartwood serves for support only, having outlived its other functions.

Conduction, the movement of water, mineral solutions, and plant food, is a life-sustaining plant function. Water and mineral solutions move upward from the roots to the leaves through tissue called xylem. Plant food manufactured in the leaves moves in any direction in the plant through tissues called *phloem*. Between the xylem and phloem is the *cambium* layer, which forms new xylem and phloem as needed. This accounts for the increase in stem diameters.

Cells in leaves and stems store starch. The starch is digested by an enzyme that changes it to simple sugar, which then moves out to feed other parts of the plant. The *cortex* and *epidermis* are the outer portions of the stem, and provide protection from outside elements.

For a plant to remain healthy, its xylem, phloem, and cambium tissues must function without interference. For example, if the stem is inadvertently *girdled*—cut all the way around by a weed whacker, or by plastic twine or wire—the food supply to the roots will be restricted, and eventually the roots, and then the entire plant, will die. (Important tip for tree planting: to prevent accidental girdling, always remove twine or wire used to tie up the tree during transport).

ROOTS

The *root* system found in many plants is fibrous, consisting of numerous multi-branched roots that are slender and filament-like with no one root more prominent than the others.

In other plants, by contrast, like in many trees, the *taproot* system consists of one main root extending straight downward, from which branch roots grow and spread. The root may either be a fleshy food-storage organ, as with a carrot, or a woody root, as found on pecan trees. Since taproots penetrate much deeper into the soil than fibrous roots do, trees (like sassafras, for example) with taproots are much more difficult to transplant unless their roots are frequently pruned. Such trees are more easily planted if they are grown in a nursery container, with holes on the sides or bottoms to allow air pruning of the taproot and the development of a fibrous root system.

Root system functions include the *absorption* of water and nutrients, as well as conduction, anchorage, and storage. Large amounts of water are absorbed by the root system and then transported to (and lost through) the leaves. Most of the water is absorbed through millions of thin-walled root hairs in close contact with soil particles. These root hairs are formed from the outer layer of cells near the tip of a growing root. They are fragile, and many may be killed or broken during transplanting. Also, their exposure to air should be limited because they dry out rapidly. Root hairs conduct both minerals and nutrients dissolved in water to a plant. The nourishment flows through the roots and stems to the leaves, where it is manufactured into usable food.

When trees and shrubs are dug for transplanting, maintaining a healthy root system is critical for their subsequent survival. Trees growing in heavy clay soils may be difficult to transplant if the root ball falls apart, because the root hairs are likely to be torn off in the clumps of clay. Once the root hairs are gone, the roots can no longer take up water until a new crop of root hairs grow.

A balance must exist between the size of the root system and the size of the stem system for a plant to be healthy. If roots are lost, either through damage during transplanting or through improper cultivation, the reduced root system may not be able to supply the required moisture and nutrients to the stem. Cutting back a portion of the plant's top growth will reduce the water lost through transpiration, but it will also reduce the plant's capacity to photosynthesize, thus slowing recovery of the root system. If a plant can be kept moist, it is best to leave the top growth intact even if the root system has been damaged.

Plants are also anchored to the soil through the spread and depth of their root system. If soil conditions allow the roots to grow deeply and spread widely through the soil, then the plant is less likely to be damaged by cultivation or wind. Urban trees planted in sidewalk pits often have little opportunity to develop an extensive anchoring root system, so landscape design that provides a larger space for groups of trees is far preferable to the standard street tree design.

The root system also stores food and water. Some plants store food from one year to the next in a modified stem below ground. True *bulbs* (daffodils and tulips), *corms* (crocus and gladiolas), *rhizomes* (canna and iris), and some *tubers* (dahlias and caladium) are actually modified stems that provide underground storage.

There are three types of growing tissue, or *meristems*, in plants. The *apical meristem* increases length at the tips of the stems and roots. The *lateral meristem* allows the sides of stems and roots to increase in diameter. The *intercalary meristem*, which provides growth at the base of nodes and stem blades, is unique to *monocots*, flowering plants (or *angiosperms*) with seeds typically containing only one embryonic leaf, or *cotyledon*. Monocots include most bulbous plants (e.g., tulips, daffodils, lilies), grains (e.g., wheat, rye, barley), and grasses (e.g., bluegrass, ryegrass, switchgrass). The intercalary meristem is what allows the leaf blades of grass to regrow after they have been cut. (Leaf blade regrowth in grasses has evolved not to survive suburban lawn mowers but to accommodate repeated grazing from herbivores.)

Understanding plant growth patterns can help us effectively manage our many landscape plants. For example, cool season turfgrasses (growing on many lawns in Delaware) grow shoots in the spring and roots in the fall. It is more advantageous to fertilize, plant, or *core aerate* (a practice that involves removing plugs of soil with a machine comprised of hollow tines and depositing the plugs on the grass surface) in the fall to promote root growth, rather than to promote extra shoot growth that must be repeatedly mowed. By core aerating in the fall, when roots are growing, grass plants recover quickly from root injury. However, it is better to prune woody plants in the winter if the goal is to produce new growth, because stored carbohydrates are available to help push the new growth. If the goal is to shape without promoting growth, then it is better to prune in the summer, when stored carbohydrates are low.

FLOWERS

The *flower* may be thought of as a kind of specialized stem, with leaves adapted for reproductive functions. Without flowers, there would be no sexual reproduction of new plants.

The following features are present in a flower that has all the characteristic parts. An outer set of green floral leaves called *sepals* enclose the other parts of the flower until they are nearly mature. Collectively, the sepals are called the *calyx*. An inner set of white or colored floral leaves called *petals* constitutes the *corolla*. In many flowers, the petals are showy and may aid in attracting the attention of insects that assist in pollination.

Within the petals are one or more sets of *stamens* (the male flowers). Stamens are composed of pollen-bearing *anthers* supported by a *filament*. When the pollen is mature, it is discharged through the ruptured anther wall. One or more *pistils* (female flowers) are at the center of the flower. Pistils consist of an ovule-bearing vase, or *ovary*, supporting an elongated region, or *style*, whose expanded tip, or surface, is called the *stigma*. The ovule gives rise to the seed. The mature ovary then becomes the fruit if pollen reaches the female flower parts and fertilizes them.

Pollination occurs when pollen grains are transferred from an anther to a stigma. Some flowers self-fertilize and can use their own pollen (or pollen from another flower on the same plant). Others can be pollinated with the help of the wind. Many other plants need pollen from an entirely different plant of the same species; this is called *cross-pollination*, and this is the process that requires pollinators. We need bees, butterflies, moths, and flies to move pollen grains from one plant to another. Some plants, such as woodland phlox, need a very specialized type of insect (like a snowberry clearwing) to reach down the long throat of the flower and get the pollen. Other plants

FIGURE 4.1. Parts of a perfect flower. (Sketch courtesy of Olivia Kirkpatrick.)

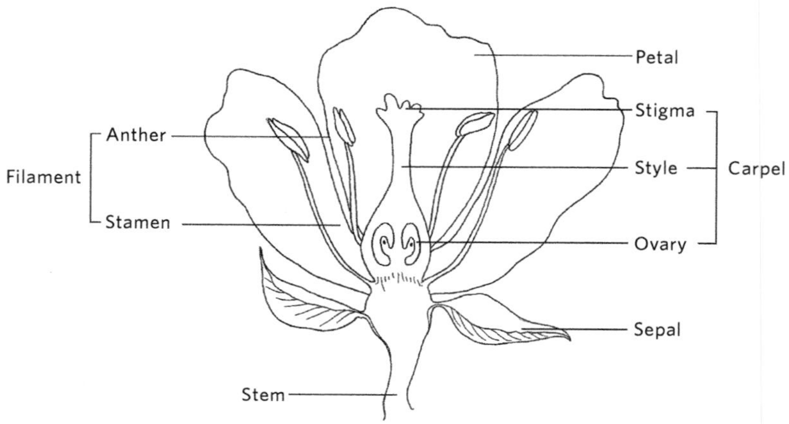

(such as hollies) have only female flower parts, and require a nearby male plant, or pollinizer, to produce fruit. Plants with flowers pollinated by bees will have fragrant flowers, while plants pollinated by flies may have putrid-smelling flowers, like the Callery pear.

Many flowers are borne in an *inflorescence*, which is comprised of individual flowers. Members of the *compositae* family (like daisies, asters, and chrysanthemums) may have hundreds (or thousands) of individual flowers in one inflorescence. They include *disk flowers* in the center and *ray flowers*, which look like petals, in the outer ring (a novice might confuse a daisy inflorescence for an individual flower). Flowers can be a showy aesthetic feature of a plant, as they are on an Eastern redbud, or they might be hidden at the top of a tree and barely visible until they fall to the ground, as they are on a tulip poplar. Flowers without petals, like those on a red maple, may still be showy when the tree is covered in small red flowers, but individually they are not particularly noticeable.

Flowering can affect the invasive potential of a plant. Callery pears in cultivation were originally all members of the common suburban landscaping cultivar known as the "Bradford." Bradford pears have an attractive lollipop form, but were subject to easy breakage due to their narrow branch angles. The trees were relatively self-sterile, so few fruits were formed, and they did not spread dramatically. When new cultivars with better branch angles were introduced, they began to cross-pollinate with Bradford pears, and the plants started producing copious fruit. This new pear has now spread throughout disturbed roadsides and abandoned fields, and has become a significant problem in the mid-Atlantic. When Callery pears bloom in early spring, the rights-of-way of roadsides throughout the region are draped in white from the dominance of this invasive tree.

Flowering time is also important when grappling with the vexing issue of invasive plants. Miscanthus cultivars, for example, bloom throughout the fall, some in September and some as late as November. Cultivars that bloom in September in Delaware have time to ripen seed, and are thus invasive, but those that bloom in November don't have time to ripen their seed in our climate and rarely become invasive.

SEEDS

A *seed* is a miniature plant in a dormant state of development. The seed is a matured ovule, although various parts of the ovary may be attached to the seed's coat. Most seeds contain a built-in food supply. All seeds require oxygen (to release ATP for growth), water (to activate the cells), and warm temperatures (for the optimal functioning of enzymes). Before they will germinate, many seeds require special conditions or treatments such as a cold period, fluctuating temperatures, light, or the removal of their seed coat. Without the necessary conditions required to begin germination, seeds can outgrow their dormant period, and no longer be viable.

Different kinds of plants can stay alive in the dormant or seed state for different time periods, varying from a few weeks to many years. Proper care of the seed lengthens its life span considerably. Vacuum-packed containers add months to the keeping quality of grass seed. "Package dating" is used by most seed firms to prevent the sale of "old" seed. In contrast, seed racks that are exposed to dampness, heat, and temperature extremes usually have a reduced length of time in which its seeds will remain viable.

Seeds present in the soil are part of what is called a *seed bank*. In an established forest or garden, existing vegetation can prevent those seeds from germinating. But when the site is disturbed, or when vegetation is removed, many of those seeds will germinate. Seed banks in the soil can produce plant communities that have both positive and negative effects on the forest as a whole. For example, changing the management strategy of a site can result in threatened or endangered plant species reestablishing themselves through the natural seed bank. But an undesired result may also happen; for instance, the seeds of Japanese stiltgrass, a problematic invasive species, are incredibly prolific in the soil in many natural areas, and will come up readily if that soil is disturbed.

FRUIT

Fruit surrounds a plant's seeds, and often provides a mechanism of dispersion and/or protection for the seed. Dry fruits (like the samaras on maple and tulip poplar trees) are spread by wind, while fleshy fruits (like blueberries and persimmon) attract birds and other animals, who eat the fruit and spread the seeds in their feces. This is great for our native persimmon, which we would love to spread more widely in natural areas, but it is also the reason you should control invasive porcelain berry or English ivy before they fruit and become attractive to and spread by birds.

Some fruits and seeds—such as those of coconut, willow, and silver birch trees—are buoyant and float, spreading through waterways. Seeds such as burdock, a common weed, can have appendages or sticky coatings that spread by attaching to animal fur. The sticky fruits of mistletoe, a parasitic plant currently spreading throughout Southern Delaware, are attractive to birds, sticking to their beaks and moving with them from one tree to the next.

FACTORS THAT INFLUENCE GROWTH

The life cycle of a seed-bearing plant is divided into two broad phases: vegetative and reproductive. The *vegetative* phase consists of two stages, the *germination* of the seed and *vegetative growth*. Although the germination stage may take only a few days, the period of vegetative growth may last for years, with repeated growth cycles. The *reproductive* phase begins with physiological changes indicating flower bud induction (when newly formed shoots are signaled to develop from resting buds), followed by flower bud initiation (the growth of those buds), followed by further development, and then flowering, leading to the subsequent production of fruit and seed.

Plant growth is influenced by many internal and external factors. Internal factors include the nature of the plant, the plant's hereditary (or genetic) potentialities, and age. Morning glories are by nature twining vines with weak stems, and although their rate of growth may be markedly influenced by variations in their external environment, they always remain weak-stemmed twiners. Some species of trees, such as poplars and willows, grow very rapidly in the presence of favorable external conditions. Pines, white oaks, and sweet gums exposed to similarly advantageous conditions grow much more slowly.

Another important internal growth factor is age. Young cells, tissues, and organs generally grow more rapidly than their older counterparts under similar environmental conditions. Some species have juvenile and mature foliage that function differently. English ivy has lobed leaves in its juvenile phase and, when it becomes mature, leaves that are entire (with no lobes). This is important because juvenile ivy will not flower, fruit, or set seeds, so it doesn't spread widely. If ivy is maintained as a low ground cover, it will never develop mature foliage. But when ivy grows up a building or tree, it develops mature foliage and produces fruit and seed that are widely distributed by birds and becomes invasive. Therefore, it is important to remove English ivy from an area before it climbs trees high enough to become mature and start fruiting.

External growth factors include light, temperature, moisture, oxygen, carbon dioxide, minerals, etc. These factors influence many physiological activities of plants. Plants are impacted, for instance, by the climates in which they live. In temperate climates, deciduous plants enter a winter dormancy period. *Perennial forbs* (herbs other than grass) die back to the ground each fall, maintaining only a root system below ground, and emerge in the spring when the weather warms. *Annuals*, plants that complete their life cycle in one growing season, can be cool-season (germinating in the fall or winter and thriving in cool temperatures) or warm-season (germinating in spring once the soil has warmed sufficiently and thriving throughout the summer until frost). As we discussed with English ivy, understanding a plant's life cycle can also help us control undesirable plants. For example, since Japanese stiltgrass is an annual, it can be controlled by preventing the next crop of seed from emerging, either by mowing off the flowers before they become seeds in late September, or by using a preemergent herbicide in the early spring.

Changes in plant development are often signaled by day length. In a warm fall, you might think you can sow grass seed later into the year and have no trouble with germination, but shorter days signal seeds to avoid germination and prevent death from an early frost.

Plants often have symbiotic relationships with other organisms. Legumes have *root nodules* that allow them to pull (or "fix") nitrogen straight from the atmosphere. *Mycorrhizal* associations between plants and fungi play an important role in plant nutrition. Some plants do not grow as well (or at all) without the presence of these symbiotic mycorrhizae. This has been found to be particularly true with pine trees. Similarly, trillium, a much-loved native forb, is difficult to plant or transplant to a new location because it is dependent on mycorrhizal associations.

Plant hormones also control some forms of growth and development. *Dicots*—flowering plants typically equipped with broad, stalked leaves with netlike veins (like daisies, hawthorns, and oaks) and an embryo that bears two seed leaves—have *auxins*, hormones that inhibit lateral buds from sprouting. When the growing point (apical meristem) is removed, the auxin no longer inhibits lateral growth and buds sprout, causing plants to fill in and become denser. This is problematic when forest edges are mowed with a vertical-boom ax mower because the growing point is removed, causing lateral buds to sprout: where one branch was growing out toward the highway, the following year, four or five branches will grow. Other hormones control stem elongation, fruit ripening, and rooting.

PLANT VARIATIONS AND HEREDITY

As with animals, one of the most striking qualities of living plants is their ability to reproduce plants of the same kind. When a species of plant or animal reproduces, its offspring are always like their parents in their fundamental characteristics. Just as robins inevitably produce more robins, snapdragons always produce seeds that grow up into another generation of snapdragon plants.

Also like animals, even though all plants are fundamentally like their parents, they usually differ from them in certain minor respects. White elms are all unmistakably white elms, yet each tree is an individual living organism that differs in some quality from all other white elms. Variations within a certain plant species are the result of three common factors—environmental modifications, mutations, and combinations.

Plants of the same species often vary from each other because they are subjected to differences in environmental conditions. Plants grown in poor soils do not grow as large or produce as much food as plants that grow in fertile soils. Plants exposed to bright sunlight manufacture more food than those growing in weak light and, therefore, have greater dry weights and larger quantities of stored food. These differences, caused by variations in available moisture, light, soil nutrients, and other environmental factors, are not inherited, meaning they are not transmitted to the offspring of the plants subjected to the varying environmental conditions.

Mutations are sudden, unpredictable genetic changes that occur in organisms and cannot be explained based on environmental modification or crossbreeding. These changes are frequently insignificant, but in some cases are so marked that offspring arising by mutation often seem to be entirely new varieties or species. Mutations occur in plants raised from seeds or develop from individual buds on a stem. In the case of a mutation that develops in a bud, the twig that grows from the bud differs in some manner from all other twigs of the plant.

Many important varieties of cultivated plants have had their origins in mutations. Among these are: navel oranges; many kinds of double-flowered plants; flowers with different colors than the rest of their species; leaves with different fall colors on them; or plants with a smaller than typical overall size. Usually the variations we select for our gardens or landscapes have a desirable ornamental characteristic. But cultural differences can occur from mutation as well. In one Virginia Tech experiment, red maple seedlings from a population adapted to wet conditions did not succumb to flood injury when exposed to excess moisture, but all the red maple seedlings grown from trees in a dry location died when inundated with excess water. This kind of environmental mutation could be extremely useful in the planned landscape, but the nursery industry has typically not been selective for these characteristics in the past.

The third cause of variation, combination, occurs from the breeding of related, though somewhat different, types of organisms. This "crossing" of two varieties of a species or of two closely related species often results in offspring with new characteristics—a *hybrid*.

A great example of plant variation and *plasticity* can be illustrated by the cabbage family. If the original wild cabbage is selected for terminal buds, we can then create the commercially familiar cabbage we know today. Selection for lateral buds produces brussels sprouts. Selection for a strong stem results in kohlrabi. If we select for leaves, we get kale; for stems and flowers, broccoli; and for flower clusters, cauliflower.

SOILS

To care for your landscape, whether it is a garden or a meadow or a forest, you must also understand what your landscape needs to remain healthy. And that requires understanding *soil*.

As you read in the Geology chapter, soils are defined as the upper layer of the Earth. Soil depth can range from a few inches to more than one hundred feet overlaying the rocks known as the Earth's crust. Soil is a heterogeneous material in which chemical, biological, and physical processes are constantly occurring. Soils come from the weathering of inorganic material in the presence of oxygen, as rocks and minerals are slowly broken down by oxidation, freeze-thaw cycles, biological activity, and chemical reactions. But the soil system is also composed of organic matter, air, water, and living microorganisms. A healthy soil system has an appropriate balance of those four components and functions as a key component of an ecological system.

Soil provides a medium for plant growth, serves as a reservoir of nutrients, and allows for gas and water exchange between it and the air. Soil helps to regulate water levels, flows, and discharges, and is a medium to cycle nutrients between plants, animals, and the atmosphere. Healthy soils are the basis for good plant growth and, therefore, it is critical to have healthy soil conditions prior to planting.

ORIGIN OF SOIL

As Tom McKenna has written elsewhere in this volume, soil formation—the development of soil from parent rock—can take thousands of years. Soils vary in their thickness, layers, quantity of organic matter, mineral materials, pore space (the volume of voids that can be filled by water or air), and contained organisms. The inorganic or mineral portion comprises approximately 45–48 percent of the soil system. This mineral component comes from the breakdown or decomposition of weathered rock that may have originally existed in situ or was transported by gravity, wind, water, or glacial action. Soils form from the interplay of five main factors: parent material, time, climate, relief, and organisms. Some soils accumulate to form *sediments*, which occur when weathered rock is transported to a new location. Sediments can be found in lakes, rivers, sand dunes, beaches, and the ocean. If sediments are buried deep enough and undergo pressurized compaction and cementation, they turn into sedimentary rocks.

Soil mineral particles contain elements important to plant nutrition, such as potassium, calcium, sodium, iron, and magnesium, as these elements are bound within the particles crystalline structures and may be released in the soil solution over time. Sand and silt are composed mainly of primary materials such as quartz, feldspar, mica, hornblende, and augite. The clay portion contains secondary materials such as kaolinite, montmorillonite, and illite. During the weathering process, elements bound within the crystalline structure are released, providing nutrients for absorption by plant roots.

SOIL TEXTURE

The term *texture* is used to express the relative size of the mineral particles in soil. It refers to the fineness or coarseness of those particles. The relative proportions of sand, silt, and clay in soil determine the texture of the soil. Soil texture governs many physical and chemical reactions in soils because it determines the amount of available surface area on which reactions can occur. Soil particles range in size from fine (clay <0.002 mm) to medium (silt 0.002 mm–0.05 mm) to large (sand 0.05 mm–2 mm).

Soil textures are named by the predominant particles they contain—for example, sandy soils, clay soils, clay loams, silt loams. A *loam* soil contains all three separates of sand, silt, and clay in approximately equal proportion. Sand particles are coarse in texture and have little effect on soil fertility because they have relatively little surface area for chemical reactions. Sandy soils have good drainage and aeration, but they are also prone to drought and are easily eroded, depending on topography. Silty soils have more moisture-holding capacity than sandy soils and are finer in texture than sandy soils. However, they are prone to crusting and wind erosion. Clay soils are comprised of the finest particle-size class, having a much higher surface area to hold nutrients. However, with their smaller pore spaces, clays tend to hold water and may become waterlogged.

Soil texture affects water movement as well as soil workability. It is implausible to change soil texture. Adding sand to a clay soil will not make it drain better (you will just make concrete!). Just as having a well-rounded education with knowledge in a variety of subjects is good for one's mind, having a good variety of sand, silt, and clay in a given soil texture is best for fertility and water movement.

SOIL STRUCTURE

Soil *structure* is determined by the arrangement of soil particles/textures within the soil. Structure can either be single-grained, such as in sands, or it can be a composite of smaller particles grouped together to form soil *aggregates*, such as blocky structures, or more defined features, such as prisms found in clayey soils. Soil structure affects both the workability of soils and water movement. Structure will vary with soil texture and depth at any given location. Platy structure is typically an indication of compaction, a problem in urban and suburban areas where soils have been (often severely) compacted during the construction process. Heavy pedestrian traffic can compact soil as well, as can over-tillage or tillage when the soil is wet. Compacted soils drain poorly and can impede a plant's formation of roots.

An ideal soil system contains approximately 50 percent soil particles, 25 percent water, and 25 percent air. This ratio is controlled by the size of the

FIGURE 4.2. Common soil aggregate shapes. (Sketch courtesy of Olivia Kirkpatrick.)

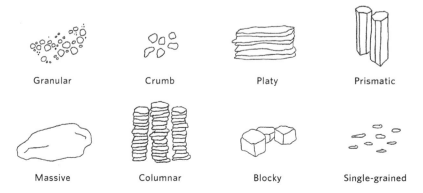

pore spaces in the soil. In heavy clay soils, there are small pore spaces. In light sandy soils, the pore spaces are large. As a result, clay soils typically hold more water than sandy soils and sandy soils typically contain more air-filled pores than water-filled pores. While you can't change soil texture, you can change soil structure to improve drainage or nutrient-holding capacity. *Organic matter* is the "magic sauce" that fills in pore spaces in a sandy soil to hold water and nutrients, and can help glue together clay particles into aggregates to improve drainage.

Organic matter typically accounts for 2 to 5 percent of a given soil, and is derived from decaying plants, animals, animal excrement (manure), and/or living organisms. It supports bacteria and other soil organisms that help provide nutrients and water to plants. Beneficial soil organisms also reduce disease and weed populations by feeding on disease-causing microbes, and provide nutrients (through the soil's composition and chemical properties) that are favorable for plant fertility. Like mineral textures, the various states of decomposition of organic matter have their own textures, typically known (from least to most decomposed) as *fibric*, *hemic*, and *sapric*.

Compost is a great source of organic matter to improve soil structure and provide soil nutrients. Yard and kitchen waste can be used to build compost at home, or compost can be purchased from large municipalities. (Funny story: once, a soils professor-turned-administrator requested assistance on landscape recommendations for a new home in Newark. After a quick look at the property, I recommended he spend his landscape allowance on getting a load of compost to till into the soil before he planted trees, shrubs, or a lawn. He was disappointed and said he had been thinking more along the lines of buying some dogwoods and azaleas. I thought, "Wow, my career is over, I can't even convince a soils professor to spend money on compost!")

Density is a measure of weight per volume of a substance. A pound of cement and a pound of feathers weigh the same, but the volume needed to obtain a pound of each material is vastly different. With soil, the density of its particles plus pore spaces is called bulk density. When the bulk density is greater than 1.6 grams/cubic cm, the soil will impede growth. The bulk density of a forest soil and a pasture is lower than that of a subdivision lawn or athletic field. These heavily trafficked environments make plant growth difficult. Core aeration and adding organic matter can help loosen the soil and improve plant growth.

Porosity or pore space is inversely related to bulk density. Porosity is calculated as a percentage of the soil volume:

- *Bulk density:* oven dry soil ÷ volume of soil
- *Percent soil porosity:* (1 − (bulk density ÷ 2.65)) × 100

Loose, porous soils have lower bulk densities and greater porosities than tightly packed soils. Porosity varies depending on particle size and aggregation.

It is greater in clayey and organic soils than in sandy soils. Aggregation also decreases porosity because more large pores are present, as compared to single clay and silt particles, which are associated with smaller pores.

SOIL MICROBIOLOGY

Living organisms exist in the soil and (in most cases) contribute to plant growth. The organisms found in soil vary from *macrofauna* such as earthworms, ants, millipedes, and termites to *microfauna* such as nematodes, rotifers, ciliates, flagellates, bacteria, viruses, algae, and fungi. A single teaspoon of rich garden soil can hold up to one billion bacteria, several yards of fungal filaments, several thousand protozoa, and scores of nematodes. Some of these organisms break down dead plant and animal material, releasing nutrients for plant uptake. Others fix nitrogen, making it available to plants; combat harmful organisms; burrow tunnels that improve aeration; and/or establish symbiotic relationships, such as fungi that supply a plant with water and nutrients by improving access to pores the roots cannot reach. More than 75 percent of all terrestrial plants form mycorrhizal associations between their roots and soil fungi. The plant roots provide carbon-containing compounds that serve as food for the fungi, and the fungi improve the roots' absorption of phosphorus and other plant nutrients.

SOIL CLASSIFICATION

All the soils in the world can be assigned to one of the twelve soil orders. The twelve orders illustrate properties and stages of soil development, soils of particular regions or climates, and soil chemistry.

In Delaware, since we are on layers of transported and deposited sediments from the weathering of the Appalachian Mountains, the soils here are mostly *ultisols*. In tidal wetlands and in some non-tidal wetlands, *histosols* are present. In some coniferous areas, *spodosols* exist. In and around floodplains, *inceptisols* and *entisols* may be found. In some places where sea level rise is occurring and changing the chemistry of the soils, utlisols are changing to *alfisols*. Recently, soil classification has expanded to include subaqueous (underwater) soils. The Delaware Inland Bays were mapped in 2010 and are one of the few places with a completed and published subaqueous soil survey.

The United States Department of Agriculture (USDA) Natural Resources Conservation Service (NRCS) has distribution maps of dominant soil orders and can be found by googling USDA soils and choosing a state within the Web Soil Survey. Starting before (and largely since) the Dust Bowl of the 1930s, the USDA NRCS has been mapping and classifying the country's soils, primarily to help farmers maintain food production (and thus maintain national food security). These soil maps are updated periodically, and are free and publicly

TABLE 4.1. SOIL ORDERS AND GENERAL DESCRIPTIONS

Type	Description	Type	Description
Entisols	Little, if any horizon development (see below)	Inceptisols	Soils showing the beginning of horizon development
Ardisols	Soils located in arid climates	Mollisols	Soft, grassland soils, thick dark surfaces
Afisols	Deciduous forest soils, basic soils (chemistry)	Spodosols	Acidic, coniferous forest soils
Ultisols	Extensively weathered soils, acidic soils (chemistry)	Oxisols	Extremely weathered, tropical soils
Gelisols	Soils containing permafrost	Histosols	Soils formed in organic material
Andisols	Soil formed in volcanic material	Vertisols	Shrinking and swelling clay soils

available. Soil survey maps serve as a guide and do not supplement a site-specific investigation. If you want to know the characteristics (texture, pH, soluble salts, nutrients, etc.) of the soil at a site, it is best to send a soil sample to a university or private lab for analysis.

SOIL DESCRIPTIONS

Soil exists in the landscape as "*horizons,*" or "layers." A soil scientist describes the soil based upon the arrangement and characteristics of soil layers. The *O horizon* is comprised of organic material, such as decomposing leaves. It is thin in some soils, thick in others, and may not be present at all in some soils. Forests provide such a good environment for diverse plant growth because of their unmatched ability to create organic layers. *Horizon A*, or topsoil, is a mineral surface horizon. In a mineral soil, this horizon typically has the most organic matter content. Topsoil is often removed during construction and may not be replaced, resulting in poor conditions for growth in many built landscape situations. *Horizon E*, the "zone of eluviation," is the zone of leaching where sand and silt particles of quartz or other resistant materials have had clay, minerals, and organic matter removed by the flow of water through the soil. The E horizon is often found in older soils. *Horizon B* is the subsoil and is a zone of accumulation. In Delaware, most B horizons are an accumulation of clays. The B horizon typically shows a soil's development by weathering and transformations over time. The *C horizon* is unconsolidated parent material that overlays bedrock, *Horizon R*. In Delaware's Piedmont, R horizons are present at various depths; however, on the Coastal Plain, R horizons are situated at deep depths below the layered sediments and are rarely if ever described.

FIGURE 4.3. Soil horizons. (Sketch courtesy of Olivia Kirkpatrick.)

COLOR

Soil color is used to identify and describe soil horizons. Color development and distribution are the results of weathering and biogeochemical processes. As rocks containing iron or manganese weather, the elements oxidize. Soils are primarily colored by that iron (yellow or red colors) and by organic matter (brown or black colors).

Soil colors are a signifier of several things, including soil parent material, drainage, a given climate, or soil age. In Delaware and in many other regions, red- or yellow-colored soils are markers of upland soils. Black and grey soils are markers of wetlands or soils with some degree of waterlogged conditions. The seasonal high water table can be identified based upon the depth and changes in soil color. This is because iron is primarily the coloring agent of subsurface horizons. When iron is found in saturated conditions, it becomes colorless and may translocate within the soil profile, creating mottled or blotchy soil coloring, or it may completely wash out.

UPLAND VS. WETLAND SOILS

Wetland soils fall into three categories: soils permanently inundated with water above the soil surface; saturated soils with the water table at or just below the soil surface; and soil where the water table is always below the surface. These protected ecosystems may be specifically defined by their location in the landscape and their tidal influence.

Wetland soils are anaerobic for some part of the year at or near the surface, while upland soils are largely aerobic at the surface. Wetland soils can be distinguished from upland soils by their physical properties, such as their higher organic matter content resulting from slowed decomposition in the absence of oxygen. Upland soils have a good balance of oxygen and water to support upland vegetation ecosystems including woodland, grassland, and shrubland habitats. Both wetland and upland habitats sequester carbon and cycle nutrients, energy, and water.

Wetlands are some of the most biologically productive ecosystems in the world. Yet for many hundreds of years, and in countries throughout the world, wetland soils have been drained or filled to support agriculture. Here in Delaware, tidal marshes were drained in the 1920s in Sussex County, some to grow strawberries.

SOIL CHEMISTRY

Soil *pH* measures the relative acidity or alkalinity of the soil, and is measured on a logarithmic scale of zero to fourteen, with seven considered neutral. The descending numbers are more acid, the ascending numbers are more alkaline or basic. Since the scale is logarithmic, the difference between a pH of four and five is much greater than the difference between six and seven. The pH level affects nutrient availability and microorganism activity. If the pH is too low, some nutrients, like manganese and aluminum, may be present in toxic quantities. If the pH is too high, some nutrients, like iron, may not be available. Most microorganisms, essential for decomposing organic matter, thrive in a pH range of about six to 6.5, which is one of the reasons that most plants thrive in that pH range. Some plants, particularly those in the *Ericaceae* (rhododendron, blueberries, mountain laurel), prefer a slightly acid pH. To increase the pH, gardeners can add lime; to decrease the pH, they can add sulfur.

Saline soils have high levels of soluble salts and usually a high pH. They develop when evaporation is more prevalent than leaching and can be characterized by a crusty white film on the soil surface. When soils have high levels of soluble salts, the plants in them have difficulty absorbing water because the water molecules are held tightly by the salt ions. *Sodic* soils have a very high pH (greater than 8.5) because of an excessive accumulation of sodium without other salts. This condition can disperse soil particles and destroy soil structure.

SOIL FERTILITY AND FERTILIZER

Nutrients are the raw materials taken from the soil by plant roots and used in manufacturing the food and energy needed in the plant growth process. Soils vary in their amount of natural fertility, primarily resulting from the parent material from which they were formed. Usually, clay soils contain more nutrients than sandy soils because nutrients are retained on their greater surface area. However, these nutrients may be bound to the clay surfaces and thus inaccessible to plants. Only soil testing by a reputable laboratory can indicate the actual amounts of nutrients in a soil sample.

Fertilizer is considered any natural or artificial material added to the soil to supply one or more of a plant's nutrients. Currently, there are sixteen known chemical elements essential for the optimum growth of plants. These elements are generally present in the form of chemical compounds in both soils and fertilizers. The compounds are divided into ions in the soil solution through various chemical reactions. The *soil solution* consists of the water contained in the soil and the chemicals dissolved in it. Of the sixteen essential elements, carbon (C), hydrogen (H), and oxygen (O) are obtained from air and water. The other thirteen are absorbed by plant roots from the soil solution. Nutrients are absorbed by the plant roots along with the solution. They move upward in the xylem tissue to the leaves and other plant parts.

It is seldom necessary to be concerned about the supply of oxygen, hydrogen, or carbon, even though large amounts of each are used in plant growth and development. The air around us usually provides all the oxygen and carbon needed by plants. Water supplies plants with both hydrogen and

TABLE 4.2. ESSENTIAL GROWTH ELEMENTS ABSORBED FROM SOIL

Primary Elements	Symbol	Micronutrients	Symbol
Nitrogen	N	Boron	B
Phosphorus	P	Manganese	Mn
Potassium	K	Copper	Cu
		Zinc	Z
Secondary Elements	**Symbol**	Iron	Fe
Calcium	Ca	Molybdenum	Mo
Magnesium	Mg	Chlorine	Cl
Sulfur	S		

oxygen, as well. But elements such as sodium that are detrimental to plant growth may also be contained in water, so a chemical analysis is often necessary to ensure a safe water source.

PRIMARY ELEMENTS

The primary elements for plant growth are nitrogen, phosphorus, and potassium (N, P, and K). You may also hear these referred to as the primary *macronutrients*. When purchasing fertilizer, these are the primary elements sold and labeled on products.

Nitrogen is important for good vegetative growth, and produces the dark green color in leaves. But nitrogen can also be used in excess; typically, more nitrogen is present than a plant can use, but excess nitrogen can cause plants to stay vegetative and reduce fruiting. Too much nitrogen in soil can produce soft succulent growth that is more susceptible to disease or insect damage.

Nitrogen is a negatively charged particle, and in Delaware's acidic soils it can be easily lost due to the deficit of cations (positively charged ions) to which nitrogen can adhere. Increased plant uptake of nitrogen is favored by coarser soil textures, cooler temperatures, and limited irrigation. Well-established root systems and high water infiltration capacity limit soils' nitrogen loss. When nitrogen leaches out of the soil system and makes its way to groundwater or to surface waters, it causes pollution by increasing algae growth and reducing the oxygen present in the water for aquatic animals (*eutrophication*). Babies who drink water too high in nitrates can become seriously ill with shortness of breath and blue-tinted skin, known as "blue baby syndrome."

Many people get confused by the difference between organic and synthetic nitrogen. Nitrogen is simply a mineral element. The real issue is whether it is quickly released by its source, as in soluble nitrogen, or slowly released, as in a sulfur-coated urea (a common type of fertilizer). There are organic products that contain readily available nitrogen, like raw poultry manure, and synthetic sources of nitrogen that are designed to release it more slowly. Many composted manures and other organic sources of nitrogen are considered "slow-release" because they must be decomposed by microorganisms for that release to take place. For most gardening projects, it is best to apply small amounts of soluble nitrogen—no more than one pound of nitrogen per thousand square feet of land—or to use a slow-release source of nitrogen.

Phosphorus is a primary element that stimulates root formation and gives plants a rapid and vigorous start. It can also stimulate flowering and improve winter hardiness. Unlike nitrogen, phosphorus is a positively charged particle, and therefore adheres to Delaware's primarily acidic, negatively charged soils. Phosphorus is usually lost by erosion or organic debris that gets washed in surface water. The best way to prevent phosphorus loss from a landscape is to minimize bare soil, which can erode, and to collect and compost organic debris.

Always test the soil prior to adding phosphorus. If there is plenty of phosphorus already in the soil, adding more will not improve plant growth. In fact, high-phosphorus soils (often resulting from the high phosphorus content in poultry manure and decades of application) are a major environmental issue in Delaware, with many agricultural fields already at their maximum limit and not able to receive more phosphorus due to nutrient management laws. Excess phosphorus flows off of land and into rivers and bays, further contributing to algae blooms and low-oxygen (or hypoxic) dead zones.

Potassium is the third primary element required by plants. It imparts vigor and disease resistance, produces strong, stiff stalks, and is essential for the formation of starches and sugars. Like phosphorus, potassium is positively charged and rarely moves in the soil.

SECONDARY ELEMENTS

Calcium, magnesium, and sulfur (Ca, Mg, and S) are considered secondary plant macronutrients, not because they are less important, but because they are required in smaller quantities. Calcium is supplied primarily in the form of limestone, and is involved in holding together a plant's cell walls. Ground limestone is very fine, with a large surface area, and it reacts quickly in the soil. Adding it to soil is the cheapest and best way to increase soil pH, but its powdery consistency makes it hard to work with. Granular limestone is easier to handle because the particles are larger, but it doesn't have as much surface area, so it reacts slowly with the soil. Pelletized lime is ground limestone that has been formed into pellets. It releases quickly and is easier to handle but is more expensive than ground limestone.

Magnesium is present in chlorophyll, the green pigment in plants. It acts as a catalyst, or activator, for certain plant processes. Magnesium can be supplied as dolomitic limestone and is present in some mixed fertilizers. It also can be supplied as magnesium sulfate.

Sulfur is present in certain plant proteins. It affects cell division and formation. It can be supplied as elemental sulfur and is rarely deficient in plants or in soil. It is usually added to soil to lower the pH.

Micronutrients are required in such small quantities by plants that the soil almost always contains enough of them for adequate plant growth. However, if the soil pH is too low, micronutrients may be present in toxic quantities, and if the pH is too high they may be unavailable altogether. Micronutrients must be added to soilless media when plants are grown in containers without actual soil. Micronutrients are added to bark mixes used to grow nursery crops, for example.

SOIL TESTING

A soil's pH, as well as the relative amounts of its primary and secondary nutrients, can be determined with a reliable soil test made from a representative soil sample. The University of Delaware soil testing lab offers soil testing at a modest price, with soil test kits available from county extension offices and the Plant and Soil Sciences office. Other soil labs are also available in the state.

Each distinct area within the landscape where crops or plants may be grown and managed should be tested. Soil test results will determine management decisions about nutrient applications. To sample, take soil from ten random locations in each area and mix the samples to generate a composite soil sample. Sample soils from at least four inches in depth and avoid incorporation of surface litter. A basic soil test will provide the pH and P, K, Ca, Mg, and S levels in the samples. An additional soil test for soluble salts is used primarily for greenhouse, nursery, and home garden soils where very high application rates of fertilizer may lead to a buildup of soluble salts. Established plants, such as ornamental shrubs, will wilt and show symptoms typical of drought in high-salt conditions. Such conditions are often found along roadsides or in parking lot medians when salt has been used for ice and snow melting over the winter.

We normally think rich, fertile soils are best for plant growth, but when trying to establish a meadow, for example, infertile soils will favor the native meadow plants and reduce weed competition. If you are lucky enough to have the opportunity to establish a planting on a former parking lot, you are likely to be faced with highly compacted soils, requiring significant inputs of compost to improve soil structure and fertility before planting.

The role of organic matter in creating and maintaining a healthy soil system cannot be overstressed. In addition to improving the soil structure of

FIGURE 4.4. Nutrient availability at varying pH levels. (Sketch courtesy of Olivia Kirkpatrick.)

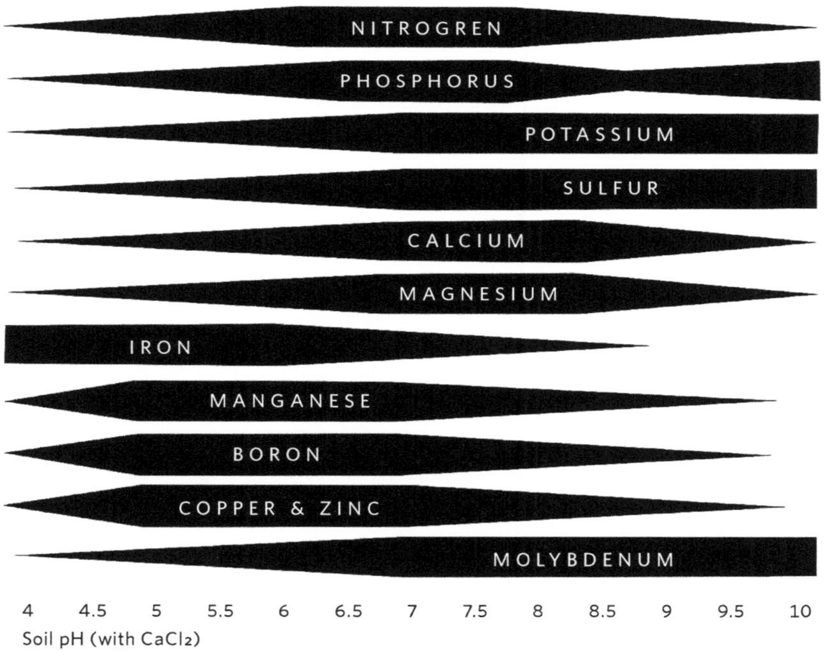

both sandy and clayey soils, when organic matter is added, biological activity and diversity is increased. That results in reduced soilborne diseases and parasitic nematodes. It also results in decomposition, which detoxifies harmful substances, releases nutrients, provides *humus* (the substance formed when leaves and other plant material are decomposed by microorganisms), and increases aggregation. Increased aggregation improves soil's pore structure, resulting in improved *tilth* (the physical condition of the soil) and water storage. All this leads to healthy plants.

WATER

Plants are made up primarily of water: 70 to 85 percent of the weight of most herbaceous plants is water, while trees may have between 45 and 50 percent fresh weight in water. Water broken down by chemical reactions becomes a part of the sugars, starches, and proteins found in plant sap and cell tissues. Minerals are dissolved in water and transported from the soil to the plant. When absorption of water by the roots does not occur as quickly as does transpiration, the plant wilts.

Water dynamics are an important part of the soil system. When it rains, water fills the porous space in the soil and air is displaced. Without enough air in the soil, gas exchange slows, CO_2 accumulates, O_2 decreases, and there is reduced root activity and shifts in microorganisms. Gravity moves water down and when the largest pores in the soil are drained the soil is at what is called *field capacity*. Field capacity is defined as the water retained from freely drained soil twenty-four hours after a thorough wetting. At field capacity, water is held in small pores and thin films around fine soil particles. Water is lost from this system as plants take up water and transpire through evaporation or when films around the soil particles become so thin that they are held too tightly for plant uptake. Only about 75 percent of water held in soils is available to plants.

When plants experience drought, they respond by wilting or leaf rolling to conserve as much water as possible. Water stress symptoms in plants include: leaf scorch, the death of leaf tissue at the edges of leaves; interveinal chlorosis, the yellowing of leaf tissue in between leaf veins; midsummer leaf drop; dead leaves remaining on the tree or shrub; crown decline, the death of branches in the top part of the tree; and suckering, the development of vigorous shoots with leaves coming from the roots or plant crown that are larger than normal for the species. The best way to prevent drought damage to plants is to select plants well-adapted to their site conditions. But, since Delaware is a *mesic* environment, normally having a well-balanced supply of moisture, we can have extremely wet conditions some years and dry conditions others.

Plants growing in the landscape may require supplemental water, but it is important to remember that overwatering can be just as detrimental as a *lack* of water. When plants are overwatered, roots die, and the plants are no longer able to take up water. The effect is wilting, which may be confused with having too little water. Always feel the soil to determine its moisture content before watering. Deep, thorough, infrequent watering encourages the development of deep root systems, and is much better for plants than frequent, shallow watering. Compacted clay soil may take up water slowly, so water will run off of it without penetrating. It is important to apply a small amount of water to the soil, let it soak in, and then apply more water until you reach field capacity.

Understanding how plants grow, the soil that supports and nourishes them, and the water necessary for plant growth and development will help you manage plants in landscape situations. It will also help you manipulate natural settings to allow for normal biological processes to occur that will improve the health of the whole ecosystem.

SOIL SCIENCE RESOURCES FOR DELAWARE NATURALISTS

TEXTBOOKS

- Ray R. Weil and Nyle C. Brady, *The Nature and Properties of Soils*, 15th ed. (Harlow, UK: Pearson Education, 2016)
- Stanley W. Buol, Randal J. Southard, Robert C. Graham, Paul A. McDaniel, *Soil Genesis, Morphology, and Classification*, 6th ed. (Hoboken, NJ: Wiley-Blackwell, 2011)
- Delaware National Resources Conservation Service (NRCS): https://www.nrcs.usda.gov/wps/portal/nrcs/detail/de/soils/?cid=nrcs144p2_024890
- Mid-Atlantic Nutrient Management Handbook: https://www.pubs.ext.vt.edu/CSES/CSES-122/CSES-122.html
- P. J. Schoeneberger, D. A. Wysocki, E. C. Benham, and Soil Survey Staff, *Field Book for Describing and Sampling Soils, Version 3.0* (Lincoln, NE: Natural Resources Conservation Service, National Soil Survey Center, 2012). Available as an online PDF at https://www.nrcs.usda.gov/wps/portal/nrcs/detail/soils/research/guide/?cid=nrcs142p2_054184
- Soil Survey Staff, *Soil Taxonomy: A Basic System of Soil Classification for Making and Interpreting Soil Surveys*, 2nd ed., U.S. Department of Agriculture Handbook 436 (Lincoln, NE: Natural Resources Conservation Service, 1999). Available as an online PDF at https://www.nrcs.usda.gov/wps/portal/nrcs/main/soils/survey/class/taxonomy/

WETLANDS

- Atlantic and Gulf Coastal Plain Region and Eastern Mountains and Piedmont supplements. Manuals for identification of wetlands, includes soil indicators at https://www.usace.army.mil/Missions/Civil-Works/Regulatory-Program-and-Permits/reg_supp/
- A Guide to Hydric Soils in the Mid-Atlantic Region: https://www.nrcs.usda.gov/Internet/FSE_DOCUMENTS/nrcs142p2_052291.pdf

WEB SOIL SURVEY

- Review, search a specific area, download soil survey reports at: https://websoilsurvey.sc.egov.usda.gov/App/HomePage.htm

SOIL SERIES DESCRIPTIONS

- Search and review approximately 20,000 different soil series descriptions at https://www.nrcs.usda.gov/wps/portal/nrcs/detail/soils/survey/geo/?cid=nrcs142p2_053587

GOOGLE EARTH SOIL LAYER ADD-IN IN "SOIL WEB"

- https://casoilresource.lawr.ucdavis.edu/soilweb-apps

NATIONAL WETLANDS INVENTORY

- https://www.fws.gov/wetlands/

GOOGLE EARTH NATIONAL WETLANDS INVENTORY LAYER ADD-IN

- https://www.fws.gov/wetlands/data/google-earth.html

STATE OF DELAWARE WETLANDS AND SUBAQUEOUS LANDS

- https://dnrec.alpha.delaware.gov/water/wetlands-subaqueous/

MID-ATLANTIC ASSOCIATION OF PROFESSIONAL SOIL SCIENTISTS

- http://www.midatlanticsoilscientists.org/

CHAPTER FIVE

Field Sketching and Nature Photography

SUSAN BARTON, JULES BRUCK, AND JON COX

INTRODUCTION

"Field sketching is more than simply planning for a future work or keeping your drawing hand in practice. It's more, even, than learning to see. It is learning, period."
—Cathy Johnson, *The Sierra Club Guide to Sketching in Nature*[1]

"A great photograph is a full expression of what one feels about what is being photographed in the deepest sense, and is, thereby, a true expression of what one feels about life in its entirety."
—Ansel Adams, "A Personal Credo"[2]

There are many reasons why Master Naturalists will want to develop skills in sketching and photography. Both practices offer powerful tools you can use to learn about and understand the natural world. Even for the beginner, taking good photographs and drawing accurate sketches are possible with simple tools—a pencil, paper, and a digital camera—and both are rewarding personally. Both mediums provide the means to more fully see an object, and therefore to document information that builds understanding. With practice, your ability to "see" the subject you wish to represent will grow and develop, allowing you to make good decisions about the most effective way to record your observations.

When you sketch from observation or compose a photograph, you form a relationship with an object or scene. What you see also affects how you think, so this relationship between sight and thought provides each of us with unique ways to draw and think creatively.[3] By observing intently enough to sketch an object or compose a good photograph of it, you will learn more about the object, and ultimately about the environment in which it is situated.

Throughout this chapter, we will emphasize field observation and ways to learn from the natural world. You will learn to use creative tools of both sketching and photography to document what you encounter outside. Sometimes a quick photo or sketch is all you need to capture the essence of what you are seeing, perhaps to share it with someone else or to simply identify the subject. Other times, you may want to use these mediums as creative tools to communicate a vision you have to a larger audience. Either way, you will find similarities between the two mediums. We will discuss ways to position yourself to take advantage of the direction of light, to analyze the color of a scene, and to compose a perspective that offers your viewer a unique way of thinking about the beauty of the natural world.

There are infinite ways to capture a subject, so we will start with the basics and remind you that building skills requires *practice*: the more you sketch or work on photography, the more you will improve. Naturally, you will want to improve your skills to produce attractive results, but don't let your concern for perfection limit you. Being critical of your own work, or your abilities generally, will stand in your way of having fun and learning. As you engage in mastering these skills, you will learn more and more about the landscape around you.

WHAT'S IN A MASTER NATURALIST'S DAYPACK?

For starters, you will want a field sketching notebook. Choosing a field notebook is fun! Some have colorful covers and pockets to place found objects. Select a notebook with unruled pages that is both sturdy and comfortable for you to carry. Balance notebook size with convenience. We have found that a five-by-seven-inch or larger book is best to capture landscape scenes; however, if a smaller notebook means that you can carry it in your back pocket and have it with you at all times, that may be a more important decision. Select a notebook that is comfortable, practical, and makes you happy. This will be the first of many you will have, so if you are not satisfied, you can always try another type. These days, many people sketch on a digital tablet. All the exercises in this chapter will help you learn to sketch no matter the surface you are using.

A simple case or bag will suffice for the rest of your supplies. Find something to hold a range of pencils, some erasers, a small ruler, sharpener, pocket knife, sunscreen, insect repellent, something to sit on (a small ground cloth or collapsible stool), knee pads, water, and some snacks. You will want a range of pencils, so purchase a sketching kit that includes 4B to 4H pencils. Soft (B) pencils are easy to use for quick sketching and composition. Hard (H) pencils allow you to add fine detail. HB, commonly known as the No. 2 pencil, is a good general-use pencil for starting a sketch.

Carry a variety of erasers; these are easy to find. You can use a white vinyl eraser or a pink pearl for a good clean job. The pink pearl is angled so it provides a sharper tip to erase in tight areas. A six- or eight-inch ruler is a great field tool. Use it to draw straight lines and to measure when you want to accurately explain an object in your field notes. You will also need a pocket knife or small pencil sharpener.

Your cell phone may be all you need to capture some amazing subjects in nature. However, if you find your cell phone no longer records images the way you envision, you may want to purchase additional lenses for your cell phone or even consider a mirrorless or digital single-lens reflex (SLR) camera with a macro lens. But remember: it is always the person behind a camera who makes a great image, not the specific camera itself.

Some additional tools that are optional but may come in handy include an erasing shield, watercolors, and black ink pens. An *erasing shield* is a thin metal object with perforations of different sizes. You can place it over a portion of a drawing you want to erase to eliminate mistaken lines without removing the part of the sketch you want to keep. To add vibrancy to your sketch, bring along some watercolor paints. Watercolor kits for field sketching are small and convenient. Watercolors can help you compose a landscape scene quickly by allowing you to lay down a wash of color above and below the horizon line. We have discussed pencils, but if you want to try another medium, waterproof black ink pens can make a wonderful addition to your sketch practice and they work well with color mediums. As people get more comfortable sketching, they often switch to sketching in ink.

A DAY IN THE FIELD

Before heading out to sketch or photograph, let someone know where you are going and the approximate time you plan to spend there. Learn the opening and closing hours of the park(s) you plan to visit and map out a plan for your day.

When planning for field photography, it is a good idea to research what you might see at that time of year and pack your bag accordingly. What camera gear will work best on the subjects you may encounter? When the oak leaves are the size of a squirrel's ear, it is the right time to search for morels. When photographing these low-growing mushrooms on the forest floor, you may want to bring a ground cloth, so you can get down and personal with your subject. On the other hand, if you are photographing giant silk moth caterpillars, you will want to wait until you hear the dog-day cicadas during the hot and humid days of August. Since silk moth caterpillars feed on the leaves of shrubs and trees, you would probably want to consider bringing a tripod along, so that you could fine-tune your composition while viewing your subject through the branches.

Interspersed throughout the chapter are sketching and photography exercises to stimulate the flow between and help you become comfortable with the two mediums. So grab your supplies, including your camera/lenses, water, and snacks, and let's start with a warm-up.

BEFORE YOU HEAD OUTSIDE—WARM UP!

Just like stretching before exercise, we warm up before sketching. Many objects are formed using a combination of basic familiar shapes—circles and squares, or triangles, as well as curving organic forms. Start your warm-up by drawing a series of these basic shapes. Using a quarter of one blank page, practice drawing circles. Use your HB pencil to start but take the opportunity to try out your full range of pencils to see how they feel and what type of lines they create. Overlap the circles on your paper to save space, and make groupings of large and small circles. Concentrate, and try to start and stop at the same point. In other words, close your circles. Do your circles look circular or like an egg? Practice drawing circles until they become more and more round.

Next, create a series of squares and triangles. Interlock large and small ones and concentrate on closing each of their corners with a slight overlap rather than leaving a gap. Do your squares have parallel vertical and horizontal lines? Practice until you begin to draw straight parallel lines. Finally, try spirals. First, visualize a spiral. We will practice spiraling out from the center and then into the center. Place your pencil tip down on the paper and begin to move it outward in a circular motion toward the left. Next, move from the center out to the right. Then, start a spiral from the outside moving into the center. Do this toward the right and toward the left. Do your spirals form neat concentric lines, or do they cross at times? Practice spirals until you can make them with concentric lines.

Now, let's get expressive with lines. Think for a minute about what a happy line might look like and then represent a happy line on your paper. Next, draw a sad line and an angry line. Different types of emotional lines will come in handy as you try to capture the essence of an object. Remember, the varied pencils you have and use them to get as expressive as possible during this warm-up exercise.

Use these warm-up exercises each time you start to sketch and eventually they will become a natural divider in your field notebook between different days spent outside sketching.

TRANSLATING 3D TO 2D

We observe the world in three dimensions (3D), but must represent it on paper or a digital screen in two dimensions (2D). The transformation from 3D to 2D can be challenging for people new to sketching and photography. One trick is to close one eye while you look at an object you want to sketch. Doing so will flatten the image by taking away depth perception. Try to think about the scene you are seeing as a photograph (a 2D object) and represent it on paper as if you were copying a photo. To sketch what you see, rather than what your brain thinks the object you are seeing is *supposed* to look like, try to break a subject down into more recognizable forms. Analyze the lines that create the outline of the object and the shapes that are created between it and another object. In doing this, we try to trick our brain to ignore our preconceived notions about the object and truly represent what we are seeing.

One way to get the hang of representing objects in 2D is to copy them from a 2D image such as a photo or a sketch from a book. Doing so, you will start to learn some tricks for how to represent three-dimensional items two-dimensionally. When you copy sketches from a book, don't think about the object you are sketching at first. Instead, just copy the lines exactly the way you see them in the image. You might be surprised as you get further into the process that your sketch looks remarkably like the one in the book.

Once you've mastered a sketch from a book, keep practicing! Perhaps your next copied sketch will be a bit more complicated or challenging. You can find good images online or at a local art museum. Remember to copy the lines and forms exactly as you see them without thinking about the subject. If you are having trouble disassociating the subject from its representation as lines or forms, try this fun exercise.

EXERCISE: COPYING AN IMAGE

Print or find a simple image to copy. Before you get started, turn the image upside down. Once it is upside down, it is no longer a picture of a recognizable object, rather it is just a bunch of lines and forms. In other words, this eliminates the meaning of the object. Carefully copy all the lines from the upside-down image to your sketchbook. It might help to tape the image onto one page (upside down) so it is secure while you sketch on the opposite page.

Figure 5.1. Copy this image of a dandelion upside down. When it is flipped you will see how close your sketch looks to the original drawing. (Sketch courtesy of Claire Ciccarone.)

This can be a tedious process if it is a complicated scene, but you will be surprised by how accurate your representation is once you flip the page right-side up. In completing this exercise, you will have accomplished the difficult task of drawing what you see instead of what you think you see, and started the hard work of seeing objects you wish to represent as just lines and forms. As you advance in your sketching practice, continue to refrain from thinking in terms of what the image is *representing*—for example, a person or a building—and instead, draw the lines and forms exactly as you *see* them. Once you learn these skills, you can apply these lessons to sketching something you are viewing in real time.

EXERCISE: EMULATING A PROFESSIONAL NATURE PHOTOGRAPHER

One of the best ways to learn about what makes a successful photograph is to observe the work of professional photographers. Try this exercise: research the work of a notable nature photographer, and evaluate their style. While working in the "style" of the selected photographer, create three images that are similar to their work. You may select any professional nature photographer you find interesting. From traditional black-and-white photographers like Clyde Butcher to digital color photographers like Shannon Wild, there is a wealth of talented individuals to explore and emulate. This is often a good place to begin when you want to find direction in your own work. Your images should look like the chosen photographer might have created them, not be an attempt to exactly reproduce their images. Eventually, as you continue your photographic journey, your work will expand and evolve into your own style.

PERSPECTIVE

Perspective has a variety of definitions. It is a point of view. It is also the way a person photographs subjects and draws objects on a two-dimensional surface to give an accurate impression of their height, width, depth, and position in relation to one another. Notice how, as your location changes, your perspective (impression of size and position) of the objects you are studying also changes. If you look into the distance straight down a street toward its end, you will notice all lines converge at a distant *vanishing point*. The lines of the built objects like the road and buildings and even the tops and bottoms of trees all point toward that distant point, but the vertical lines of the doors and windows of the buildings, and the tree trunks, remain vertical. Also, the objects closer to you appear larger and the objects further away appear smaller. You see more detail on closer objects and less detail on distant objects. Finally, the objects in the distance lose vibrancy and may even appear dull, as if you are viewing them through a light mist. Objects in the foreground (up close to the viewer) retain vibrancy in line, color, and texture.

An awareness of these rules will help you to create a more accurate sketch and photograph.

If you come across an outdoor scene and all the lines of the buildings and roads are not converging to a single point, then you may be looking at a two-point perspective scene. This is the case if you are looking at the corner of a building and can see down a street or pathway in two directions. If that is what you see, observe where the lines are converging to the right and left of the scene along two different vanishing points, and notice that all the verticals in the scene still remain vertical.

Now consider perspective as it refers to point of view. You can look at an object from above (a high perspective) or below (a low perspective) or simply head-on. A change of point of view to high or low creates a more dramatic scene than one that is head-on. It also makes for a more complicated sketch, so it may be best as a beginner to use high and low perspective to compose a photo rather than to sketch. Either way, it is good practice to recognize and try different perspective compositions.

EXERCISE: RECORDING PERSPECTIVE

Take a walk outside. As you walk, alternate bringing your attention to an element close by, and to something far away in the distance. Do this several times before selecting an object close by to focus on for a while. You will then be able to compose a photo of the object from a more interesting perspective. Consider how you could compose a photograph from a high or low point of view to capture a dramatic image. Think of this in terms of a worm's-eye view (low point) looking up at your subject or a bird's-eye view (high point) looking down at your subject.

Once you complete that exercise, look into the distance. Find a scene that you would like to sketch. It should have at least one building and a road or pathway. Look for the vanishing point along the horizon line (the place where the sky and ground meet). Do all the lines converge at one vanishing point or are there multiple points? (Hint, if you are looking down a road, it should be one). Practice sketching the scene, remembering to focus all lines on the vanishing point except the vertical lines (which always remain vertical). Also, look to see, and then represent, objects that are closer to you as larger and distant objects as smaller (compare where they are starting and stopping to the horizon line). Lastly, think about the value of the colors. Are objects in the distance more or less vibrant? How can you soften the look of distant colors? Make a note in your field notebook about all your observations. Once again, find an object close by and compose a photograph from the opposite point of view of your last attempt. Finally, complete one more sketch of a distant view. Challenge yourself to find a different perspective. So, if you looked down a pathway, look at the corner of a building and try to create a two-point perspective.

RELATIVE SCALE

When you draw a scene, it is important to have all the objects the right size in relation to one another. Taking care to draw the scene with closer objects being larger and more distant objects being smaller will make the scene look accurate.

To determine relative scale, you can use a simple object you should always have with you: your pencil. Hold the pencil at arm's length so you know you are always measuring from the same distance. Use your pencil to determine the length or height of an object. It helps to close one eye when using your pencil to measure. Depending on the size of the object and the size of the paper in your sketchbook, you may be able to represent your object at the exact proportion of your pencil, or you may need to draw everything in half-pencil or quarter-pencil lengths. Measure the full extent of the scene you plan to sketch and see which scale will fit on your paper, then consistently use that scale throughout the drawing. For example (at arm's length while closing one eye), is a bench one pencil long? Is the height of the bench equal to a quarter of your pencil? If all you are sketching is the bench and the pencil is smaller than the size of the page, then you can measure in multiple pencil lengths.

EXERCISE: RELATIVE SCALE SKETCH

To begin, draw a horizontal and a vertical guideline lightly through the center of your paper. To test your ability to draw what you see, try drawing a simple object. Holding your pencil at arm's length and squinting, measure the relative distances as you proceed. Notice how each part of the object you are sketching relates in terms of scale to the other parts.

FIGURE 5.2. Pavilion sketched by using a pencil to determine relative scale. (Sketch courtesy of Claire Ciccarone.)

Next, you are ready to tackle a more complicated scene. You view the scene in three dimensions, but you must represent it in two dimensions in your sketchbook. If you sketch a pavilion (like Figure 5.2), you will notice the posts closer to you are longer than the posts further away. By getting the relative heights of the posts correct, you will be able to accurately sketch the pavilion and make it look three dimensional.

VIEWFINDER

Renaissance artists like Claude Lorrain used instruments to frame their view. A *viewfinder* is a simple device that allows you to isolate or "crop" a scene within a rectangular area. You can adjust the viewfinder back and forth, left and right, and up and down, looking for the most dramatic and engaging composition. The viewfinder helps to organize the composition and represent perspective and can be used in both photography and sketching. Walk around your subject, change your angle, perspective, and distance until you find a compelling composition, and then use your tool of choice to capture the scene.

Before you pick up a pencil or a camera, you can use an empty slide frame (for those of you who remember slides) or a four-by-six-inch picture frame mat to compose your next scene. Any rectangle will work, but an old slide is a handy size to frame your view. Push the slide film out of the holder. To make the viewfinder even more useful, glue a string across the middle of the viewfinder horizontally and another string through the vertical middle. If that is too difficult, you can simply draw a mark at the center point on the top and bottom of the slide or mat and two more marks at the center point on the right and left sections. This will help you make sure that you are locating the objects properly in the scene.

EXERCISE: VIEWFINDER SKETCH

Locate an empty slide frame or make a four-by-six-inch frame out of stiff paper. Mark the center points along the horizontal and vertical edges of your viewfinder. Draw a rectangle the same shape as your viewfinder on a page in your sketchbook. Outline the main shapes of the objects you see through the viewfinder. Use the relative scale technique you just learned with your pencil to get the sizes and relative locations of the objects correct. Once the objects are on paper, use a light line to map the shadows. Fill in the shadows with tone to show high and low lights of the major forms in your composition.

FUNDAMENTALS OF LIGHT

Before we get much farther, let's discuss the fundamentals of light, including how the quality and direction of light affects a subject. As you study, start by looking at the quality of the natural sunlight. When you start out in sketching or photography, it is crucial to understand how natural light affects a subject. In this section, a variety of types of light are discussed, including hard, diffused, frontal, back, top, and side light.

To more easily explain light quality, we can break the subject into two main categories: *hard* and *soft* (or *diffused*) light. Hard light is light directly hitting a subject from an unfiltered source, often found on a bright sunny day. It helps give the subject shape and form and often exaggerates the texture. Diffused light, on the other hand, can be seen on a cloudy, rainy, or snowy day, and also in partial or total shade. Clouds act as a giant diffuser and reduce or even eliminate shadows. When viewing a subject in diffused light, the viewer often has trouble knowing where the direction of the light is coming from. The soft light given off by a cloudy day is perfect for capturing soft textures and subtle details.

When you're in the field looking at a subject to photograph or sketch, ask yourself why you are drawn to the subject. What qualities does your subject possess? Once you consider these questions, you will be able to decide what the best type of light to capture or exaggerate those qualities is. For example, if you are photographing or sketching a pink lady's slipper orchid, you might consider using a diffused light source to highlight the soft and delicate texture of the flower. Even though you are drawn to a particular subject, make sure you look around and see what possibilities—angles, light qualities, or even other similar subjects—you can shoot or draw.

Frontal lighting occurs when the light source is in front of the subject. This means that the light is positioned to the artist's back. Frontal light exaggerates the color of a subject and reduces or eliminates the texture, shape, and form. If you're drawn to the color of a scene, then frontal lighting may be just the right type of light to capture your subject. Remember, to capture a frontal lighting situation, the sun will be at your back.

The opposite of frontal lighting is *backlighting*. Backlighting occurs when your light source is directly behind the subject. There are several effects that may occur when using backlighting. Your subject may be silhouetted, or appear translucent, or you may see an overexposed rim of light around the subject's edge. Some scenes may even contain several of these characteristics. Backlighting is one of the best types of light to create a dramatic effect in your final work. Look for this spectacular type of light right after sunrise and right before sunset.

Toplighting occurs during the time of the day when the sun is at its highest peak, around 11 A.M. to 2 P.M. Toplighting is usually not considered the best type of light because it is often harsh, providing too much contrast and creating very dark shadows and very bright highlights. However, if you take your time, you can often find the right perspective to make toplighting work for you. For example, toplighting often occurs at the perfect time of day when insects are most active, and works well when photographing butterflies sipping nectar from flowers.

Sidelighting occurs when the light source is coming from either the right or left side of the subject. This is one of the richest types of light to give a subject shape and form, and to accentuate texture. Sidelighting creates a combination of shadows and highlights that give a subject a three-dimensional look.

Whether you are getting ready to sketch or to compose a photograph, always look at your subject to analyze how the light is affecting its appearance. The goal is to capitalize on the direction and quality of light to create just the right representation of the subject.

EXERCISE: CAPTURING LIGHTING TYPES

Start to become aware of the different types of light. Use your camera to capture each of the light types discussed in the paragraphs above. For each one, practice composing a photograph and then take the time to write in your field notebook how the quality and direction of light affects your subject. Afterward, attempt a quick sketch of the subject highlighting the different light types. You can sketch directly from the photograph or from the scene in the field. Which type of light produces the most dramatic sketch, or provides the mood or atmosphere you are trying to capture? As you study light in photographs, you may start to see strong contrasts. A scene captured under hard lighting conditions will most likely have darker shadows and brighter highlights than a scene captured under soft lighting conditions. You can practice representing the contrasts between light and dark to create realistic sketches.

TONE

Tone is the representation of light and dark values. Using pencils, you can represent a wide range of values to represent dark and light tones that will help make your sketches read more realistically.

To start with tone, we will create a *tone line*. Use a hard pencil and your ruler to draw a rectangle on your page that is one inch high by six inches wide. Create equally spaced vertical lines to break the rectangle into eight equal parts. Directly above the rectangle, label the first box zero. Label the rest of the boxes in order one through seven. Next, select a sharpened soft pencil, like a 2B or 4B, and use the side of it to move back and forth with even, very light pressure to represent a light value in the box labeled one. This value should be just one time darker than the previous value, zero.

Continue as you work down the line. One is one time darker than zero, but seven is seven times darker than zero. Seven will be the darkest value on this line. You will use dark values to represent dark colors (hues), including the parts of your sketch that you see as black, and you will also use it to form shadows. It is good practice to assign a value to each part of your sketch before you get started. This value mapping allows you to accurately portray the darkest parts of your drawing as sixes and sevens, the mid-tones as threes to fives, and your lightest parts as zeros through twos. White or off-white portions of a scene can be represented by the absence of lead or by zero on your tone line.

Create several different tone lines for practice. Try another one that you create by hatching or crosshatching using a harder pencil. Represent lighter tones with hatch lines further apart and dark tones with hatch lines close together. No matter what technique you use, the idea is to match the value of the light you see in an object and its surroundings with a box on your value line to get the most accurate representation.

EXERCISE: TONE MAPPING

After making a tone line, try working on an object out in nature to practice mapping values including hues and shadows. Do this by looking for a subject that has a lot of contrast between light and dark and a reflected shadow—perhaps a shadow that is visible on a surface like a sidewalk. Using a hard pencil (2H), lightly sketch obvious forms, followed by the *contour* (outline) of your subject. Next, assign a number to various parts of your subject related to a value on your tone line. Start by applying mid-range values and then proceed to the darkest values. Use your eraser to remove value to make sure you achieve a distinction between each element of your sketch. Practice with smooth tones created by the side of a sharp pencil and also experiment with some hatching.

EXERCISE: PHOTOGRAPHING TONE

Find a subject close to home that you can photograph on multiple days and at different times of the day. Try and photograph the exact same subject using the exact same composition under hard lighting conditions and under soft lighting conditions. Then, change the images to black-and-white and compare the tonal value of each image. Are there darker shadows and brighter highlights in one of the images?

THE PRACTICE OF COMPOSITION

Practice composing a photo with your camera in a certain type of light, use your pencil and field notebook to record your results, and then use the same tools to produce a quick sketch. Now, incorporate in *field documentation*. Field documentation is about observing, composing, sketching, and writing. You will often see many words in a field notebook. Words are used to reflect on a scene, to describe or document the dimensions of an object, and to note the date, time, and weather. Get accustomed to using your notebook to record many different things. You can even jot down your "to-do" list in your field notebook.

The next exercise requires you to practice the full range of field documentation. You will observe for an extended period before writing, sketching, and composing a photograph of an object in the landscape.

EXERCISE: OBSERVE/WRITE/SKETCH/COMPOSE

To begin, walk around outside until you find a subject that captures your interest. It might be a natural item like a flower or a small branch of leaves, or a built object like a wall or a springhouse. Study the object of your choice thoroughly and prepare yourself for a new way of interacting with the objects around you and the natural world.

Sit comfortably about six feet away from the object. Completely focus on it, purposefully ignoring everything else around you. Do not talk to anyone, listen to music, or become occupied with anything other than your object for the duration of the exercise.

Observe quietly for ten minutes. Ask yourself and answer at least ten questions about the object as you sit and observe. For example:

- How does the light hit the object?
- What type of light am I observing?
- What shapes do the shadows make?
- What is the primary line of the object—vertical, horizontal, curving, straight, etc.? What is its form? Its color? Its texture?

After you have completed your observations, open your field notebook and take five minutes to write a thorough description of the object you have observed. Write the answers to the questions you asked and anything else that comes to mind about what you have seen.

Finally, draw what you have seen with a quick sketch, taking only about two to three minutes. Before you finish, select several views and compose several different photographs.

After you finish these four steps, reflect on your experience. Were you able to look at your object with uninterrupted concentration for ten minutes? Very few people can do that. If your mind strayed, what happened when you brought yourself back to the object? Most people find that they observe something new when they refocus.

Which aspect of the experience did you enjoy the most? Some people like the peaceful, relaxing phase of observation, while others are dying to get to the writing stage so they can capture all the thoughts floating around in their mind. Others find writing a description out longhand is cumbersome and would prefer to capture the details of a subject in a sketch.

Once you spent time studying your subject, were you better able to compose a good photograph? A picture is worth a thousand words! Did your extended observation give you any ideas for a creative composition?

This exercise shows you that observation requires a process. That does not mean every sketch or photograph must follow this progression, but it is useful to understand the process that works best for you and, if you are feeling stuck creatively, to come back to this process for inspiration. Field sketching never starts with putting pencil to paper before some period of observation takes place, and composing good photographs requires study before snapping.

POSITIVE/NEGATIVE SPACE[4]

Close your eyes and picture a kitchen colander—something you would use to drain pasta after it is cooked. The tool has many perforations large enough to only allow water, not pasta, to pass through. As we discuss positive and negative space, it is helpful for you to think of this object. *Positive space* refers to the areas on an object that are solid. *Negative space* is the area that is not solid (the perforated holes in the colander), as well as the area surrounding the object. In any scene you observe, you will find both positive and negative space. Look around at some nearby objects. A coffee mug is a solid form with a large oval negative space formed by joining the cup and handle. Put your keys on the table and look for the parts that are solid (positive space) and parts that are void (negative space). Once you start to see positive and negative space around you, you will always notice it when you are trying to draw and to compose photographs and you will start to rely on

FIGURE 5.3. Photograph showing positive and negative space. (Photo courtesy of Jon Cox.)

FIGURE 5.4. Positive space is white and negative space is shaded in these three sketches. (Sketch courtesy of Claire Ciccarone.)

"Tomatoes" Pos.Neg. Space.   "Bird feather" Pos.Neg. Space.   "Landscape Plants" Pos Neg. Space

it to help you represent accurate forms. If you are having trouble seeing the two types of space, try converting a color image to black and white. In looking at the image of the tulips taken from a low perspective in Figure 5.3, see how the bright white portion is all the negative space in the photo. If you want to capture a dramatic sketch of this scene, try representing just the negative space in a dark tone, as in Figure 5.4. Eliminating any details on the flowers and working to just represent the shapes between the flowers and the edge of the composition will be excellent practice in seeing negative space.

Use positive and negative space exercises to improve your concentration and to breakdown your preconceived notions of the objects you are drawing. You know a lot about the objects you are trying to represent and that can often get it your way. However, you most likely know very little about the random spaces around and between the objects. Changing your preconception about that space will help you record more accurately what you are seeing in your drawing in the same way that flipping an image upside down forces you to just copy lines and forms.

Before you begin a positive/negative space exercise, let's consider composition. You have several choices to make when you embark on a photo or sketch, including whether the format will be vertical, horizontal, or square. Select the format that accentuates the form of the object.

The *rule of thirds* is a compositional approach that artists apply to their work to draw the viewer into the frame. What the rule does is place the viewer's interest at one of four points in the image. Imagine a tic-tac-toe grid being drawn on a rectangle, then that grid being placed over the image that's being captured. Many cameras and cell phones have this feature built directly into the viewfinder and when sketching you can simply draw light lines in this fashion across the page. If the rule of thirds is followed, one of the four points where the lines of the grid cross will be placed over a key element of the image, or a key element will lie directly on one of the lines of the grid. It might be the center of a flower, the eye of a frog, or a focal spot in a landscape; because of the subject matter and the way humans see things, it is the point where the typical viewer's eye will first fall.

FIGURE 5.5. Rule of thirds grid. (Courtesy of Jon Cox.)

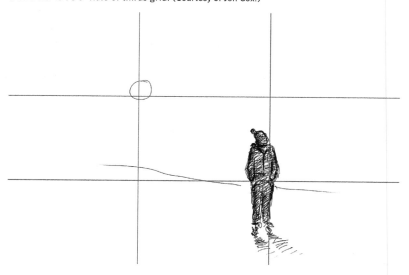

Another method of composing is the use of the *bull's-eye*, where the subject is placed squarely in the center of the image. While it is the least popular way of handling composition because it tends to produce an image that is far less dynamic than that produced by the rule of thirds, in some cases, such as in capturing the center of a flower or in landscape photography, the bull's-eye approach can be highly effective. For subjects that are static, such as trees, statues, or large elements of flora, implying movement may not make sense, so centering the image may not be a problem.

It is important to also consider *balance* when creating your compositions. We can look at two types of balance: *symmetrical* and *asymmetrical*. Symmetrical balance is when both sides or the top and bottom of a scene are essentially the same. You will most likely encounter this type of symmetry when

FIGURE 5.6. Symmetrical balance (left) and asymmetrical balance (right). (Courtesy of Jon Cox.)

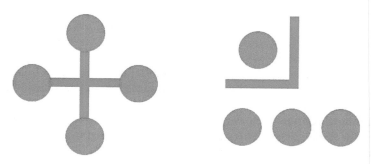

working with macro (close-up) subjects. Asymmetrical balance means that the top and bottom and/or sides of a scene are not the same. But this doesn't mean the scene is not balanced. Arranging the scene so the elements align to create a visual balance through asymmetry is what you will most likely have to do in most of the scenes in nature. It will be up to you to create and evaluate how well you have achieved balance through your composition, perspective, and lighting techniques.

EXERCISE: SKETCH NEGATIVE SPACE

Choose an object/scene that has interior spaces. In other words, the structure of the object connects with itself in a way that produces voids between its parts (such as space between the slats of a chair, space between the branches of a house plant, or elements that are composed so that they overlap each other, forming closed shapes of negative space). If you are having trouble distinguishing between positive and negative space, remember to compose a photo of the object first, and then convert it to black and white. Black-and-white images accentuate the shape and form of an object. Also, removing color may eliminate a distraction that was keeping you from seeing the voids.

Focus on the voids between the mass of the object you are drawing. This will help you develop concentration and hand-eye coordination. Rather than draw the object, draw the void spaces within it that form the angles and the shapes that surround and fall within your composition. Fill the voids dark and leave the mass white. Then, reverse this technique with the same object. Draw the angles and shapes, filling the mass dark and leaving the voids white.

If you use very intense observation skills to draw only the negative spaces of the still life, then the drawing will be remarkably accurate. You will see how important the negative space is when you are finished. With drawing nothing but the negative spaces, the positive forms (the objects themselves) emerge and are created with accuracy. Your concentration should be on "seeing" and only representing the space inside of and between objects.

RECORDING WHAT YOU SEE

As an observer of the natural world, it is important to record what you see on a regular basis. When you notice something interesting in the world around you, use one of the mediums we have been discussing to record the details of that object or scene.

An architect from Brazil who speaks to our Landscape Exploration of Brazil study abroad class carries a small sketchbook with her wherever she goes. She records details of the lighting fixture over the bar in a restaurant while she is waiting for her dinner to arrive. She captures the lines of a particularly attractive chair so she can find it for later use. Her sketchbooks have become a valuable tool to inspire design ideas and remind her to look up details she wants to access.

We all carry cell phones with us that contain powerful digital cameras. We have become accustomed to snapping photographs of menus, bus schedules, or anything we may need for future reference. Learn to use such tools with some thoughtfulness to hone your ability to observe the world around you.

EXERCISE: SKETCHING ON A HIKE

Plan for a time when you can take two similar hikes. They do not need to be to the same location, but they should both be nature trail hikes, city hikes, or whatever type of hike you prefer. On the first hike, bring your camera/phone. Record what you find interesting, taking as many shots as you need to capture

FIGURE 5.7. A page of field sketches from a hike in the Amazon region of Brazil. Note quick sketches with field notes. (Courtesy of Claire Ciccarone.)

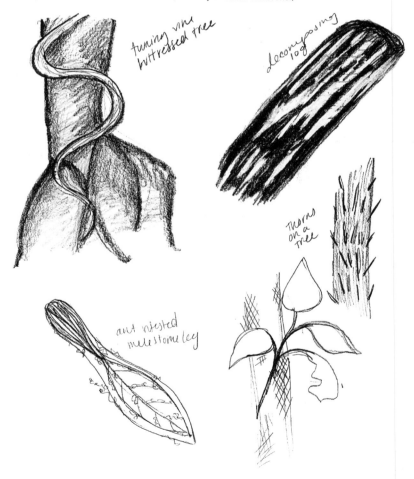

your hiking experience. After the hike, think about what you observed and how it helps you better understand the terrain you covered.

For the next hike (same-day or another day), bring your field notebook and your pencil case. Fill at least three pages with quick sketches of objects or scenes that catch your attention along the hike. Take enough time to capture details of the objects to help you remember what you observe, but don't worry about drawing beautiful sketches. Label the sketches as needed to communicate details you can't capture in a quick sketch, like color. Sketch at whatever size helps you complete the drawing easily. Don't worry about the scale of the sketches in relation to one another. When the hike is over, without looking at your field notebook, think about what you observed and how that observation informs your understanding.

Consider which recording medium—camera or field notebook—allowed you to more accurately remember the objects and scenes that caught your interest along the hikes. Did one medium help you understand the location better than the other? What insights did you gain? Now, go back and look at your sketchbook and your photos and see how a review of your work informs your understanding.

Most people find that they remember and learn more about an object, scene, or location by taking the time to observe and sketch it, or by carefully composing a photograph before just snapping the shot, forcing themselves to see more details than they might observe in an uncomposed photo or in a few words and lines jotted in their field book. Learn to understand the benefits of each medium and determine situations when one medium is more appropriate.

TEXTURE

Texture is the appearance of how an object feels. People have strong sensory reactions to different textures, so if you want to evoke emotion in your photo or sketch, pay attention to different textures. There are a variety of techniques for rendering texture in a sketch that range from literal depictions of patterns to abstract representations of how a texture makes you feel. Just like in your warm-up exercises when you created lines with emotions, you can warm up by producing textures that represent a word. Select one soft pencil (2B) and one hard pencil (2H). Warm up by copying the textures in the boxes described below into your field book. As you copy each texture, try using both pencils and imagine a word to describe the texture or an object that might contain that texture.

Think about the following textures and use both pencils to represent them in your field book inside a small box: bumpy, fuzzy, and rough. Think about the last fruit you ate and draw its texture in a box. Do not try to draw the shape of the fruit; rather, fill the box with a representative texture. For example, I last ate an orange, so I would try to represent the small grainy dimples you can find on the surface of an orange peel.

Here are some other fun ideas. Grab an object that is close to you and rub it on a page of your field notebook to produce a texture. Describe the resulting texture. Step on a page in your field notebook—what do you notice about the resulting texture? Find something outdoors or indoors with an interesting texture and represent it in another box on a page in your field book. Review all the textures you have sketched. Which ones most accurately evoke the texture they are supposed to represent? How might you use these interesting textures in a sketch? When two very different textures are next to each other they form a contrast, making each element easy to read. As we discussed in the section on tone, when composing a landscape scene, you should map light and dark tones. You should do the same for textures. Map the fine versus coarse textures and practice achieving a strong contrast to make the distinction legible.

Now that you have thoroughly explored texture and how you might represent a variety of textures, try switching your medium and move on to photography!

EXERCISE: REPRESENT TREE BARK

Tree bark has a wide variety of textures, from long peeling strips (shagbark hickory) to rough chunks (native dogwood) to smooth muscular surfaces (beech or ironwood), and provides a great opportunity to explore different textures.

Stand back from a tree about six feet so you can see the bark and some of the primary branching patterns. Draw the trunk of the tree in contour first. Be careful to fully observe the lines of the tree and try to disassociate yourself from the actual object (forget it is a tree!). You have probably drawn a tree many times in your life, but this time look at the shape of the trunk and how the branches split to make an accurate representation of the outline. Next, start to fill in the texture you see from the tree bark, using some of the techniques you practiced earlier with abstract textures. Find another tree and sketch it as well. Now, photograph the trees you sketched. Notice how the direction and quality of light will impact the texture of the tree trunk in your photographs. Does light also impact the texture you saw when you sketched the tree?

ZOOM IN/ZOOM OUT

Now that you have worked on representing the texture of the tree trunk, you can tackle the entire tree. It is time-consuming to sketch an entire tree with accurate detail of every leaf. *Realism* (in the arts sometimes called *naturalism*) is generally the attempt to represent subject matter truthfully, without artificiality and avoiding artistic conventions, or implausible, exotic, and supernatural elements. Great artists in this genre take a lot of time to represent exactly what they see with a high level of detail. *Impressionists* capture the image of an object as someone would see it if they just caught a glimpse of it. Field sketches are somewhere between these two genres. Field sketching requires you to quickly represent the objects you see, so you don't have time to sketch every leaf on a tree.

EXERCISE: DRAW A TREE

Sketching a tree accurately, but with less detail, requires a process you can teach yourself. Start by selecting a tree you want to sketch. Next, pick a leaf from the tree and do a detailed sketch of the leaf including tone for shading. Once you have accurately sketched a single leaf, try to *abstract* that leaf and sketch a branch with multiple leaves and fewer details. Next, sketch several branches of the tree with even less detail. Finally, sketch the entire tree. Sketch the trunk and then start placing branches on the tree using less detail but keeping what you learned from your detail sketches in mind. With sketching, the process requires you to zoom out to achieve the abstraction you need to sketch an entire tree without depicting each leaf perfectly. With photography, the process is often reversed: we start by looking at the overall subject and compose an inclusive photo. Picture yourself trying to capture a bird you've spotted in a tree. You would compose a photo far enough from the subject to capture the overall scene before stepping forward and zooming closer and closer.

FIGURE 5.8. Accurate sketch of an individual leaf (left), multiple leaves on a branch (middle), and a full tree sketched with less detail (right). (Courtesy of Claire Ciccarone.)

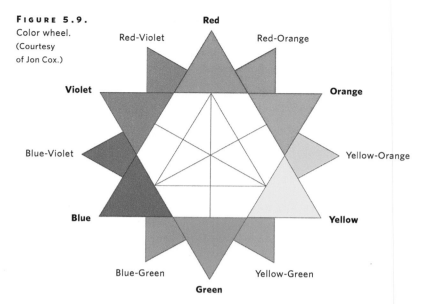

FIGURE 5.9. Color wheel. (Courtesy of Jon Cox.)

COLOR

Whether you are viewing a painting, a pastel, or a photograph, color can evoke a strong emotional response. Now that you are beginning to train your eye and create a visual vocabulary, we will build additional visual skills by concentrating on color. In this section, you will learn how to observe and capture the differences created by color. Specifically, we will look at capturing monochromatic, complementary, and analogous colors to add feeling and create a specific mood within your images. We will examine how and when to change a color image into a black-and-white photograph. We will also use the basic color wheel to help you understand the relationship between different colors and how individual colors can be combined to create a variety of moods in a scene.

An *analogous* scheme is made up of colors that are adjacent to each other on the color wheel. For example, adjacent to yellow we find yellow-green and yellow-orange. Adjacent to red we find red-orange and red-violet. And finally, adjacent to blue we find blue-green and blue-violet. Analogous color schemes are often pleasing to the viewer because they are frequently found in our natural environment. Including analogous color schemes in your garden images can create a harmonious and peaceful atmosphere to draw the viewer into the scene. When looking for analogous colors, you may need to search for the proper background. This can often require you to change your perspective by lying on your stomach or climbing to a higher location for a different vantage point.

A *complementary* color scheme is made up of colors located opposite one another on the color wheel. Examples include red and green, orange and blue, and violet and yellow. Complementary colors create high contrast in a photograph and are especially good to use in small doses to make the subject stand out. In addition to searching for complementary colors, you should continue to pay attention to composition and frame the subject to highlight its specific features. A vertical format accentuates, for example, the vertical structure of the reproductive parts of the crocus in Figure 5.10b. The crocus's complimentary yellow-orange and blue-violet colors produce a dramatic effect. You probably won't have to search hard to find complementary colors in nature because people are naturally attracted to this scheme.

When we think of *monochromatic* images, black-and-white ones often come to mind, but there are many more examples of this scheme to explore.

A

B

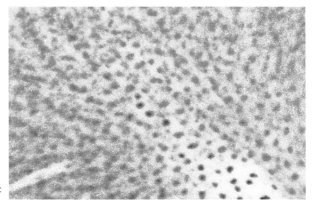

C

FIGURES
5.10A–C.
(A) Analogous color scheme;
(B) Complimentary color scheme;
(C) Monochromatic color scheme.
(Photos courtesy of Jon Cox.)

Think about the different seasons of the year and how the time of day can influence the color and mood. Winter is an obvious choice for capturing a monochromatic scene, like a snow-covered landscape. Fall can be a good time of year for capturing the monochromatic haze associated with atmospheric fog, especially in the morning light. Early summer is the perfect time of year to capitalize on the lush monochromatic shades of green foliage. Consider shooting with different angles, perspectives, or camera settings, such as shooting on a macro (close-up) setting, to capture something unique.

To start photographing monochromatic subjects, you may want to begin by zooming in closely on your subject to isolate individual colors in the scene. Subtle details and carefully framed compositions are crucial elements to bring attention to a monochromatic photograph. When looking for monochromatic images, you should push yourself to search for images that elicit an emotional response.

Why would you shoot a scene in color if you knew you wanted to turn it into a black-and-white image? You have a much larger tonal range to work with in an editing program if you shoot the image in color. Scenes that have texture and shadows can make excellent candidates to turn into black-and-white images. Black-and-white images accentuate the shape and form of the subject. Changing a color image to a black-and-white one automatically changes the mood. By removing the color, you remove the viewer's emotional response to the different colors.

To add color to your field notebook, use watercolors. You will need to supplement your field sketching supplies with:

- Watercolor paper or your regular field notebook paper (but be aware that as the watercolor dries on regular paper it may buckle a bit)
- Watercolor pan set (big or small, depending on the situation)
- Two or three brushes of various size
- Small water container
- Paper towel

Watercolor is a great medium for quick sketching and can help you more accurately share a feeling or impression of a scene. To start adding watercolor, use a pencil to draw a rectangle frame for your sketch in horizontal or vertical format depending on your subject. Use a 2B pencil to lightly sketch the shapes of the primary subjects in the scene you are viewing through your viewfinder. Next, dilute your selected watercolor hue with enough water to make a very light color on the brush. Use this light color to create a wash across the entire sheet, changing colors to paint the major elements as they meet along the horizon line. For example, paint a wash for the sky in a light blue and a wash for the ground in a light green. Allow the two colors to meet at the horizon line. Next, use your paintbrush to add the primary features in a color of your choice. Remember that objects in the foreground are more vibrant, so it is helpful to build your sketch from the background to the foreground. The background objects will be lighter and more faded and those in the foreground will be more vibrant and also include more detail. Optionally, use pencil or ink on top of the watercolor background to emphasize edges. Or you can work entirely in watercolor. To get fresh, vibrant watercolor scenes, a blog author named Anne-Laure offers the following tips:[5]

1. Keep enough white. Don't add color everywhere. Letting the white paper show through adds highlights and sparkle to your sketches.
2. Use plenty of paint and plenty of water. The water will play with the paint and you'll get a fresher look.
3. Be bold. Allow colors to run into each other, add some splatters and allow color variations to be strong enough to really show. Paint big shapes.
4. Apply light watery washes and then add dark colors on top. Don't wait for one color to dry before you add more color. The colors will blend and look more natural.
5. Use strong, bright colors. For tree trunks, use orange or dark red, rather than brown. Roads and roofs look grey? How about using blue and purple? Watercolors dry lighter and duller than when first applied.
6. Contrast is important. Add some small dark strokes to your sketch. To some dry areas of the sketch, add a few strokes of dark paint for details or shadows.
7. Get to know your paper. Different types of paper respond differently to watercolors. Know what you are working with and how it will respond. Does the paper soak up color or will the color spread quickly across the paper?
8. Stop soon enough. Don't keep adding more color after more color. Usually simpler is better.

EXERCISE: COLOR

Photograph a scene that highlights the following examples:

1. Monochromatic Color
2. Complementary Color
3. Analogous Color
4. Black and White

After photographing the scenes, represent each one in watercolor.

CONCLUSION

There are many ways a Master Naturalist can use the field sketching and photography tools and techniques described in this chapter. You may find that you will use these tools exclusively as a method to help you learn. As you start to spend more time outside, it is natural to want to document and describe what you see so you can refer to your sketches, notes, and photos as a way to build your knowledge and understanding of nature's elements. Over time, you may gain a level of creative confidence that drives you to use these tools to capture the incredible beauty of what you see and experience outside. Either way, if you work through the exercises, you will notice that you will start to see differently. You may become more aware of the way the light hits an object to highlight it, bring out its texture, or cast a dramatic shadow. You may become drawn to looking at the spaces between objects (the negative space) and find yourself thinking about how you would represent shapes on paper. You may become more aware of the colors you see in the natural world and, in looking at a scene, try to identify the color scheme you are seeing. Since sketching and photography are both predicated on careful observation, mastery of observation is a prerequisite to accurate sketching and successful photography. But, learning to take time to observe can also be considered a result of learning to sketch and photograph, so honing all of these skills will be critical in helping you capture your environment.

TOOLS OF THE TRADE

- *Field sketching notebook*—bound notebook with unlined pages, usually about 5" × 7".
- *Ground cloth*—a cloth placed on the ground for protection from soil or moisture when outdoors.
- *Macro lens*—a camera lens designed for photographing small subjects at very close distances. It can focus much more than a normal lens, allowing you to fill the frame with your subject and capture more detail.
- *Mirrorless SLR camera*—a mirrorless camera is one that doesn't require a reflex mirror, a key component of DSLR cameras. With a mirrorless camera, you get a preview of the image on-screen.
- *Pink pearl*—soft and pliable eraser that always gives smudge-free erasures; beveled on the ends for better control.
- *White vinyl eraser*—softer and a bit stickier eraser made of vinyl that doesn't abrade the paper's surface.

NOTES

1. Cathy Johnson, *The Sierra Club Guide to Sketching in Nature* (San Francisco: Sierra Club Books, 1997), 83
2. Ansel Adams, "A Personal Credo," *American Annual of Photography* (1944).
3. Paul Laseau, *Freehand Sketching: An Introduction* (New York: W. W. Norton & Co, 2004).
4. Drawing and Painting Lessons with Edward A. Burke, http://www.drawingandpaintinglessons.com/Drawing-Lessons/Negative-Space-2.cfm.
5. Anne-Laure, "10 tips for nice and fresh watercolors," *Watercolor Sketching: Sketching Adventures and Watercolor Exploration*, http://www.watercolorsketching.com/10-tips-for-nice-and-fresh-watercolors/ (retrieved January 15, 2020).

CHAPTER SIX

Plant Identification and Taxonomy

SUSAN BARTON AND ANNA WIK

A CRITICAL SKILL for understanding, managing, and restoring the landscape is learning the language of plants. To become fluent in this language, you must understand the systematic code based on the *binomial system* developed by the Swedish botanist Carl Linnaeus in 1752.

In the binomial system, each plant has two names: a *genus* name and a *species* name. The Latin binomial is usually enough to denote plants that occur anywhere in the world, either in your garden or in the wild. However, an additional name called the *cultivar* name (cv.) is often needed for cultivated plants that arise through *mutations* or *hybridization*. Cultivar names, which are the official names of plants that can be used by anyone, have become complicated. Some nurseries that develop and promote new plants use complicated cultivar names and then designate catchy trademark names for those plants that can be used only by that nursery for ten years, renewable every ten years. This is done for marketing purposes, so the gardening public ends up referring to plants by their trademarked name rather than their cultivar name.

In addition to the scientific name, most plants have *common* names that may be easier to remember but are not definitive for a plant and may differ from region to region or even state to state. For example, sour gum, black gum, yellow gum, tupelo, and pepperidge are all common names for *Nyssa sylvatica*.

Plant names are controlled by the International Code of Nomenclature. To see how it works, let's use the Delaware native red maple as an example. The genus name for maple is *Acer*. All maples—the sugar maple, silver maple, striped maple, Japanese maple, and all other maples—have the same genus name. The species name *rubrum* denotes our native red maple. No other maple will have the same species name and no other plant will have the same combination of names as *Acer rubrum*.

But in the nursery trade, there are many selections of red maples propagated asexually that have special characteristics. Thus, we have *Acer rubrum* "October Glory," *Acer rubrum* "Red Sunset," and *Acer rubrum* "Bowhall." October Glories and Red Sunsets have better fall color than most red maples. Bowhalls have a narrow, pyramidal shape.

When buying or selling a plant, the scientific name and the cultivar names help ensure you are getting the plant you desire. But common names may be easier to use when referring to plants locally if everyone is using the same common name.

Linnaeus organized plants in the binomial system based on the number and arrangement of their flower parts, which is still the basis for classification and identification of plants in scientific texts and in taxonomies. Plant leaves, however, offer a more practical (and familiar) means of identification for people working with plants in all stages. Leaves of plants differ in size, shape, and arrangement on the stem, and in their margins, tips, and bases. Many people are familiar with the typical shape of the leaves of oaks, maples, willows, and tulip poplars. With careful study, most plants commonly found can be identified by their leaves. Follow-up clues, such as habit of growth, buds, bark color, flowers, fruit, thorns, and venation, can all be learned to help identify our plants, flowers, and trees.

The first step is to identify the genus of a plant. For instance, is it an oak, maple, or dogwood? Once the genus is known, reference texts or the Internet can be used to determine the species and cultivar. You might recognize certain characteristics, like flowering dogwood buds that look like little buns; this alone can trigger recognition whenever you see these trees. But to learn new plants, you will need a systematic approach to fix characteristics in your mind.

Start identifying a new plant by eliminating as many plants as possible that it can't be so that you can concentrate on a smaller number of choices. Trees, shrubs, and vines are logical groups to use as starting points. Is the tree an *evergreen*, or is it *deciduous*, dropping its leaves in winter? Evergreen plants are further divided into conifers (cone-bearing) or needle evergreens and broadleaf evergreens that keep their leaves in winter but produce flowers and fruit. Try to identify the group to which a plant belongs, then systematically narrow the possibilities within the group until you can make a final identification. Then fix the characteristic that you associate with the plant in your mind.

CONIFER IDENTIFICATION

The *conifers* or the needle evergreens (although a few are deciduous) are a distinct group of plants, easily recognized by their cones and the shape, color, and texture of their needles (leaves). Conifer leaves fall into four categories:

1. Long, narrow needles fastened together at the base in *fascicles* (bundles of needles typical of all pines).

2. Flat or angular needles fastened to the stem by a narrowed constriction (fir, spruce, hemlock, yew).
3. Short, sharp, awl-shaped needles, overlapping and sharp to the touch (some junipers, some false cypress).
4. Scale-like needles closely overlapping each other like shingles on a roof (arborvitae, some junipers, some false cypress).

The following are genera of conifers, or groups of genera, with similar characteristics followed by distinguishing characteristics of each genus:

Pinus, Pine—distinguished from all other evergreens by the arrangement of needles in bundles (usually two, three, or five needles) spirally along the stem. Pine cones are *pendulous* (that is they hang from the branches).

Taxus, Yew; *Tsuga*, Hemlock; *Picea*, Spruce; *Pseudotsuga*, Douglas fir; and *Abies*, Fir—all have needles arranged singly, spirally along the stem.
- *Taxus*, Yew—distinguished from the others of this group by the light yellow-green color on the underside of the needles; the others tend to be *glaucous* (blue; powdery bloom). The *Taxus* seed is surrounded by a red gelatinous cup and called an *aril*.
- *Tsuga*, Hemlock—distinguished from *Picea*, *Pseudotsuga*, and *Abies* by the short, flat, blunt needle, which is more or less two-ranked (with one row of needles on opposite sides of the stem), and the constriction of the needle at its base. Cones are terminal and drooping.
- *Picea*, Spruce—characterized by pendulous cones and usually four angled needles, which are very stiff and sharp, and distinguished from *Pseudotsuga* and *Abies* by the peg remaining on the branch after needles are shed. When you rip off a needle, a part of the needle that comes off is attached to the stem, so it looks like you are ripping off part of the stem. Cones are pendulous.
- *Pseudotsuga*, Douglas fir—distinguished most readily by the distinct, pointed bud. Needles are flat and soft and have slightly raised leaf-scars. Stems are smooth, and cones are pendulous. Woody female conifer cones are made up of a series of cone scales. In the case of Douglas fir, there is a fish-tail-like bract scale in between each cone scale.
- *Abies*, Fir—distinguished from all of the above by the usually smooth stem with round leaf scars that can be depressed. Needles are usually flat with prominent glaucous bands underneath. Cones are upright.

Juniperus, Juniper; *Chamaecyparis*, False cypress; and *Thuja*, Arborvitae—have small, opposite, either scale-like and needle-like or awl-shaped leaves. Junipers sometimes have short, needle-like leaves in whorls of three.
- *Juniperus*, Juniper—branches are round and leaves are both scale-like and needle-like, frequently on the same plant. Glaucous bands, if present, are on the upper surface. Juvenile leaves are awl-shaped. The fruit is blue-grey and berry-like.

- *Chamaecyparis*, False cypress—branches are slightly flattened or round. White markings can be found on the undersides of needles. Cones are round, pea-sized, upright, and contain few seeds.
- *Thuja*, Arborvitae—branches are flattened and frond-like. There are no white markings on leaves and foliage is light green on the underside. There is a characteristic odor coming from glades at the base of each leaf. Cones are small (½"), oval with few scales, and born upright.

Cupressocyparis, e.g., Leyland cypress—branchlets are sharp-angled and plumy. Scale-like leaves. Glaucous markings may be evident.

Cedrus, Cedar—needles are in bunch-like whorls, one-inch to two-inches long, on short shoots along branches. Cones are upright and require two years to mature.

Metasequoia, Dawn redwood; *Taxodium*, Bald cypress—deciduous conifers. Needles are thin, soft, and linear and are found in two flattened ranks.
- *Metasequoia*, Dawn redwood—opposite leaf arrangement, more loosely crowned.
- *Taxodium*, Bald cypress—alternate leaf arrangement with branchlets arranged spirally.
- *Cryptomeria*, e.g., Japanese cedar—spirally arranged, bright green to bluish green, awl-shaped, four-ranked leaves that curve inward. First leaves of the year are shorter than the later ones.

BROAD-LEAVED EVERGREEN IDENTIFICATION

Broad-leaved evergreens hold onto their leaves during the winter but have larger leaves than the needle evergreens. American holly, inkberry holly, bayberry, wax myrtle, rosebay rhododendron, and mountain laurel are native broad-leaved evergreens found in Delaware. For broad-leaved evergreen identification, follow the strategies suggested for deciduous trees and shrubs.

DECIDUOUS TREE AND SHRUB IDENTIFICATION

For deciduous trees and shrubs, first look at the arrangement of leaves on the stem. Most leaves will be either *opposite* each other or *alternate* with each other at the point of attachment to the stem (Figure 6.1). A few exceptions occur where leaves are not quite opposite (*sub-opposite*) or where three or more leaves are attached at the same point (*whorled*). When identifying a tree in winter, you can look at leaf buds to determine if a plant has alternate or opposite leaves.

Next, look at the leaf and determine if it is a *single leaf*, as is the case for most trees and shrubs, or a *compound leaf* made up of several leaflets attached to a single *petiole*, or leafstalk (Figure 6.2). To determine if a leaf is simple or compound, look for the bud in the *axil* of the leaf where the petiole attaches

FIGURE 6.1 Leaf arrangements—alternate and opposite. (Sketch courtesy of Olivia Kirkpatrick.)

FIGURE 6.2. Single and compound leaves (pinnate and palmate). (Sketch courtesy of Olivia Kirkpatrick.)

FIGURE 6.3. Oval, heart-shaped, round, and lanceolate leaf shapes. (Sketch courtesy of Olivia Kirkpatrick.)

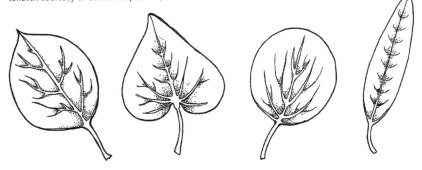

to the main stem. In both simple and compound leaves, buds will only be found where the main stem of the leaf is attached. You will not find buds in the axil of leaflets of a compound leaf. Compound leaves can be arranged *pinnately* (with leaflets arranged like a feather) or *palmately* (with leaves arranged like the palm of your hand with leaves as fingers).

Next, look at the shape of the leaf. Some leaves are *round*, others are *heart-shaped*, *elliptical*, *oval*, etc. (Figure 6.3). Then look at the margin of the leaf to determine whether it is *smooth*, *toothed*, or *lobed*. Oaks and maples are examples of common trees with lobed leaves. You can separate species of maples by the number of lobes on their leaves. Red maples have three lobes and sugar

FIGURE 6.4. Acute, acuminate, bristle-tipped, truncate, or obtuse leaf tips. (Sketch courtesy of Olivia Kirkpatrick.)

maples have five. Sweet gums have five lobes, but they are arranged in a starlike pattern. All the oaks in the white oak group have rounded lobes, while red oaks have pointed lobes.

Additional features, such the shape of the top of the leaf and of the base of the leaf, can be checked in a reference text to complete the identification. Common top shapes include *acute, acuminate, bristle-tipped, truncate,* or *obtuse* (Figure 6.4). Leaf bases can be *cuneate, attenuate, rounded, cordate, truncate, sagitate, hastate, auriculate,* or *oblique* (if one side is different from the other).

When a plant is flowering, there are even more identification clues. Flowers can be borne singly or in an inflorescence (that is, with multiple flowers in a structure). Common flower structures include *racemes, spikes, corymbs, umbels, panicles, cymes,* or a *capitulum (composite flowers)* (see Figure 6.5).

The *pedicel* is the main flower stalk. A *raceme* has flowers borne on short stalks arranged alternately along the pedicel. A *spike* has flowers borne directly on the pedicel. A *corymb* is like a raceme except the flower stalks are different lengths, so all the flowers form a flat-topped structure. An *umbel*

FIGURE 6.5. Spike, raceme, umbel, panicle, and composite flower structures. (Sketch courtesy of Olivia Kirkpatrick.)

results in the same flat-topped structure, but each flower stalk originates from the same point on the pedicel. A *panicle* is a multi-branched raceme and a *cyme* is multi-branched, but all the flowers are borne at about the same height, forming a round-topped inflorescence.

Composite flowers look to the beginner like a single flower, but they are really hundreds or thousands of individual flowers in one inflorescence. Their outer flowers are called *ray flowers* and look like standard petals. The inner flowers or *disk flowers* are small and make up the center of the flower. A daisy is an example of a composite flower.

Bark can be a distinguishing characteristic any time of year but is especially important for winter identification. Some plants (like maples) have smooth bark, others (like birch) have peeling or exfoliating bark. Some bark (like on walnut or dogwood) is chunky and in a fractured pattern. Cherries often have reddish bark and have horizontal *lenticels* (raised pores on the stem for gas exchange). Carpinus or ironwood has a sinuous bark that looks like strong muscles rippling below the skin of a shark.

Branch structure is also a defining characteristic of trees. Flowering dogwood have a horizontal branching pattern. Pin oaks have their upper branches ascending, middle branches horizontal, and their lower branches descending. Sassafras have branches in horizontal clusters that look like horizontal clouds from a distance when they flower in early spring. Birch trees have multiple stems with ascending branches.

Overall *habit* or plant *form* is another useful characteristic for identifying plants. We can look at the general outline of the tree for this (see Figure 6.6).

Pyramidal (conical) trees are primarily needle-bearing evergreens, but American holly is a broad-leaved evergreen that has a pyramidal form, and katsura is a deciduous tree that is often pyramidal. Trees can be *oval*, with the

Figure 6.6. Pyramidal, columnar, oval, obovoid, and weeping tree habits. (Sketch courtesy of Olivia Kirkpatrick.)

lower branches wider than the top branches; or *obovoid*, with the top branches wider than the lower branches. Some trees are more *rounded*, and others are *elliptical*. *Columnar* trees don't usually occur naturally, but cultivars have been bred to have a narrow, columnar habit to fit into tight spaces in the landscape.

TABLE 6.1. DECIDUOUS TREE KEY

Leaf type	Leaf margin	Leaf arrangement	Genera		
Simple leaves	Smooth margin	Leaves alternate	Asimina Cercis Diospyros	Elaeagnus Maclura Nyssa	Quercus* Salix* Sassafras**
		Leaves opposite	Catalpa Chionanthus	Cornus Paulownia	Syringa
	Toothed margin	Leaves alternate	Alnus Amelanchier Betula* Carpinus Castanea Celtis Crataegus* Fagus*	Franklinia Halesia Malus* Morus* Ostrya Oxydendron Populus Prunus	Pyrus Quercus* Salix* Stewartia Styrax Tilia Ulmus Zelkova
		Leaves opposite	Acer*	Cercidiphyllum	
	Lobed margin (can also be toothed)	Leaves alternate	Betula* Crataegus* Fagus* Ginkgo	Liquidambar Liriodendron Malus Morus**	Plantanus Populus* Quercus* Sassafras**
		Leaves opposite	Acer*		
Compound leaves		Leaves alternate	Ailanthus Albizzia Aralia Carya Cladrastis	Gleditsia Gymnocladus Juglans Koelreuteria Maackia	Robinia Sophora Sorbus
		Leaves opposite	Acer* Aesculus***	Fraxinus	Phellodendron

* genus has species with different leaf forms
** more than one leaf form appears on the same tree
*** palmately compound

TABLE 6.2. SHRUB KEY

Leaf type	Leaf margin or type of compound leaf	Leaf arrangement	Genera		
Simple leaves	Smooth margin	Leaves alternate	Arctostaphylos (E) Azalea (Rhododendron) (E) Buddleia** Cornus** Cotinus	Cotoneaster (E) Daphne (E) Elaeagnus Gaylussacia (E) Kalmia (E) Lindera	Lyonia Rhamnus* Rhododendron (E) Spiraea* Vaccinium (E)
		Leaves opposite	Buxus (E) Calycanthus Cornus Hypericum	Kalmia (E) Ligustrum (E) Lonicera (E) Mitchella (E)	Syringa Viburnum* (E)
	Toothed margin	Leaves alternate	Alnus Aronia Berberis (E) Chaenomeles Ceonanthus Corylopis Clethra Fothergilla	Gaultheria (E) Hamamelis Ilex* (E) Kerria Leucothoe (E) Myrica (E) Pieris (E) Prunus (E)	Pyracantha (E) Rhamnus* Salix Sorbus Spiraea* Vaccinium (E)
		Leaves opposite	Abelia Buddleia Callicarpa Deutzia	Diervilla Euonymus Forsythia Hydrangea*	Rhamnus* Viburnum* (E) Weigelia
	Lobed margin	Leaves alternate	Hedera* (V) (E) Hibiscus	Physocarpus Ribes	Rubus
		Leaves opposite	Hydrangea Viburnum* (E)		
Compound leaves	Trifoliate or palmately compound	Leaves alternate	Akebia (V)*** Cytisus (T) Lespedeza (T)	Parthenocissus (V)***	
		Leaves opposite	Clematis (V) (T) Jasminum (T)	Vitex***	
	Pinnately compound	Leaves alternate	Aralia Maackia Mahonia (E) Paeonia	Potentilla Rosa Rhus Sorbaria	Sorbus* Robinia Xanthorhiza
		Leaves opposite	Campsis (V) Sambucus	Staphylea	

(E) genus is evergreen or contains some evergreen species
(V) vine
(T) trifoliate (compound leaves with three leaflets)
* genus has species with different leaf forms
** more than one leaf form appears on the same shrub
*** palmately compound

Weeping habits occur naturally in some plants like the weeping willow, but many cultivars have been bred to weep, providing a graceful component in the landscape. Plants with an arching, *vase-shaped* habit are essentially an upside-down pyramid. This habit is particularly pleasing when trees are planted in an allée on either side of a path.

On the University of Delaware Green, the allée of American elms long made a grand statement inviting students to Memorial Hall. When most of those elms succumbed to Dutch elm disease, they were replaced with zelkova, an Asian tree with a vase-shaped habit. But the zelkova has a much stiffer habit, doesn't grow as tall, and does not have branches that recurve at the top of the tree, so the effect was far from graceful. Fortunately, new Dutch elm disease-resistant cultivars of American elm are now available, and the university is slowing replacing dying elms and zelkovas with these new cultivars.

Leaf buds can be distinctive and useful for winter identification. Saucer magnolias have large, hairy buds, but sweet bay magnolias have smooth, pointy buds. Beech trees have extremely long, pointy buds. Flowering dogwood have biscuit-shaped buds, while kousa dogwood have pointed buds. On both dogwoods, the buds are arranged like fingertips at the end of an upturned palm.

Tables 6.1 and 6.2 on pages 140–41 separate different types of woody plants by their leaf characteristics. These are common genera found in either the planted or natural landscapes of Delaware.

IDENTIFYING PLANTS BY REGION

The following section presents common plants found in different habitats in the state. The contiguous United States are divided into physiographic provinces according to their *geomorphology*, which refers to the physical features and processes of landforms, and their relation to geologic structures. The climate, underlying geology, and geologic history of an area affect the modern topography. Some areas have been scraped flat by glaciers; others are dominated by towering mountains; and still others are subject to changing sea levels and coastal processes. Delaware contains two physiographic provinces: the Piedmont (located in the upper northwest corner of the state) and the Coastal Plain, which comprises the rest of the state. For people familiar with New Castle County, Kirkwood Highway is approximately the dividing line between them.

As you have read elsewhere in this volume, the Piedmont in Delaware is characterized by low, rolling hills—remnants of several ancient mountain chains that eroded away. Soils here are clay-like and the *climax vegetation* (the community of plants found after a cleared area has undergone a series of developmental stages until a steady state is achieved) is an oak-hickory forest.

The Coastal Plain in Delaware is lowland, with few hills. There are many wet areas here, including rivers, marshes, and swamplands. Soils become sandier the closer one gets to the coast. Pine-dominated uplands give way to grass-dominated meadows near the coast.

In addition to natural sites, Delaware has many planned landscapes that humans have planted or managed. Even so-called natural areas, however, have been impacted by human activity throughout the state.

Plants included in this chapter can be found in Delaware's planned and natural landscapes. Common plants in the following categories are described below with key characteristics to help in their identification: Ornamental, Woodland, Woody Oldfield/Meadow, Coastal, or Invasive. Plants native to Delaware are designated by (N). As you read this Master Naturalist handbook, you will find many good reasons for planting native plants in the planned landscape. Including some non-native (but non-invasive) species in ornamental landscapes is fine, but we encourage planting native plants whenever possible.

ORNAMENTAL

- *Betula nigra*, River birch—exfoliating bark is cinnamon-colored on the species and can be white on cultivars ("Heritage"). Triangular-shaped, alternate leaves are toothed. (N)
- *Buxus sempervirens*, Boxwood—opposite leaves distinguish boxwood from Japanese holly, a similar shrub. An unpleasant odor is also characteristic. This evergreen shrub is usually grown as a dense hedge or foundation plant.
- *Chionanthus virginicus*, Fringetree—elongated, opposite leaves have midribs (veins along the midline of the leaf) that disappear near the end of the leaf. Flowers are fragrant, fleecy, white, and bloom before leaves emerge. Individual petals are small and strap-like. Females have bluish, grape-like fruit. (N)
- *Cercidiphyllum japonicum*, Katsura tree—blue-green foliage is heart-shaped, turning yellow in the fall and bark is shaggy.
- *Ginkgo biloba*, Maidenhair tree—fan-shaped leaves are bright green and yellow in fall. Narrow upright habit and ability to withstand tough conditions make ginkgo a popular landscape tree in urban sites.
- *Gleditsia triacanthos*, Honeylocust—bipinnately compound leaves turn yellow in fall. Often has thorns on the trunk, but thornless cultivars are available. Fruit is long pods that turn dark brown.
- *Hydrangea macrophylla*, Bigleaf hydrangea—toothed, with opposite leaves, and lustrous green. Flowers borne in large corymbs and can be pink or blue depending on the soil pH.
- *Ilex crenata*, Japanese holly—alternate cup-shaped leaves are evergreen. Many cultivars in different habits and sizes are available.
- *Malus* sp., Crabapple—grown for its early spring flowers in white, pink, and red, depending on cultivar. Red or yellow fruits in fall. Alternate leaves are toothed. Many cultivars of this popular landscape tree are available.
- *Pieris japonica*, Andromeda—dense, rosette-like evergreen foliage, opens bronzy in spring, changing to lustrous dark green. Creamy white pendulous clusters of

urn-shaped flowers in early spring. So susceptible to the lace bug that the foliage is often stippled with white dots.
- *Prunus serrulata*, Kwanzan cherry—probably the most commonly planted landscape cherry. Double deep-pink flowers bloom as bronze new leaves are emerging. This tree has a stiff vase-shaped habit.
- *Quercus palustris*, Pin oak—alternate lobed leaves have deep sinuses with pointed lobes (red oak group) turning reddish-brown in the fall with leaves hanging on into winter. Lower branches are pendulous. (N)
- *Quercus phellos*, Willow oak—alternate simple leaves are not lobed, like the leaves of most oaks. Narrow, strap-like leaves turn yellow in fall and hang onto their trees into winter. (N)
- *Rhododendron* sp., Azalea—evergreen azaleas have small, often glossy green leaves. Flowers are in smaller clusters than rhododendrons. Early spring blooms usually cover the plant.
- *Rhododendron* sp., Rhododendron—large, oblong evergreen leaves are sometimes pubescent on the underside depending on the cultivar. Large flower clusters bloom in spring in a wide variety of colors.
- *Viburnum carlesii*, Korean spice viburnum—dull, grey-green to blue-green pubescent leaves. Fragrant flowers open pink and turn white before the leaves appear in early April.

WOODLAND

- *Acer saccharum*, Sugar maple—five-lobed leaves turn a yellow/orange color in fall. Can be confused with Norway maple, but there is no white sap present when leaves are removed. (N)
- *Amelanchier canadensis*, Serviceberry—alternate, finely toothed leaves turn orange in fall. White flowers are borne very early in spring and this tree is usually found in woodland edges. (N)
- *Carpinus caroliniana*, Ironwood—alternate, toothed leaves have prominent ridged veins on the leaf undersurface. Muscle-like blue-grey bark is distinctive. (N)
- *Carya ovata*, Shagbark hickory—pinnately compound leaves usually have five leaflets. Bark peels off in long strips. (N)
- *Clethra alnifolia*, Summersweet—alternate, dark-green leaves with toothed margins turning yellow in fall. Small fragrant white flowers in racemes in summer. This shrub is multi-stemmed and spreads from a single plant into a broader clump from smaller crowns produced near the parent stem. (N)
- *Cornus florida*, Flowering dogwood—opposite leaves with veins that appear to run parallel to leaf margins. Red fall color, horizontal branching, bun-shaped buds and white flowers comprised of large bracts characterize this tree. (N)
- *Ilex opaca*, American holly—sharply toothed evergreen leaves are dull green in the species. Red berries are born in fall and the habit is pyramidal. (N)
- *Juglans nigra*, Black walnut—pinnately compound leaves are up to two feet long with five to eleven pairs of leaflets. Bark is dark and deeply furrowed. (N)

- *Kalmia latifolia*, Mountain laurel—elliptical, leathery, glossy evergreen leaves are dark green above and yellow-green below. Cup-shaped white flowers with darker markings are borne in inflorescences. The habit is usually gnarled in the woods. (N)
- *Liriodendron tulipifera*, Tulip poplar—alternate leaves are truncate at the top (shaped like a tulip flower), turning yellow in the fall. Tall, straight trunks characterize this tree in the woods. Flowers are very showy and also look like tulips, but are borne high in the tree and may not be noticed. (N)
- *Quercus alba*, White oak—alternate leaves have five to seven rounded lobes and turn brown to red in fall. This oak has a thick trunk, wide-spreading branches, and oblong acorns. (N)
- *Nyssa sylvatica*, Sour gum, Tupelo—simple leaves with entire margins are nondescript. Easy to identify in the fall due to its early orange-red fall color. Local wisdom says if you can't identify a tree in the woods, it is likely to be a nyssa! (N)

WOODY OLDFIELD/MEADOW

- *Acer rubrum*, Red maple—red flowers, red leaf petioles, and red fall color typify this maple. Leaves are three-lobed and opposite. (N)
- *Cercis canadensis*, Redbud—heart-shaped, alternate leaves turn yellow in the fall. Distinguishing characteristics for redbuds include dark grey bark and purple flowers borne directly in the stem in early spring. (N)
- *Diospyros virginiana*, Persimmon—elliptical, alternate, glossy green leaves. Bark is chunky and edible fruit matures to an orange-reddish color. (N)
- *Liquidambar styraciflua*, Sweetgum—alternate leaves are five-lobed in a star-like pattern, turning yellow, red, and purple on the same tree in fall. Fruit is brown, spiked balls. (N)
- *Myrica pensylvanica*, Bayberry—semi-evergreen leaves are one to two inches long and often twisted. They are fragrant when crushed and often stickly. Fruit is a wrinkled berry with a blue-purple waxy coating. (N)
- *Prunus serotina*, Black cherry—finely toothed leaves are shiny on the upper surface with a long pointy tip. White flowers are borne in drooping racemes. (N)
- *Rhus typhina*, Staghorn sumac—pinnately compound leaves are two feet long and contain thirteen to twenty-seven pairs of toothed leaflets, turning orange-red in fall. Fruit are showy, dark red *drupes* (fleshy fruit with a thin skin surrounding a seed) in terminal pyramidal clusters. This colonizing shrub often forms large thickets. (N)
- *Sassafras albidum*, Sassafras—three different leaf forms appear on each plant (entire leaf, leaf with two asymmetrical lobes—resembling a mitten, and three-lobed leaves). Small yellow flowers borne in profusion in early spring can be showy. Fall color is a spectacular bright orange. (N)

COASTAL

- *Baccharis halimifolia*, Groundsel bush—coarsely toothed, grey-green foliage on brittle twiggy branches. Fleecy flowers cover the plant in fall. (N)

- *Ilex glabra*, Inkberry holly—leaves are flat, alternate, and larger than Japanese holly or boxwood. Black berries are present in fall and compact cultivars are available. (N)
- *Magnolia virginiana*, Sweetbay magnolia—semi-evergreen elliptical leaves have silvery undersides and creamy white fragrant flowers are borne after the foliage in late spring. (N)
- *Myrica cerifera*, Wax myrtle—semi-evergreen, olive-green, waxy leaves are coated with yellow resin glands providing fragrance to this colonizing shrub. Clusters of blue-grey fruit mature in fall and persist into winter, each covered with a waxy coating. (N)
- *Rhododendron viscosum*, Swamp azalea—one of many species of native deciduous azaleas, this species has lustrous green leaves clustered at the ends of branches. White fragrant flowers with long corollas are covered in a sticky substance. (N)

INVASIVE PLANTS

Many invasive plants "escape" from suburban neighborhoods and establish themselves—sometimes in shocking density—in the forest patches around our state. These plants are included here so you can learn to identify them, but they should NOT be planted in the landscape. In fact, they should be removed from planned landscapes whenever possible:

- *Acer platanoides*, Norway maple—shorter in stature than most native maples, this five-lobed-leaved maple can be differentiated from sugar maple by the presence of a white sticky substance at the end of the petiole when a leaf is removed. Yellow fall color emerges later than color does in most native trees.
- *Ailanthus altissima*, Tree-of-heaven—able to grow almost anywhere, this rangy tree has one- to four-foot pinnately compound leaves with ten to forty-one pairs of leaflets. While sometimes confused with native sumac, young walnut trees, or hickory, the leaflets are easily identified by the glandular, notched base on each leaflet. Large clusters of yellow flowers are followed by tan clusters of winged fruit.
- *Berberis thunbergii*, Japanese barberry—spiny shrub with obovate, green leaves. Cultivars sold in the nursery trade have red leaves, but seedlings that escape usually revert to green. Small yellow flowers are followed by fruit, which is spread by birds.
- *Buddleja davidii*, Butterfly bush—grey-green, four- to ten-inch-long leaves can be hairy on the undersurface. Four- to ten-inch terminal flower panicles are fragrant, and come in a variety of colors from white to pink and purple.
- *Elaeagnus umbellata*, Autumn olive—alternate leaves are bright green to grey-green above and covered with silvery scales below. Extremely fragrant flowers occur in clusters near the stem in early spring.
- *Euonymus alatus*, Burning bush—dark green leaves taper at the tips and turn a brilliant red in fall. Two to four corky wings often appear on young stems. Fruit is reddish capsules that split to reveal orange seeds.

- *Paulownia tomentosa*, Empress tree—large, hairy, heart-shaped leaves are sometimes mistaken for the native catalpa. Large, fragrant purple flowers are borne in upright clusters before the leaves emerge. Twigs are speckled with white dots (lenticels).
- *Pyrus calleryana,* Callery pear—alternate, shiny leaves have wavy, slightly toothed margins and turn a reddish-bronze fall color. Showy, white, malodorous flowers (pollinated by flies) appear in spring before foliage.
- *Rosa multiflora*, Multiflora rose—pinnately compound leaves with five to seven leaflets have toothed edges and fringed petioles (unlike other roses). Fragrant, white, five-petaled flowers are borne in clusters in early spring. Stems remain green throughout the winter.

IDENTIFYING HERBACEOUS PLANTS

Based on the skills you are developing by looking at woody landscape plants, you can now begin to explore a whole world of interesting herbaceous plants. By definition, an *herb* is a plant that does not have woody tissue that causes it to persist through the winter and therefore dies down to the ground when the temperature gets cold. We also use the term herb to describe plants used for culinary or medicinal purposes, but the definition used in this book refers to all plants with *herbaceous* or *soft* tissue.

There are many clues that can be used to differentiate herbs from one another, and there are a variety of useful identification keys available (see below for a recommended booklist). Some of these use flower color or the number of flower parts to identify herbs, and others use families. Understanding the difference between plant families can be an extremely helpful way to narrow down the possibilities for an unknown plant you are trying to identify in the field.

It is important to know flowering plant parts (or basic plant morphology) to differentiate plant families and species from one another.

Some tricks you can use to remember flowering plant parts are that the male plant parts, or *stamen* (pronounced "STAY MEN") make pollen, and there are more of these than the female plant part (called the *pistil*), which is comprised of the *stigma*, the *style*, and the *ovary*.

Petals are pretty recognizable, although when the old Disney cartoon shows Oswald the Lucky Rabbit saying "she loves me, she loves me not" as he plucks the "petals" of a daisy, they are not actually petals. These are the ray flowers, or *ligules*, of the Asteraceae family, which, together with disk flowers, make up the *composite head*.

Soft petals combine to create the *corolla*, which provides protection for the reproductive organs of the flowering plant, and to attract pollinators (and humans) with their beautiful colors, shapes, and smells. At the base of their inflorescence, the leaf-like whorl of sepals forms the *calyx*. Together, the corolla and the calyx create the *perianth*.

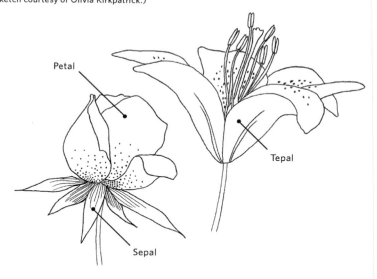

Figure 6.7. Petals, sepals and tepals on typical flowers. (Sketch courtesy of Olivia Kirkpatrick.)

Figure 6.8. Mint, composite, mustard, and parsley family typical structures. (Sketch courtesy of Olivia Kirkpatrick.)

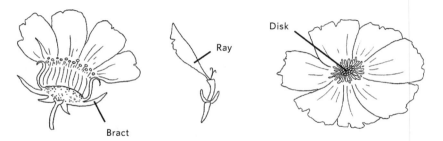

Figure 6.9. Ray and disk flowers of the composite family with bracts at the base of the inflorescence. (Sketch courtesy of Olivia Kirkpatrick.)

In the case of some flowers, like *Tulipa* and *Lilium*, the petals and sepals are indistinguishable, and are known as *tepals*, the flowers of which are comprised in multiples of three.

Here are some examples of some common plant families and keys to their identification. (A great resource if you want to learn more about plant families is the book *Botany in a Day*).

- *Lamiaceae*, Mint—the mint family is one of the easiest to identify. Plants in the mint family have square stems and opposite leaves and are often aromatic. The perianths of mint family flowers are always made up of five united sepals and five united petals. You might be familiar with the brightly colored annual, coleus, or common herbs such as peppermint, oregano, and sage. These are all members of the mint family!
- *Asteraceae*, Aster/Composite—the aster family is far-reaching and includes many different plants. It is the second largest plant family. It does not play by the rules as far as its flowers go, as the most helpful clues in its identification focus on its non-conformity to the morphology of the standard inflorescence. In most cases, inflorescences in this family are made up of ray and disk flowers. Sometimes a species only has one or the other. The base of the flower head is cupped by a series of *bracts*, which are modified leaves. Plants you may recognize from the Asteraceae family include sunflowers (*Helianthus*), coneflowers (*Echinacea, Rudbeckia*), and daisies (*Leucanthemum*).
- *Brassicaceae*, Mustard—the mustard family is tough and widespread, with the recognizable traits of radial arrangement of seedpods around the stem in a raceme, and an inflorescence with four petals, four sepals, two short stamens, and four tall stamens. Plants you may recognize from the mustard (Brassicaceae or Cruciferae) family include kale, broccoli, and other cruciferous veggies, as well as the annual alyssum and weeds such as the bittercress and pepperweed.
- *Apiaceae*, Parsley—plants in the parsley family are notable for their compound *umbels* (or umbrella-shaped inflorescence), which are comprised of many tiny flowers with five petals and five stamens. They also generally have hollow stems. Common edible plants like carrots, fennel, and parsley are in the parsley family.

DICOTS VS. MONOCOTS

While the members of the other families reviewed are *dicots*, meaning they have two embryonic leaves, members of the grass, lily, sedge, and rush families are *monocots*, and have a single embryonic leaf. Instead of netted veins, monocots have parallel veins. Their inflorescences are comprised of multiples of three parts, rather than four to five, as is typical for dicot families. Two monocot families are described in a bit more detail below.

- *Liliaceae*, Lily—the lily family is recognizable by its showy inflorescence, with its modified petals and sepals known as tepals. The lily family has been further

FIGURE 6.10. Common monocot families—lily and grass. (Sketch courtesy of Olivia Kirkpatrick.)

divided into multiple subfamilies, but for learning purposes, if you see a plant with parallel veins that is not grass-like, and it has showy flowers in parts of three, then you can be sure it is related to the lily family. Common plants you may be familiar with, such as tulips, daylilies, crocus, trillium, asparagus, and onion, are all in subfamilies of the lily family.

- *Poaceae*, Grass—the grass family has a less showy inflorescence than the lily family. Hallmarks of the grass family include hollow stems interrupted by joints or nodes and inflorescences that are wind-pollinated, and thus not very showy. Most of the grains we consume (such as wheat, oats, corn, and barley) come from the grass family, as do ornamental meadow grasses like switchgrass and bluestem, and of course the quintessential American lawn, which is comprised of fescue and Kentucky bluegrass.

Grasses are divided into a few subcategories that will be helpful to understand. The first are *clumping* and *running* grasses. Clumping grasses, also called bunch or tufted grass, increase in girth as they grow, and thus have a more predictable mature size, while running grasses increase via *stolons* or *rhizomes*, which are modified stems that spread either above or below ground, and can quickly fill an area. The latter can be good for soil stabilization, but, depending on the plant, can get out of hand if not carefully monitored. The second division of grasses that aids in identification is *warm* versus *cool* season grasses. Warm season grasses start to grow only when the soil and air are warm enough, flower in the late summer, and go dormant and change into gorgeous colors in colder temperatures. Conversely, cool season grasses grow the most in late winter and early spring, and again in the fall, they bloom early, and go dormant when warmer midsummer temperatures and dry conditions set in. By understanding these categories, you can narrow down your grasses when practicing identification.

COMMON HERBACEOUS PLANTS OF DELAWARE

In the following pages, we describe several species you are likely to encounter throughout the state of Delaware. While we encourage the planting of native species when possible, many excellent non-native ornamentals are common in the trade and useful in the landscape. Native plants for various sites are also included. These are listed by their cultural conditions (i.e., whether they prefer sun or shade, dry or wet conditions) and include mature size—Tall (T) and Wide (W)—and family. Refer to any of the excellent books or other resources listed at the end of this chapter for images and additional information about the plants.

ORNAMENTAL NON-NATIVES

In need of full sun:

- *Achillea millefolium (naturalized),* Common yarrow (1'-3' T × 1'-3' W) (Asteraceae)—this tough member of the aster family is found throughout the world, and was brought to the U.S. so long ago that it is considered to have naturalized. Recognizable by its fern-like clumps of basal foliage, and composite corymbs made up of teeny tiny disk and ray flowers, these hardy plants can survive pretty much anything you put them up against. While the species (or true version) of the plant is off-white, many cultivars have been selected in colors of orange, pink, rose, and magenta. This plant is tolerant of salt and coastal conditions.
- *Crocosmia × crocosmiiflora*, Crocosmia or montbretia (1.5'-3' T × 1' W) (Iridaceae)—related to the iris, crocosmia is another bright flower borne upon a *scape*, or leafless stalk above a clump of sword-like foliage. A grouping of 1.5-inch tubular flowers face one direction from a zig-zag stem. The flower at the tip of the stem blooms last. It has many hybrid cultivars with varied and rich shades of orange, yellow, and red have been bred.
- *Hemerocallis hybrids*, Daylilies (1'-3' T × 1'-5' W) (Asphodelaceae)—this ubiquitous plant is recognizable by its strap-like foliage, with parallel veins and six parted flowers made of tepals. It can come in virtually any color as it has been so cultivated and hybridized, but is often seen in orange and yellow hues. Some of the parentage for these cultivars comes from a species called *Hemerocallis fulva*, which can be seen along roadsides and at field edges across the country. Note that *H. fulva* is considered by many to be invasive and can be found in large monocultures in the Delaware Valley.
- *Kniphofia uvaria*, Red hot poker or torch lily (Asphodelaceae) (2'-4' T × 3' W)—this eye-catching plant is native to South Africa. Mostly evergreen clumps of sword-shaped leaves support bright reddish-orange flowers on dense racemes, which turn yellowish-green as they mature, providing a bi-color appearance. These flowers cluster on the upper six to ten inches of the scape. Early summer flowers must be cut after blooming for the plant to provide continuous color throughout the season. This plant is tolerant of salt and coastal conditions.

- *Nepeta* × *faassenii*, Catmint (Lamiaceae) (10"-24" T × 12"-24" W)—a member of the mint family, catmint is a low-growing (between one- and two-feet high), tough perennial that can take full sun and tolerates drought and salt. It has scalloped, pubescent grey-green leaves and light purple flowers borne in racemes. Its aromatic leaves and stems are not as beloved by cats as its cousin, catnip, but its long flowering season is enjoyed by many pollinators. Opposite leaves, tubular flowers, and ease of culture make this a winner in the Delaware landscape.
- *Perovskia atriplicifolia*, Russian sage (Lamiaceae) (1-3' T × 1-5' W)—another member of the mint family, Russian sage is a perennial favorite in the sunny garden. The toothed leaves are grey-green and pubescent, which indicates this plant is drought tolerant, as these light fuzzy leaves can reflect sunlight and protect the plant from water loss. Long panicles of blue-purple tubular flowers waft above the loose foliage. While beautiful in well-drained and sunny conditions, the humid Delaware summer can be brutal for this long-blooming plant.

Shade tolerant:
- *Astilbe* × *arendsii*, False spirea or astilbe (2'-4' T × 2' W) (Saxifragaceae)—a popular early summer bloomer, the six-inch pink, red, white, or purple panicles of the tiny flowers of false spirea appear above clumps of fernlike, ternately compound leaves (leaves in multiples of three) with toothed leaflets. Native to China, Japan, and Korea, a wide variety of cultivars provide the gardener with lots of color and size choices. Adequate moisture and part- to full shade will ensure full growth.
- *Anemone* × *hybrida*, Windflower or Japanese anemone (2'-4' T × 2' W) (Ranunculaceae)—this lovely late summer and fall blooming plant from the buttercup family boasts white to pink two- to three-inch flowers on graceful stems above a mound of trifoliate (three-part) leaves. The flowers generally have six to nine indistinguishable tepals around a grouping of yellow central stamens.
- *Galium odoratum*, Sweet woodruff (6"-8" T × 1'-1.5' W) (Rubiaceae)—whorled leaves in groupings of six to eight clasp the stem of this delicate ground cover. When crushed, leaves smell like newly mown hay. In late spring, white cymes of tiny flowers cover the foliage. Can be aggressive when given proper moisture.
- *Helleborus orientalis*, Lenten rose (1'-1.5' T × 1'-1.5' W) (Ranunculaceae)—this early blooming tough plant is a great choice in shady environments that need all-season interest, due to its evergreen leaves and showy, three-inch to four-inch-wide, cup-shaped flowers in varied colors of white to peach to purple. The nodding flowers, which have five sepals and surround contrasting yellow stamens, bloom in late winter and can last up to ten weeks.
- *Heuchera villosa*, Hairy alum root (1.5'-2.5' T × 1'-2' W) (Saxifragaceae)—a good plant for moist shade, the seven- to nine-lobed leaves of hairy alum root and its stems are covered in brown hairs, hence the common name. This species provides the parentage for many of the well-known commercially available cultivars, which attributes to their tolerance of heat and clay soils. White-pink

flowers raise above clumps of foliage in August and September on three-foot-tall panicles.
- *Hosta species*, Plantain lily (variable size) (Liliaceae)—the shade garden would not be complete without the hosta, its recognizable large basal leaves with long petioles clumped around a graceful white or purple trumpet-shaped grouping of flowers on a scape. A member of the lily family, this plant is recognizable because of its parallel venation. Cultivated for the variety to be found among its foliage, there is a hosta for every need.

Grasses and sedges:
- *Carex morrowii*, Japanese sedge (1'–1.5' T × 1.5'–2' W) (Cyperaceae)—a great evergreen choice for the shade garden, Japanese sedge is a clumping ground cover that can fill in a moist area of the garden. Cultivars such as "Aurea Variegata" and "Ice Dance" provide variegated foliage and can lighten up a shady corner of the garden. Brown, wind-pollinated flowers appear in late to early summer and are not conspicuous.
- *Calamagrostis × acutiflora*, Feather reed grass (3'–5' T × 1'–2' W) (Poaceae)—this handsome European grass is popular in trade, especially the cultivar "Karl Foerster," and not without good reason. It is a cool season grass that grows most in the spring and fall, rather than in the heat of summer. Feathery flower spikes rise above 1.5- to 3-feet-tall sharply pointed foliage; seeds persist into winter, providing four-season interest. Feather reed grass tolerates heavy clay, thus is a great choice for Delaware landscapes. Full sun to part shade.
- *Hakonechloa macra*, Japanese forest grass or hakone grass (1'–1.5' T × 1'–2' W) (Poaceae)—the arching bright green leaves of this graceful, cool season grass brighten shady corners. Some cultivars have variegation on the linear leaves, providing even more accent to the shade garden. Native to the mountains of Japan, it blooms inconspicuous flowers in mid-late summer. In warmer microclimates, may be semi-evergreen, but plants benefit from a late winter cut back for best effect.
- *Festuca glauca*, Blue fescue (.75'–1' T × 1' W) (Poaceae)—found in full sun with good drainage, clump-forming, evergreen blue fescue is popular on roof gardens and in traffic islands. Tufts of fine foliage range from true blue to grey, sometimes tinged with purple, and panicles of flowers appear in early summer.

NATIVE PLANTS FOR SPECIFIC LOCATIONS

Woodland (shade to part-shade natives):
- *Eurybia divaricata*, White wood aster (1'–2.5' T × 1.5'–2.5' W) (Asteraceae) (N)—an excellent plant for the woodland garden, white wood aster will spread and fill under trees and can tolerate competition for moisture from roots as well as hold its own against invasive plants. White ray flowers surround red or yellow disk flowers on the abundant composite inflorescences that bloom from late summer through fall. Heart-shaped, coarsely toothed leaves distinguish this member of the Asteraceae family.

- *Asarum canadense*, Wild ginger (.5'–1' T × 1'–1.5' W) (Aristolochiaceae) (N)—this excellent, low ground cover for the native garden has two leaves covered in fine soft hair that emerge in spring. It forms large colonies in shady, moist conditions, and blooms in early spring with an unmistakable "little brown jug" that hangs below the leaves to attract fly or beetle pollinators.
- *Polygonatum biflorum*, Small Solomon's seal (2'–3' W × 1'–2' T) (Liliaceae) (N)—arching stems dangle pairs or trios of .5- to 1-inch greenish flowers beneath alternate leaves. This spreading plant is in the Liliaceae family and has parallel veins as typical of monocots. Blue-black berries replace the flowers in the fall and the leaves turn a cheerful yellow, providing three seasons of interest.
- *Polemonium reptans*, Jacob's ladder (1'–1.5' T × 1'–1.5' W) (Polemoniaceae) (N)—this plant has leaflets of pinnately compound leaves arranged alternately in a ladder-like fashion along the loose and sprawling stems. This shade tolerant woodland plant has true blue, bell-shaped flowers in late spring and early summer.
- *Pachysandra procumbens*, Allegheny spurge (.5'–1' T × 1'–2' W) (Buxaceae) (N)—this semi-evergreen shade-loving ground cover is related to the invasive *Pachysandra terminalis* (Japanese spurge) but does not have its aggressive qualities and is native to the Southeast. New growth is mottled green and purple in whorls with toothed apices and entire bases. While leaves can survive our winters in Delaware, removal of the previous year's growth highlights attractive new leaves in spring. Fragrant white flowers in short (two- to four-inch) terminal spikes emerge before new growth in spring.

Ferns:
Unlike plants from the other families, ferns do not flower and set fruit or seed in the traditional sense. Instead, they go through an amazing life cycle in which the mature fertile *fronds* release *spores*, which reproduce asexually to create a gametophyte, which then releases male and female cells to sexually create a new fern plant. Ferns can be identified by the shape of their overall leaf, or frond, which includes the *blade*, or leafy portion of the plant, usually subdivided into leaflets and subleaflets known as *pinnae* and *pinnules*. The stem of a fern is known as the *stalk* below the leaf blade, and as *rachis* within the blade. Often spores are covered by protective shells called *sori*, which burst or disintegrate when it is time to begin the reproductive cycle.
- *Adiantum pedatum*, Northern maidenhair fern (1'–2' T × 2'W) (Pteridaceae) (N)—this lovely deciduous fern holds its fronds out like a flat, circular fan or bird's foot. The delicate stem is a dark reddish brown and contrasts beautifully with the lime-green pinnae. It is said that this dark stem was once used to weave into baskets. Requires shade and good soil moisture to avoid browning out in late summer.
- *Dryopteris marginalis*, Marginal wood fern (1.5'–2' T × 1.5'–2' W) (Dryopteridaceae) (N)—this easy-to-grow, evergreen fern shoots up a flush of thick leathery leaves in spring. These fronds are intended to last all year and help identify the plant as they display prominent large round sori on the back side of the leaves, organized along the margins of the pinnules. Can tolerate more drought than other members of the genus.

- *Matteuccia struthiopteris*, Ostrich fern (2'-4' T × 3'-6' W) (Onocleaceae)—another deciduous species, the ostrich fern is the source of the culinary curiosity of *fiddleheads*, which many foragers look for in early spring. These are the tightly curled precursors to the fern's elegant and upright fronds, which resemble the plumage of an ostrich. The green or vegetative fronds lose their attractiveness by late summer, but sterile fronds linger through winter and extend the interest of this plant for another season.
- *Onoclea sensibilis*, Sensitive fern (1'-2' T × 2'-3' W) (Dryopteridaceae) (N)—instantly recognizable by its round-lobed, chartreuse leaves and ability to form large patches in sunny to shady conditions, the sensitive fern is easy to grow and is widespread in the area. Even better, it does not seem to mind the variable soils found in our region, from rocky Piedmont outcrops to clay flats in the White Clay Creek region. What it does mind is the first frost, and this deciduous fern essentially melts into the landscape at the first sign of true winter in the late fall.
- *Polystichum acrostichoides*, Christmas fern (.5'-1.5' T × 1'-2' W) (Dryopteridaceae) (N)—tough and a good selection for a variety of conditions, this evergreen fern gains its common name because of its Christmas stocking-shaped pinnae, as well as the fact that it can be used as holiday decoration. Clumping, and not particularly quick to spread, the Christmas fern is still a great choice for the Delaware Valley due to its tolerance of drought conditions as well as of clay soils.

Woody oldfield/meadow (sun natives):
- *Asclepias tuberosa*, Butterfly milkweed (1'-2.5' T × 1'-1.5' W) (Asclepiadaceae/Apocynaceae) (N)—part of the genus named for the Greek god of medicine, Asklepios, this species of milkweed, along with *A. incarnata* and *A. speciosa*, support the larvae of the monarch butterfly. Narrow, linear leaves along hairy stems support umbels of orange clusters of flowers. Ornamental seedpods follow the long bloom period.
- *Baptisia australis*, False indigo (3'-4' T × 3'-4' W) (Fabaceae)—bluish green, trifoliate, oval foliage on shrub-like plants supports pea family flowers in lovely indigo blue racemes up to a foot long. Cultivars have been developed in yellow, purple, and white. Blooms fade to persistent, black seedpods, which rattle when shaken and provide winter interest. This plant is tolerant of salt and coastal conditions.
- *Liatris spicata*, Blazing star (2'-4' T × .75'-1.5' W) (Asteraceae) (N)—the upright structure, fine, grass-like leaves, and long purple spikes of the blazing star provide an architecturally interesting punctuation in the native meadow garden and other full sun locations. While this plant is a member of the Asteraceae family, note that it has no ray flowers, only disk flowers. Likes moisture, but the *corm* (underground stem) that produces the plant cannot tolerate sitting in wet conditions all winter long.
- *Solidago caesia*, Blue stem goldenrod (1.5'-3' T × 1.5'-3' W) (Asteraceae) (N)—some of our native goldenrods, particularly *Solidago canadensis* and *Solidago rugosa*, while great in meadows and other large-scale plantings, can be quite aggressive and out of scale for the home landscape. While more diminutive,

blue stem goldenrod is a low-maintenance plant that tolerates both clay soil and drought conditions, and sports bright yellow, tiny clusters of daisy-shaped blooms in late summer and fall, which appear along the length of the stems at the base of its linear, alternate leaves. Blue stem goldenrod can also take some shade, though has better bloom in full sun.
- *Symphyotrichum novae-angliae*, New England aster (3'-6' T × 2'-3' W) (Asteraceae) (N)—robust plants with typical Asteraceae blooms, with purple ray flowers surrounding cheerful yellow disk flowers, New England asters are stalwarts of the fall garden. Their linear, hairy leaves are *sessile*, meaning they clasp stems of the plant, which are hairy too. Can tolerate clay soil.

Raingarden plants:
- *Eutrochium purpureum*, Joe-pye weed (5'-7' T × 2'-4' W) (Asteraceae) (N)—recognizable for its whorled, serrated leaves, which appear at purplish leaf nodes along tall stems, this moisture-loving plant thrives in roadside ditches and the open wet woodlands of the mid-Atlantic. Large compound inflorescences of pale mauve flowers top the plants in mid-late summer.
- *Hibiscus moscheutos*, Swamp rose mallow (3'-7' T × 2'-4' W) (Malvaceae) (N)—masses of large (four- to six-inch) white to pink flowers appear on the swamp rose mallow in late summer and early fall atop the statuesque plant, which is native to wetlands including marshes, swamps, and riverbanks. Sometimes lobed, its oval leaves are arranged alternately along its hairy stems.
- *Iris versicolor*, Blue flag iris (2'-2.5' T × 2'-2.5' W) (Iridaceae) (N)—a great architectural feature for the rain garden, the arching, sword-like leaves of blue flag iris are instantly recognizable. In late spring and early summer, purple-blue flowering stalks emerge. This swamp and marsh native can tolerate wet feet and shallow standing water.
- *Lobelia cardinalis*, Cardinal flower (2'-4' T × 1'-2' W) (Campanulaceae) (N)—easily recognized by its bright red racemes of tubular flowers that bloom in late summer and early fall, cardinal flower can survive short periods of flooding. From a clump of evergreen basal leaves emerge alternate leafed stalks that support the bloom. Though individual plants are short-lived, they can self-seed in the garden. Also of note is the comparable species *Lobelia siphilitica*, which has a similar habit, but purple flowers.
- *Packera aureus*, Golden ragwort (.5'-2.5' T × .5'-1.5' W) (Asteraceae) (N)—this moisture-loving native thrives in wet and shady areas. Evergreen, oval basal leaves provide winter protection, out of which a stalk topped by a corymb of bright golden flowers, typical of the Asteraceae family, appears in spring. Stem leaves are deeply lobed.

Grasses and sedges (native):
- *Andropogon gerardii*, Big bluestem (warm) (3'-9' T × 3'-5' W) (Poaceae) (N)—big bluestem was the tallest and most common grass in the tallgrass prairie and is still worthy to include in many landscapes, with its rusty orange and red foliage

contrasting against the blue autumn sky. Many cultivars are available in the trade, making for great selection for height and colorways. Easily recognizable for its "turkey foot" bloom, which turns to seed and persists for winter interest.
- *Carex pensylvanica*, Pennsylvania sedge (.5'–1' T × 1'–1.5' W) (Poaceae) (N)—this popular woodland ground cover is a great choice for dry shade. It has a fine texture and is slow growing, but once a stand is established, it is a lovely native understory in a woodland setting. Reliably evergreen, but may need a trim to refresh in late winter.
- *Panicum virgatum*, Switchgrass (warm) (3'–6' T × 2'–5' W) (Poaceae) (N)—upright form and tolerance of clay, drought, and occasional inundation are hallmarks of this often-planted species. Many cultivars provide variation in height, habit, and color. Recognizable for its graceful pinkish panicles that rise over the foliage in mid-late summer and convert to lovely seed heads to provide winter interest.
- *Schizachyrium scoparium*, Little bluestem (warm) (1'–4' T × 1.5'–2' W) (Poaceae) (N)—its smaller stature and reliable performance make this a great choice in the Delaware landscape. Tiny wisps of "eyelash" seed heads identify this grass, known for its blue grey to bronze coloration, maturing to a ruddy orange foliage in fall. Many excellent cultivars are available.
- *Sorghastrum nutans*, Indian grass (warm) (3'–5' T × 2'–3' W) (Poaceae) (N)—another dominant plant of the tallgrass prairie, Indian grass is well suited to the poor clay soils of Delaware. Highly recognizable for its beautiful yellow-orange color in fall that lines the roadways in Northern Delaware.

INVASIVE HERBACEOUS PLANTS

- *Alliaria petiolata*, Garlic mustard (Brassicaceae)—in the first year, a small clump of heart-shaped, toothed leaves up to six inches long is produced by this biennial herbaceous plant. In the second year, a one-foot- to four-foot-tall flowering stalk bolts from the center of the clump. Early spring clusters of tiny white four-petaled and four-sepaled flowers, typical of the Brassicaceae family, are followed by green pods, which proliferate and spread this highly shade tolerant species. In addition to forming large stands in our wild lands, the plants produce *allelopathic* compounds that inhibit other species. It is easily recognizable by the garlic smell of all plant parts when crushed.
- *Vinca minor*, Periwinkle (Apocynaceae)—this popular, trailing, evergreen ground cover is often used as an ornamental planting and is still widely sold. Notable and beloved for its five-petaled, lavender flowers in April and May, periwinkle has escaped cultivation and is taking over our natural and wild areas. Oppositely arranged, oval, thick leaves tolerate shade, and slightly woody stems root at leaf nodes, propagating vegetatively throughout the region.
- *Hedera helix*, English ivy (Araliaceae)—long planted and maintained for its waxy, evergreen leaves and vining and trailing habit, English ivy has escaped cultivation and completely taken over natural forests, forest fragments, and wildlands throughout the United States. Slightly woody, this plant proliferates through its woody stems and

can climb and destroy a tree through inhibiting photosynthesis, choking or girdling its trunk, or pulling it down in storms due to the additional weight.
- *Microstegium vimineum,* Japanese stiltgrass (Poaceae)—this prolific invader of moist woodlands and shady ditches has alternate light green leaves with a prominent silver center line dividing the blades. The leaves are slightly pubescent and the wiry yellowish stems can root at nodes, further increasing Japanese stiltgrass's ability to take over an area in a short time. Raceme flowers appear in late summer through mid-fall and are followed by grain-like seeds that persist through winter and are dispersed via waterways and by humans and other animals. Shade tolerant and allelopathic.
- *Pachysandra terminalis,* Japanese spurge (Buxaceae)—this aggressive evergreen ground cover is related to the native *Pachysandra procumbens* (Allegheny spurge) and has been extensively planted in the North American landscape. Shiny green leaves in whorls will quickly take over an area. White flowers in three-inch terminal spikes emerge before new growth in spring.

LEARNING RESOURCES

There are many great books out there to help with plant identification, both at home and in the field. Listed below is a representation of some of these additional resources.

FOR NOMENCLATURE AND IDENTIFICATION

- L. H. Bailey, *Manual of Cultivated Plants* (New York: The MacMillan Co., 1949).
- L. H. Bailey, E. Z. Bailey, Staff of Liberty Hyde Bailey Hortorium, *Hortus Third* (New York: MacMillan Publishing Co., 1987).

PLANT FAMILIES

- T. Elpel, *Botany in a Day: The Patterns Method of Plant Identification, An Herbal Field Guide to Plant Families of North America,* 6th Edition (Pony, MT: Hops Press, 2013).

FOR IDENTIFICATION AND GENERAL CHARACTERISTICS

- A. M. Armitage, *Herbaceous Perennial Plants: A Treatise on their Identification, Culture, and Garden Attributes, third ed.* (Athens: University of Georgia, 2008).
- C. F. Brockman, *Trees of North America* (New York: Golden Press, 2001).
- S. Carter, C. Becker, and B. Lilly, *Perennials: The Gardeners' Reference* (Portland, OR: Timber Press, 2008).
- R. Clausen and T. Christopher, *Essential Perennials* (Portland, OR: Timber Press, 2014).
- W. Cullina, *Native Ferns, Moss & Grasses* (New York: New England Wildflower Society and Houghton Mifflin, 2008).
- P. Del Tredici, *Wild Urban Plants of the Northeast* (Ithaca, NY: Comstock Publishing, 2010).
- M. A. Dirr, *Manual of Woody Landscape Plants* (Champaign, IL: Stipes Publishing Co., 2009).
- M. A. Godrey, *A Sierra Club Naturalist Guide to The Piedmont* (San Francisco: Sierra Club Books, 1980).

- C. E. Phillips, *Woody Vines, Shrubs and Trees of Delaware and the Eastern Shore: A Guide to Identification in the Summer* (Newark: Plant Science Dept., University of Delaware, 1974).
- S. Still, *Manual of Herbaceous Ornamental Plants* (Champlain, IL: Stipes Publishing, 1994).
- W. S. Tabor, *Delaware Trees* (Dover: Delaware State Forestry Department, 1960).
- A. T. Viertel, *Trees, Shrubs and Vines: A Pictorial Guide to the Ornamental Woody Plants of the Northern United States Exclusive of Evergreens* (Syracuse, NY: Syracuse University Press, 1970).

GENERAL REFERENCE FOR LANDSCAPE CHARACTERISTICS AND USE

- D. C. Peattie, *A Natural History of North American Trees* (New York: Bonanza Books, 2013).

FIELD GUIDES

- National Audubon Society, *Field Guide to North American Trees: Eastern Region* (New York: Alfred A. Knopf Publishers, 2000).
- L. Newcomb, *Newcomb's Wildflower Guide* (Boston: Little, Brown, and Co., 1977).
- R. T. Peterson and M. McKenny, *A Field Guide to Wildflowers of Northeastern and Northcentral North America*, Peterson Field Guides (New York: Houghton Mifflin, 1968).
- G. Petrides, *Peterson Field Guides to Eastern Trees*, Peterson Field Guides (New York: Houghton Mifflin, 1998).

FIGURE 7.1 An insect's body plan includes a head, thorax, and abdomen, two sets of wings, and three pairs of legs. (Photo courtesy of Doug Tallamy.)

CHAPTER SEVEN

Insects: The Little Things That Run the World

DOUG TALLAMY

MANY OF US SUFFER under the misconception that human societies are sustained by human ingenuity, technology, ever-growing economies, and the (occasional) Einsteins among us. Not so. We humans (and most other terrestrial species on earth) are sustained by insects! The famed entomologist E. O. Wilson has called insects "the little things that run the world" because of the many essential ecological roles they play every day.[1] It is insects that pollinate 87.5 percent of all plants, and 90 percent of all flowering plants;[2] and it is those plants that turn energy from the sun into the food that we and an unimaginable diversity of birds, mammals, reptiles, amphibians, and freshwater fish need to exist. Insects are also the primary means by which the food created by plants is delivered to these animals. Most vertebrates do not eat plants directly; far more often, they eat insects that have converted plant sugars and carbohydrates into the vital proteins and fats that fuel complex *food webs*.

And so, it is insects that sustain Earth's ecosystems by sustaining the plants and animals that run those ecosystems. And the more plants and animals, the better: ecosystems with many interacting species are more stable, more productive, and better able to support huge human populations than depauperate ecosystems with few species. Insects also provide much of the planet's pest control in the form of millions of species of predators and parasitoids (tiny wasps that develop in and kill their hosts) that keep food webs in balance. It is insects that rapidly decompose dead plants, releasing the nutrients they contain for use by new plant life. And, by keeping the planet well vegetated, it is insects that maintain the watersheds in which we all live, providing us with clean water and minimizing the frequency and severity of floods. As if all of that were not enough, the plants that insects pollinate sequester enormous amounts of carbon within their bodies and within the soil around their roots, carbon that would otherwise be in the atmosphere, wreaking havoc on the Earth's climate.

Indeed, humans would last only a few months if insects were to disappear from Earth. It is remarkable, then, that our cultural relationship with insects is built on disgust and animosity rather than on awe and appreciation. We have created a culture in which insects and their arthropod relatives are routinely maligned and worse, exterminated in the name of protecting crops and fighting a few disease vectors. We have declared war on all insects and we kill as many of them as possible, whenever and wherever we can. We have sponsored "National Insect Killing Week" and taught our children to fear them, rather than to respect those few that might sting to defend their nest, and admire the rest.[3]

And so we are winning our undeclared war against insects—but at our peril. Precipitous declines in populations of the European honey bee, the four thousand species of bees native to North America, and beautiful butterflies like the monarch and Karner blue have gotten our attention, but many other insects are disappearing utterly without notice. We have already driven three North American species of bumblebees to extinction,[4] and some 30 percent of the grasshoppers, crickets, and katydids of Europe are facing extinction.[5] Flying insects in Germany have declined 79 percent in abundance and diversity since 1989, and forty-six species of butterflies and moths have disappeared from German soil altogether.[6] Similar statistics are coming to light in England and other parts of Europe and the American tropics. By killing insects, we are biting the hand that feeds us, and that has led to the most alarming statistic of all: invertebrate abundance (read: "insect population") has been reduced 45 percent globally since 1974.[7]

It is obvious that we must form a new and more positive relationship with insects, and we must do it now. The good news is, you do not need to be an entomologist to help in this regard. As Master Naturalists, a cursory knowledge of the most diverse and numerous multicellular organisms on earth is required but, in my view, it is more important that you understand the ecological value of insects and learn how to lead the charge in saving them.

WHAT IS AN INSECT?

All insects are members of the phylum Arthropoda: organisms with a segmented body, an exoskeleton, and jointed legs. Insects can be distinguished from other arthropods like spiders, crustaceans, and centipedes by having three primary body parts—head, thorax, and abdomen—and only three pairs of legs. The head bears the eyes, antennae, and mouth of the insect, the thorax is the locomotion engine and houses the wings, the legs, and the muscles that work them, and the abdomen is the home of the digestive, renal, and reproductive systems. Insects are the only invertebrates that have evolved the ability to fly, and this, along with their small size, has enabled them to occupy nearly every niche on earth and diversify into more than four million species—some four hundred times the number of bird species.

And so, a crucial question: how are we going to coexist with insects in human-dominated landscapes?

STEP ONE: RESTORE THE PLANTS ON WHICH INSECTS DEPEND

Thank goodness it is "little things" that run the world. If we needed to share our neighborhoods with big things like tigers, elephants, bison, and giraffes, we would be challenged, indeed. Not only is it easy to create a world in which insects can coexist with humans, but it is easy to create landscapes in which they can actually flourish. All we need is more of the right plants in our landscape designs. But which are the right plants? To answer that question, we need to decide which insects we want to help.

There are a lot of insect species in the world; three million to four million by most estimates, 164,000 of which have been described in the U.S. (a great many species remain undescribed!). What they all have in common is that they are directly or indirectly tied to plants, either by eating some part of a living plant, existing solely on dead plant tissues like fallen leaves or rotting logs, developing on the muck dead plants create when they fall into water, eating other insects that have developed directly on plants, eating insects that have eaten insects that have developed on plants, or being a parasite on a mammal that has eaten plants, etc. You get the picture: insects need plants. But the class Insecta is so large and so broad that trying to create a habitat for all of them in a given space is not only impractical, it is impossible.

I suggest we narrow our focus a bit by enhancing populations of the two insect groups that arguably have the greatest impact on terrestrial ecosystems: the first includes caterpillars, the larvae of moths, butterflies, and skippers, as well as the group of Hymenoptera we call sawflies; and the second includes the four thousand species of bees native to North America that are responsible for most of the pollination required by plants.

SPOTLIGHT ON CATERPILLARS

"The early bird catches the worm," the saying goes, but that doesn't really capture it. Caterpillars are not worms, but they are in fact the mainstay of most bird diets in North America, particularly when birds are rearing their young. Only thrushes like the American robin regularly feed their babies earthworms, but most thrush nestlings still feed on caterpillars.

So birds need caterpillars to survive, and this means we ought to landscape in a way that builds the populations of such an important group of insects. For one thing, there are many types of caterpillars to work with; estimates of the number of species of Lepidoptera (moths and butterflies) in North America top fourteen thousand. This is fewer than the number (twenty-five thousand) of beetle species in North America, but unlike most caterpillars, beetles are hard prey for birds to find and eat, and therefore do not contribute as much to local food webs. They spend most of their lives hidden underground, within seedpods, or tunneled deep in wood. Beetles also sport much thicker exoskeletons than caterpillars, particularly as adults, and often have spiny, stiff legs that make them difficult for creatures like birds to eat and digest. Caterpillars, by contrast, are typically exposed on vegetation and their exoskeleton is thin and flexible, making most of a caterpillar digestible food. Think of caterpillars as little nutritious sausages and you will understand why their texture may be one of their most important attributes. If you have ever watched a bird feed its nestling, then you know it is not always a gentle process; many adult birds forcibly stuff food down their nestling's throat, using their beak as a plunger. Insects with sharp edges can injure delicate little nestlings during such feeding bouts.

Most caterpillars are also relatively large compared to other kinds of insects. It takes two hundred aphids, for example, to equal the weight of a single medium-sized caterpillar. If you are a bird looking for insects, would you choose to hunt and handle two hundred aphids, or find one caterpillar?

Figure 7.2. It is important that food for baby birds be free of sharp edges because parent birds use their beaks to stuff insects down their offspring's throat like a plunger. (Photo courtesy of Doug Tallamy.)

Also, caterpillars are more nutritious than most other insects. They are high in protein and fats, and are the best source of carotenoids for birds, particularly during the breeding season, when few carotenoid-rich berries are available.[8] It follows, then, that if birds need protein-rich prey that are high in carotenoids to raise healthy young (they do), and if caterpillars provide the best and most easily obtained source of such nutrition during the breeding season (they do), then caterpillars may not be optional components of breeding bird diets. It is also likely that they are essential to successful reproduction. As with all parts of nature, there are exceptions to this generality. A few bird lineages like finches, doves, and crossbills, can rear young on a milky substance they make from seeds, and raptors for the most part feed their young on mammals, fish, or other birds. But some 96 percent of North America's terrestrial bird species rear their young on insects rather than on seeds and berries,[9] and for most of those species, the majority of those insects are caterpillars or adult moths.[10] Caterpillars are so important to breeding birds that many species may not be able to breed at all in habitats that do not contain enough caterpillars.[11]

How many caterpillars are "enough"? That, of course, depends on which bird species we are talking about. There are enough caterpillars in a habitat when parent birds can find caterpillars fast enough to enable three to six nestlings to grow from the egg to slightly larger than an adult in under two weeks for most cup nesters, and a little longer for cavity nesters. This is an astonishingly fast growth rate. To achieve such rates of growth, nestlings must eat often. You and I eat three to four times a day (maybe five if we include snacks). A typical nestling, in comparison, eats a full meal thirty to forty times a day! That means a parent (or couple) raising five chicks must bring food to the nest about 150 times a day. Most birds forage primarily within a well-defined territory surrounding the nest. For Carolina chickadees, this is about fifty meters in all directions from the nest, or an area approximately two acres in size. Thus, nesting territories must contain a lot of food concentrated in a relatively small area or the nest will fail.

So how many caterpillars is that? Few people have actually sat and counted all of the prey items birds bring to their nest, but those who have had the patience to do this have recorded astounding figures. Robert Stewart, for example, made detailed records of a Wilson's warbler pair while they were feeding their young in 1973. Female readers will likely not be surprised to learn that he found substantial differences in how hard the male and female worked at this endeavor. The male was no slouch, carrying food to the nest 241 times in a single day, but the female put him to shame; on that same day, she fed the nestlings 571 times! This rate was maintained over the five days Stewart watched the nest. He did not count the actual number of caterpillars the pair brought to the nest; feeding was rapid and often a parent carried more than one caterpillar in its beak at a time. Yet even if only one caterpillar was brought to the nest each trip, the pair would have brought in 812 caterpillars per day, or 4,060 caterpillars in the five days Stewart watched the nest. The chicks he observed stayed in the nest eight days before they fledged.[12]

These observations are not exceptional. Bobolinks bring food to their nests an average of 840 times a day for ten days in a row.[13] Sapsuckers feed their young 4,260 times, downy woodpeckers 4,095 times, and hairy woodpeckers 2,325 times.[14] All of these species regularly bring in multiple prey items per trip. Perhaps the most complete records of feeding rates were made by Richard Brewer, who counted the caterpillars that Carolina chickadees brought to their nests throughout their nesting period.[15] Brewer found feeding rates of 350 to 570 caterpillars per day, depending on the number of chicks in the nest. Over the course of a typical nesting period (sixteen days on average), that totals six thousand to nine thousand caterpillars required to bring one nest of a tiny bird (three ounces) to fledging. A parent's job is not over, however, when the chicks leave the nest. Chickadee parents, for example, continue to feed their young for up to twenty-one days after fledging. No one knows how many additional caterpillars are required before young chickadees no longer depend on their parents for food.

Now let's think about an ideal neighborhood that contains not just one pair of one bird species, but thriving populations of many species. I, for one, am greedy when it comes to enjoying nature. I want Carolina chickadees in my yard, but I also want cardinals, and titmice, and blue jays, and Carolina wrens. I want red-bellied and downy woodpeckers, white breasted nuthatches, yellow warblers, Kentucky warblers, robins, wood thrushes, ovenbirds, and indigo buntings. And I am so greedy that I also want bluebirds, catbirds, common yellowthroats, great-crested flycatchers, yellow-billed cuckoos, mockingbirds, eastern kingbirds, field sparrows, chipping sparrows, and grasshopper sparrows—all birds that used to be common in Delaware. If each pair of these species requires thousands of caterpillars to successfully breed, imagine the number of caterpillars my yard would need to produce to support stable populations of all of these birds!

WHICH PLANTS SHOULD WE USE?

It is clear that if we need to provide as many caterpillars in our yards as possible, then we need to use plants that serve as hosts for the most caterpillar species. But which plants are those? Assembling this information is not a trivial task. There are some 2,112 native plant genera in the lower forty-eight states and most of them contain species that serve as host plants for one or more species of caterpillars. Records of these host associations have been made over the past century by naturalists, ecologists, and particularly by Lepidoptera taxonomists, and these are scattered in their writings throughout thousands of papers and books. Needless to say, finding and categorizing this information requires a combination of old-fashioned library work plus the handy search tools of the digital age.

With financial support from the U.S. Forest Service, Kimberley Shropshire, a Research Assistant at the University of Delaware, created a mammoth database that has become the basis of a search tool developed by the National Wildlife Federation called Native Plant Finder.[16] Shropshire has ranked plant genera that occur in every county of the U.S. in terms of their ability to host caterpillars. Now, simply by entering your zip code, you can find out which woody and herbaceous plant genera native to your area are best at hosting caterpillars. This tool has removed one of the biggest obstacles to homeowner restorations. We no longer have to wonder what plants we should add to our landscape to enhance its ecological productivity.

KEYSTONE PLANTS

Shropshire's work revealed a striking pattern: wherever one looks—be it in the North, South, East, or West; the plains, deserts, forests, or mountains—just a few plant genera are producing most of the Lepidoptera so important to food webs. She and I knew from our previous work in the mid-Atlantic states that not only are native plants far superior to introduced species in their ability to generate caterpillars, but also they vary by orders of magnitude in their production of caterpillars.[17] Some Delaware genera like oaks, cherries, and willows host hundreds of caterpillar species, while for others like Virginia sweetspire, there are no records at all of caterpillars using them. This is interesting in itself, but when Shropshire assembled data for each county, we saw that this pattern held everywhere and we could quantify it. Wherever we looked, about 5 percent of the local plant genera hosted 70–75 percent of the local Lepidoptera species.

I like to call such hyperproductive plants *"keystone plants"* because they so closely fit the meaning of Robert Paine's classical terminology;[18] keystone plants enable other species in the ecosystem to coexist. Like the center stones in ancient Roman arches, remove the keystone and the structure falls down. Keystone plants function in the same way: they are unique components of local food webs that are essential to the participation of most other taxa in those food webs. Without keystone plants, the food web all but falls apart. And without some minimal number of keystone plants in a landscape, the diversity and abundance of its many insectivores—the birds and bats, for example, that depend on caterpillars and moths for food—are predicted to suffer.

The implications of this phenomenon for homeowners, land managers, restoration ecologists, and conservation biologists are enormous. To create the most productive landscapes possible—that is, landscapes in which the most plant matter sustains edible insects—we have to include keystone plant species. This is a nuanced but incredibly important extension of our knowledge about how native plants contribute to ecosystem function. Before discovering the existence of keystone genera, many overestimated the degree to which most native plants contribute to food webs and assumed that if a plant was native then it contributed a lot. We now know that a few native genera contribute so much more than most others that we cannot ignore them if we are to produce complex, stable food webs. A landscape without keystone plants will support 70–75 percent fewer caterpillar species than a landscape with keystone plants, even though it may contain 95 percent of the native plant genera in the area.[19]

LET IT BE AN OAK

Oaks are aptly placed in the genus *Quercus*, a name derived from the Celtic "quer" meaning fine, and "cuez" meaning tree, and oaks are indeed fine trees. With hundreds of species globally (taxonomists argue about the exact number, with estimates ranging from four hundred to six hundred species) and over ninety species in the U.S., oaks occur in and often dominate all forest ecosystems in North America except the great coniferous forests of the North and the driest deserts of the Southwest. Ecologically, oaks are superior plants and it would be easy to make a convincing case that they deliver more ecosystem services than any other tree genus. Many species are massive and sequester tons of carbon in their wood and roots, and pump tons more into the soil they grow in. They are long-lived, with some species achieving nine hundred years of age if you include periods of growth, stasis, and decline. Thus, the carbon they pull from the atmosphere is locked within their tissues for nearly a thousand years! In many ecosystems, oaks are also superior at

FIGURE 7.3. The spun glass slug caterpillar, one of the 557 species of caterpillars in the mid-Atlantic states that develop successfully on oaks. (Photo courtesy of Doug Tallamy.)

stalling rainfall's rush to the sea. Their huge canopies break the force of pounding rain before it can compact soil, and their massive root systems prevent soil erosion and create underground channels that encourage infiltration instead of runoff. Lignin-rich oak leaves are slow to break down once they fall from the tree and create excellent, long-lived *leaf litter* habitats for hundreds of species of soil arthropods, nematodes, and other invertebrates. For me, though, all of these contributions to ecosystem function pale before the contribution oaks make to food webs.

My early work with Shropshire showed that oaks in Delaware support hundreds of caterpillar species, and at least 934 species nationwide, making them by far the best plants to support food webs. To put this level of productivity in perspective, most other common trees in Delaware are slackers in comparison. Tulip poplar (*Liriodendron tulipifera*), for example, supports only twenty caterpillar species, black gum (*Nyssa sylvatica*) thirty-six, sycamore (*Platanus occidentalis*) forty-six, persimmon (*Diospyros virginiana*) forty-six, and sweetgum (*Liquidambar styraciflua*) thirty-five. Like oaks, native willows and cherries are also highly productive, but they do not surpass oaks in Delaware. You don't need to understand precisely why oaks help food webs better than other plants; you just need to know that they do and that we should use them accordingly in our landscapes and restorations.

COMPLETING THE LIFE CYCLE

Regardless of which trees, shrubs, or perennials we employ to increase the abundance and diversity of caterpillars on our properties, we have to use them in our landscapes in ways that enable the caterpillars they support to complete their life cycle. Caterpillars undergo what is termed *complete metamorphosis*, a type of development that is comprised of four distinct stages: egg, larva, pupa, and adult. What is relevant here is that, for most caterpillar species, only two of these life stages, the egg and larval stages, are completed on the host plant. Most caterpillars crawl off of their host plant before molting to their pupal stage. A few of the 511 species of caterpillars in New Castle County, Delaware, such as the Polyphemus moth, spin their cocoon on the host tree itself after they have eaten their fill of oak leaves. But 480 species, some 94 percent, fall to the ground when fully grown, and either burrow into the soil to pupate underground, or spin a cocoon in the leaf litter under the tree.

The exodus most caterpillars make from their host plant before they pupate is not just something oak-eaters undertake. Monarch caterpillars almost never form their chrysalis on milkweed; they crawl off to some other structure, often yards away from the milkweed plant they developed on, causing many monarch watchers to think a bird has gotten them. The pipevine swallowtails in my yard really take a hike when fully grown. I have found their chrysalides halfway up my oak trees, attached to the side of my house,

FIGURE 7.4. This is the cocoon of the polyphemus moth, one few moth species that completes its life cycle on its host tree. (Photo courtesy of Doug Tallamy.)

and one was even hanging from one of my picture frames in my living room! Some of these caterpillars have crawled more than twenty-five yards from their pipevine hosts. Experts think the evolutionary motivation for such trekking is for caterpillars to put distance between themselves and their host plant before they enter the defenseless pupal stage. Lepidopteran pupae are not only prey for hungry birds; they are also targets for numerous species of other predators, as well as Hymenopteran and Dipteran parasitoids, many of which go to a host plant to search for food. The longer a caterpillar stays on its host, the greater the chances it will be discovered and attacked by one of its enemies. By crawling some distance from its host plant before it pupates, the caterpillar decentralizes the effective search zone for its enemies. The caterpillar forces its enemies to search thousands of square feet instead of being able to efficiently search just a few square feet for their prey, making it a low-return task that is usually not worth the time or energy involved.

This survival mechanism is very effective at reducing mortality in the pupal stage, but it forces us to think beyond the needs of caterpillars when we landscape. Not only do we have to provide food for developing caterpillars, but we also must provide the microhabitats that their pupae require to survive. Your yard may contain an oak tree that can feed hundreds of caterpillars, but more often than not, that oak will be surrounded by mowed lawn growing in compacted soil. When your caterpillars drop from the tree, they will find no leaf litter in which to spin their cocoon, because each year you neaten up after the leaves fall. If they are a species that tunnels into the soil, they will have to search far and wide to find soil loose enough to permit burrowing, and the longer they have to search, the greater the odds of being pulverized by your lawn service, or squashed on your driveway or street. These challenges are even greater in urban environments where trees are often surrounded by cement.

Fortunately, providing safe pupation sites in our landscapes is not an insurmountable problem. In fact, it can be a new and satisfying gardening goal. The needs of our caterpillars can be easily met when we replace some of our lawn with three-dimensional plantings: annual or perennial beds, spring ephemeral showcases, ground covers, or shrub plantings of species appropriate for either Delaware's Coastal Plain or Piedmont, depending on where you live. Any portion of your landscape that is not regularly trammeled by feet and lawnmower wheels will quickly develop a thick layer of loose organic matter perfect for pupating Lepidoptera, as long as you don't rake away your leaf litter during your spring and fall cleanups. To the surprise of many, some caterpillar species, such as the beautiful wood-nymph, the greater oak dagger moth, and Harris's three spot, tunnel into soft wood to pupate. Adding a decaying log to your garden (artfully placed of course) would allow such species to complete their development.

There is one final safety precaution we need to take to help our caterpillars complete their life cycle. Once they emerge as adult moths from their pupae, they have to survive long enough to find a mate and for the females to locate their host plants and lay eggs. A few species like the white-marked tussock moth accomplish this in just a few hours, but most require several days to a few weeks to do so. During this time, those species with mouthparts need to eat to maintain their energy, and most of the time they are eating nectar from nocturnal flowers. A recent study by Eva Knop at the University of Bern suggests that our propensity to light up the night sky is not helping moths in this regard.[20] Using night vision goggles, Knop counted insects that visit flowers in areas with no artificial lights, and then added lights to those same areas. She found that when lights were on, moth visits declined 62 percent. Either the moths simply avoided spaces that were well-lit, or they were fatally attracted to the lights as if the lights were Sirens.

Don't ask why insects are so drawn to lights; two centuries of research have not produced a satisfying answer to this puzzling question. The point is, insects do fly into light sources by the millions, and each night we light up nearly the entire world to their detriment. Lights reduce insect populations in several ways. A light can kill insects directly in their repeated collisions with the bulb. Or the frenetic flight about the bulb can fatally exhaust them by burning up their energy reserves. Those insects that don't beat themselves to death or die of exhaustion are nevertheless waylaid from their normal activities of seeking host plants or mates. That is, lights waste precious time in the short adult lives of insects. Finally, an insect drawn to a light becomes an easy target for hunting bats, or, if it is sitting near a light, for arthropod predators like daddy long-legs, carabid beetles, hanging scorpionflies, damsel bugs, ants, assassin bugs, mantids, and spiders. Those insects that make it to dawn are often picked off by birds that quickly learn that lights provide an easy place to find breakfast. Whether you use lights at night so you can see your way around your property, or to discourage bad guys, please consider putting motion sensors on each one. That way, instead of illuminating your property all night every night, the lights only turn on when you really are out and about, or when the bad guy actually comes. That simple act will make an enormous difference for our insect populations, but particularly for night-flying moths and, thus, for caterpillars.

RESTORING NATIVE BEES

The second group we should be sure to support in our landscapes is native bees, not because bees are important components of food webs (they are not; other than a number of specialized parasitoids, very few animals eat bees), but rather because, as *pollinators*, they maintain a diverse plant base for terrestrial food webs. Though many types of insects are credited with important roles in pollination, including moths, butterflies, beetles, wasps, flies, and ants, it is members of the Apoidea, our bees, that perform the lion's share of pollination duties.

When most people think of bees, an image of the domesticated European honey bee comes to mind. Honey bees were brought to North America with the earliest colonists because they could be managed and deployed where they were most needed, and because they were generalist pollinators that were so good at pollinating the many Old World crop plants the colonists also brought with them. Before we imported honey bees, however, all of the animal pollination in North America (13 percent of our plants are wind-pollinated) was accomplished by native pollinators, primarily the nearly four thousand species of bees native to North America.

FIGURE 7.5. Contrary to popular misconception, butterflies like this Canadian tiger swallowtail do very little pollination compared to bees. (Photo courtesy of Doug Tallamy.)

FIGURES 7.6A-B. Native bees like this sweat bee (A) and bumblebee (B) are collectively our most important group of pollinators. (Photos courtesy of Doug Tallamy.)

Like so many of our insects, honey bees are in trouble. Although 10–15 percent of bee colonies have always been lost during the winter months, dramatic and in some cases sudden declines in honey bee populations became apparent across the country and, indeed, globally between 2003–2007. A suite of ills, from mites, to viruses and bacteria, to abusive pollinating demands have been blamed for these declines, now called collectively "colony collapse disorder." The threat to agriculture from the loss of honey bees was so obvious that even our politicians noticed and, today, saving pollinators has become a politically correct mantra. The press is full of articles telling us that we have to save bees because they pollinate one-third of our crops. True enough, but as a sole motivation for saving bees—indeed, all of our pollinators—this is short-sighted in the extreme. There is an even more compelling reason beyond protecting a few species of human crop plants to save pollinators from local or global extinction. In addition to pollinating one-third of our crops, recall that animals are responsible for pollinating 87 percent of *all* plants and 90 percent of *all* angiosperms.[21] That's right, if pollinators were to disappear, 87–90 percent of the plants on planet Earth would also disappear. Not only would such a loss be a fatal blow to humans, but it would also take most other multicellular species with it as well. Saving pollinators from human environmental aggression goes well beyond maintaining a diversity of fruits and vegetables in our supermarkets; it is essential to life as we know it on this planet.

One positive result of colony collapse disorder in European honey bees is that it has drawn long-overdue attention to our North American bee species. Although only a handful of our native bee species have been studied, it is no surprise that most have been found to be in steep decline. Fifty percent of Midwestern native bee species have disappeared from their historic ranges in the last century.[22] Four species of bumblebees have declined 96 percent just in the last twenty years[23] and 25 percent of our bumblebee species are at risk of extinction.[24] It is not a stretch to assume the thousands of species not yet examined are similarly challenged in today's world of Roundup-ready corn, soybeans, and lawns, and deadly roads. It is clear that we must all act quickly to save our pollinators, but to do so effectively we have to understand who our pollinators are, and what they need to thrive in our yards.

WHAT IS A POLLINATOR?

It is logical to assume that all animals that go to flowers for pollen and/or nectar actually pollinate those flowers. Yet, the opposite is true; most animals that go to flowers do not end up pollinating them, even if they successfully remove pollen and nectar from them. It is more accurate to call animals going to flowers "visitors" and reserve the term "pollinator" for animals that actually transfer pollen from the male stamens to the female pistol. Butterflies, for example, get lots of credit for being great pollinators because they spend so much time nectaring at flowers. But this credit is not deserved, as most butterflies take from flowers without giving much back in return. Generally speaking, butterflies do not have a body shape conducive to transferring pollen. Even bees that have specialized adaptations for pollinating a particular flower genus may visit other flower genera without transferring any pollen. Because pollen and nectar are costly for flowers to produce, many flower genera have developed elaborate shapes such as extremely long corolla tubes, very narrow corollas, or closed petals that make access to their nectar difficult. The evolutionary idea in these cases is to prevent *generalist* pollinators from taking their pollen and allow only *specialist* pollinators access to the pollen because they are more likely to deliver it to another flower of the same species. Specialized interactions between flowers and their pollinators are largely responsible for the myriad sizes and shapes of flowers and bees that we find in nature.

One more thing: to make supporting pollinators at one's home part of mainstream culture there is an educational hurdle that many people must leap. Unfortunately, the thought of thriving bee populations in one's front- and backyard is too often a non-starter. After all, where there are bees, there are bee stings, or so many people think. Bees do sting (at least the females do,

FIGURE 7.7. Bumblebees are one of the few groups of native bees that are social and create small nests of daughters. (Photo courtesy of Doug Tallamy.)

but males have no stingers, which is a modified egg-laying device), but only in self-defense or in defense of their hive. This is important point number one. Out of all four thousand species of native bees, only bumblebees, a mere forty-six species, have what we might call "hives," though they are tiny compared to honey bee hives. The rest of the species are solitary, and never aggressively defend a home space. When people are stung by a bee, the perpetrator is nearly always a honey bee that has been stepped on with bare feet (that was my first sting as a toddler), or that is defending its hive. Important point number two is that, while foraging at flowers, bees are not aggressive at all. They are focused solely on gathering as much pollen and nectar as possible. You can prove this to yourself by petting the next bee you see at a flower. The bee might fly off, but it won't sting you. The passive nature of foraging bees means that we can walk among flowers crawling with bees with no fear of being stung.

Before you send me an angry email listing all the times you *have* been stung at home, read on. The most common misconception about the source of painful stings stems from sloppy taxonomy: people frequently mistake yellow jackets for bees. Yellow jackets are predatory wasps, not bees and not pollinators. But like honey bees, they are social species that construct large hives in the ground or in trees that they defend as aggressively as they can. I include bald-faced hornets in this group; even though they are black and white instead of yellow and black, bald-faced hornets are close relatives of yellow jackets and behave just like them. And unlike honey bees, yellow jackets and hornets can sting repeatedly, making any close encounter with them an unpleasant one. The good news is, there is no national movement to encourage people to have yellow jacket nests in their yards. The solitary bees we need to share our yards and parks with are harmless.

MAKING BEES FEEL AT HOME

Like most creatures, native bees need the basics to exist: a place to live, food, and water. Let's deal with living quarters first. Most native bee species nest in the ground, within wood or pithy plant stems, or in any nook or cranny of the appropriate size. Bumblebees, our only native bees that are always social, favor shallow holes in the ground that are protected from rain. Abandoned mouse nests within a rock wall are ideal real estate for bumblebees because they are protected from rain and also from digging predators such as possums, raccoons, and foxes that love to eat bumblebee larvae. If you don't happen to have an abandoned mouse nest or a rock wall on your property, you can simulate a nest by burying a roll of toilet paper about three-quarters of its length in the ground (center hole facing up) in a site totally protected from rain or runoff. If you are motivated, you can build a small three-sided wooden house with an inch-wide hole drilled in one side to protect the roll from rain. Bumblebee queens fly around each spring evaluating every hole they find as

FIGURE 7.8. Many species of native bees like this *Colletes* bee, nest harmlessly in the ground. (Photo courtesy of Doug Tallamy.)

a potential nest site. Chances are good that a queen will enter your makeshift nest box and chew her way into its center to set up a cozy house.

Ground-nesting bees, some 70 percent of our native bee species, are easy to accommodate, as long as you have soil that is loose enough for bees to excavate; that is, almost all non-compacted soil types except hard-packed clay. Ground nesters prefer bare patches of dry soil with a slight southern slope, so if you have such spaces, avoid walking on them as that compacts the soil beyond use for the bees. We are not talking about huge areas. Two square feet of bare soil can provide housing for a number of reproducing female bees. Please, do not use lawn fertilizer near nests either, as lawn products usually contain pesticides that do not make life easier for our bees.

Species of native bees, particularly in the families Colletidae, Halictidae, and Andrenidae, and some Apidae like long-horned bees and squash bees, will construct nests in the ground all season long, but such nests are most evident in early spring before they are obscured by vegetation. Look for holes in the ground surrounded by small mounds of excavated soil. Active nests usually have a bee leaving or entering the hole every few minutes. The holes lead to tunnels several inches deep with side-shafts here and there, each containing a ball of pollen and a developing bee larva. If you have a site on your property that is ideal for ground-nesting bees, many individuals may nest within that small space. This can result in lots of bees frenetically coming and going, a spectacle that may be a bit scary at first, but these bees are busy rearing their young and will do their very best to ignore you. You can sit and watch them for hours with no ill effects.

Pithy stem nesters are a bit more particular in where they nest. Stems of many herbaceous plants like goldenrod, blackberries, giant ragweed, or native hydrangeas are essentially hollow except for a loose fibrous pith that bees can easily remove. These cavities make perfect nesting sites. Mason bees, small carpenter bees, and small resin bees will tunnel into the stem, remove the pith from a section several inches long, and then construct a sequence of cells starting at the end of the cavity farthest from the entrance hole. Each cell is packed with pollen upon which a bee lays a single egg. It then seals off that particular cell and starts to provision the neighboring cell. In this way, the stem contains several developing larvae at once, each one a day or two younger than the previous larva. When development is complete, the larva pupates within the cell. If it is early summer, the young adult will emerge from its cell by chewing a hole directly to the outside and then start its own family. If the larva matures at the end of the season, the resulting prepupa will stay within its cell all winter, complete its development early in the spring, and emerge as an adult as soon as there are flowering plants in bloom.

Woody stem nesters like carpenter bees and some species of mason bees behave almost identically to pithy stem nesters except they build their nests in soft wood rather than in soft stems. Soft wood could be in the form of a downed log or branch, or, just as often, a dead branch still attached to a tree. Dead elderberry branches are good examples of suitable nest sites for these bees because they are so easily excavated. Finally, species such as *Osmia* mason bees, yellow-faced bees, and leafcutter bees often choose existing cavities for their nests. I can't tell you how many times I have found the snout of my watering can or even my outdoor water spigot plugged with leafcutter bee nests!

In terms of bee conservation, there is a common theme here. Bees cannot nest or overwinter in our yards unless we provide what they need to do so. Most people do have open patches of ground, plants with pithy stems, easily excavated wood, and nooks and crannies somewhere on their property, but many of us work hard to eliminate these valuable resources. My fall cleanup is particularly hard on bee populations; the senescing stems of my black-eyed Susans, penstemons, sunflowers, and all of the other perennials many of us are so anxious to cut back after they have bloomed are where pithy stem nesters are hoping to spend the winter. Similarly, that dead elderberry branch I feel compelled to prune off and the large elm branch that fell during a summer thunderstorm are now homes for bees that favor soft wood.

The social edict to neaten up is often in direct conflict with the needs of our native bees. There are opportunities for compromise, however. Maybe we can gently cut off our goldenrod stems near the ground, but rather than mulching them, tie them together like a decorative bundle of corn stalks and stand them up for the winter somewhere out of public view. The bees and katydid eggs within them should be able to make it through the winter just as if the stems were left standing in your garden. Another solution is to learn to appreciate the winter structure of perennials. Coneflower seed heads look great with a cap of snow, milkweed pods are fun to look at all winter, and birds certainly appreciate the seed heads we have left behind.

In recent years, we have discovered how easy it is to attract many species of stem- and wood-nesting bees, as well as species that nest in nooks and crannies, using commercial or homemade bee hotels. There is scarcely a trade show these days that does not offer a variety of bee hotels for sale and we humans love them. Hang one in a dry space in your yard at the end of winter, and in short order the bees will be busy rearing their young. The number of bees you attract is often a simple function of the size of your bee hotel, and so each year I make my hotels a little bigger. Bee hotels are the perfect solution

FIGURE 7.9. Scattering several small hotels like this one about your property will make it more difficult for bee enemies to find their prey. (Photo courtesy of Doug Tallamy.)

to increasing the nesting capacity of our yards. Or so we have thought. What should have been obvious from the start is that by concentrating nesting opportunities in one place, we have made it very convenient for bee predators, parasitoids, and diseases to wipe out our bees.[25] Find one bee and you have found them all! This doesn't mean that bee hotels are useless, but it does mean that we need to make them much smaller, comprised of just a few cells each, and scatter many throughout our yards. In this way, we won't be putting all of our bees in one basket. If a bee enemy finds one hotel, it will not have easy access to all the bees on your property.

MEET THE NEEDS OF OUR SPECIALISTS!

Now that we have provided housing for our bees, we need to feed them; that is, we need to meet the nutritional needs of both adult bees and their developing larvae. Adult bees eat pollen and nectar while larval bees develop exclusively on pollen. That seems simple enough, but there are some essential particulars here that we need to pay attention to. First, we have to think about timing. Bees, like most other multicellular organisms, must eat every day. Because the pollen and nectar they need comes from flowers, we need to have blooming plants in our landscape throughout the season. And bee communities are active most of the year in most parts of the country. In Delaware, there are native bee species on the wing from early March through the end of October.[26] The need for a continuous sequence of flowering plants in our landscapes is not a trivial challenge. It requires plant choices that are choreographed to enter each ecological stage one after another. Moreover, most native bee species do not forage over large areas as do honey bees. This means that, to provide good habitats for native bees at home, we need flowering plants there. There is no problem in having more than one species of flowering plant blooming at once—in fact, that is desirable and gives bees nutritional options. But a landscape that goes through a two- or three-week period with no available blooms is deadly to bees.

Blooming phenology is only one thing to consider when landscaping for bees. We also have to recognize that many bee species require pollen from particular plant genera in order to reproduce. That is, like most of the caterpillars discussed earlier, many bee species have become host plant specialists over evolutionary time. In fact, nearly 30 percent of the native bees in the mid-Atlantic region are host plant specialists.[27] And it is no wonder that natural selection has favored bee specialization. Plants differ from each other in so many ways, so some bees can reach their nutritional goals more easily if they develop the specialized adaptations that allow them to find, gather, transport, and digest the pollen of particular plants efficiently. Think about it; plants differ in when they flower, how long they flower, and the size, shape, and color of their flowers. They also differ in their pollen morphology and the types of amino acids, lipids, proteins, starches, sterols, and chemical defenses contained within the pollen. Some generalist bees are good at dealing with all of this variation and can do well on a variety of pollen species. But others become finely tuned to the characteristics of particular plant genera; if we don't include those genera in our plantings, those specialist bees cannot survive in our yards.

How do we support both our specialists and our generalists? Sam Droege of the Patuxent Wildlife Research Center says we should meet the needs of our specialists and the generalists will follow. For example, in Delaware, there are thirty-two species of bees that can rear their larvae only on the pollen of native asters. Violets support twenty-six more species, goldenrods eleven species, native willows sixteen species, evening primroses (*Oenothera* spp) seventeen, and sunflowers eighteen.[28] If you are trying to help native bees but you plant butterfly bush, xenias, and impatiens, you will see generalist bees and you may bask in a false sense of accomplishment. But without goldenrods, asters, evening primroses, sunflowers, violets, and native willows, 120 species of specialist bees that might have been able to use your yard as a refuge will be absent. Droege has said many important things during his career as a native bee authority, but in my view, this is the most important thing he has ever said: saving our bee specialists by planting what they have specialized in is the key to saving diverse bee communities around the country.

NOTES

1. Edward O. Wilson, "The Little Things that Run the World: The Importance and Conservation of Invertebrates," *Conservation Biology* 1, no. 4 (May 1987): 344–46.
2. Jeff Ollerton, Rachael Winfree, and Sam Tarrant, "How Many Flowering Plants Are Pollinated by Animals?" *Oikos* 120, no. 3 (February 2011): 321–26.
3. Roz Weis, "National Insect Killing Week," *Decorah Journal*, July 16, 2009.
4. Gwen Pearson, "You're Worrying about the Wrong Bees," *Wired*, April 29, 2015.
5. Adrian Burton, "Crickets in Crisis," *Frontiers in Ecology and the Environment* 15, no. 3 (April 2017): 121.
6. Caspar A. Hallmann, et al., "More Than 75 Percent Decline over 27 Years in Total Flying Insect Biomass in Protected Areas," *PLoS ONE* 12 no, 10 (October 2017): e0185809.
7. Rodolfo Dirzo, et al., "Defaunation in the Anthropocene," *Science* 345 (July 2014): 401–6.
8. Tapio Eeva, et al., "Carotenoid Composition of Invertebrates Consumed by Two Insectivorous Bird Species," *Journal of Chemical Ecology* 36, no. 6 (June 2010): 608–13.
9. Roger Tory Peterson, *Eastern Birds* (Boston: Houghton Mifflin Company, 1980); Roger Tory Peterson, *A Field Guide to Western Birds* (Boston: Houghton Mifflin Company, 1990).
10. Ashley Kennedy, "Examining Breeding Bird Diets To Improve Avian Conservation Efforts" (PhD diss., University of Delaware, 2019).
11. Desirée L. Narango, Douglas W. Tallamy, and Peter P. Marra, "Nonnative Plants Reduce Population Growth of an Insectivorous Bird," *Proceedings of the National Academy of Sciences* 115, no. 45 (November 2018): 11549–54.

12. R M. Stewart, "Breeding Behavior and Life History of the Wilson's Warbler," *The Wilson Bulletin* (March 1973): 21–30.
13. Stephen G. Martin, "Polygyny in the Bobolink: Habitat Quality and the Adaptive Complex" (PhD diss., Oregon State University, 1971).
14. Louise De Kiriline Lawrence, "A Comparative Life-History Study of Four Species of Woodpeckers," *Ornithological Monographs* No. 5 (1967): 1–156.
15. Richard Brewer, "Comparative Notes on the Life History of the Carolina Chickadee," *The Wilson Bulletin* 73, No. 4 (December 1961): 348–73.
16. "Native Plant Finder," National Wildlife Federation, 2015–2019, https://www.nwf.org/NativePlantFinder/Plants.
17. Douglas W. Tallamy and Kimberley Shropshire, "Ranking Lepidopteran Use of Native Versus Introduced Plants," *Conservation Biology* 23, no. 4 (July 2009): 941–47.
18. Robert T. Paine, "A Note on Trophic Complexity and Community Stability," *American Naturalist* 103, no. 929 (January-February 1969): 91–93.
19. Narango, Tallamy, and Marra, "Nonnative Plants," 11549–54.
20. Eva Knop, et al., "Artificial Light at Night as a New Threat to Pollination," *Nature* 548, no. 7666 (August 2017): 206–9.
21. Ollerton, Winfree, and Tarrant, "Flowering Plants," 321–26.
22. Laura A. Burkle, John C. Marlin, and Tiffany M. Knight, "Plant-Pollinator Interactions over 120 Years: Loss of Species, Co-occurrence, and Function," *Science* 339, no. 6127 (March 2013): 1611–15.
23. Sydney A. Cameron, et al., "Patterns of Widespread Decline in North American Bumble Bees," *Proceedings of the National Academy of Sciences* 108, no. 2 (January 2011): 662–67.
24. Paul Williams and Sarina Jepsen, (eds.), *Bumblebee Specialist Group Report 2014.* International Union for Conservation of Nature, March 2015, https://www.iucn.org/ssc-groups/invertebrates/bumblebee-specialist-group.
25. J. Scott MacIvor and Laurence Packer, "'Bee hotels' as Tools for Native Pollinator Conservation: A Premature Verdict?" *PloS One* 10, no. 3 (March 2015): e0122126.
26. Jarrod Fowler, "Specialist Bees of the Northeast: Host Plants and Habitat Conservation," *Northeastern Naturalist* 23, no. 2 (June 2016): 305–20.
27. Ibid.
28. Ibid.

CHAPTER EIGHT

Amphibians and Reptiles

JIM WHITE AND AMY WHITE

AMPHIBIANS AND REPTILES are often (if artificially) studied together in the same scientific discipline called *herpetology*. However, although both groups are cold-blooded, they are actually not closely related. In fact, amphibians are more closely related to fish, and reptiles to birds, than they are to each other. The reason for this seemingly incorrect grouping can be traced way back to the early days of classifying animals and to none other than Carl Linnaeus, the creator of the modern system of classification of living things. Apparently, he was not too fond of amphibians and reptiles, even referring to them as "fowl and loathsome creatures." This disdain may have clouded his otherwise keen sense of animal relationships. The linking of these two groups continues today, and both amphibians and reptiles are informally referred to as "herpetiles" or "herps."

Delaware is home to a relatively large diversity of herpetiles, especially considering the state's small size. There are twenty-eight species of amphibians and thirty-seven species of reptiles found within, or just offshore of, the First State. The state's mild climate and abundance of wetlands provide habitats and conditions favorable to a variety of herpetiles. Finding and observing these animals in the wild can be challenging though, as most species are small and/or secretive. However, with a little knowledge of their life histories and the habitats in which they live, one can find, enjoy, and—importantly for Master Naturalists—help protect these fascinating creatures.

AMPHIBIAN AND REPTILE HABITATS

Since the first European settlers arrived in Delaware in the 1600s, as McKay Jenkins points out elsewhere in this volume, the natural landscape of the state has been drastically altered. Most of the original forests have been lost to agriculture or commercial, industrial, or residential development, and all of the existing woodlands have been cut at least once. In addition, streams have been dammed or channelized, ponds constructed, swamps drained, and marshes diked or filled. Most of these alterations have been detrimental to herpetile populations, although some species have probably benefited from the creation of certain altered habitats (e.g., pond-dwelling frogs and turtles may be more abundant now because of the construction of millponds and sediment retention basins). Fortunately, a variety of natural habitats supportive of amphibians and reptiles remain in public and private land holdings, although their total acreage in Delaware continues to decrease. In other words, the hardiness and prevalence of these animals, which have been around for a very long time, continue to be challenged by our rather narrow ecological view of our landscape.

AMPHIBIANS (CLASS AMPHIBIA)

Amphibians first appeared on Earth in the Devonian Period about 350 million years ago, evolving from fishlike vertebrates. The word *amphibian* is derived from the Greek words *amphi-* (double, both) and *bios* (life), referring to the fact that most amphibians live a "double life": first as an aquatic larva with gills and then, following metamorphosis, as a terrestrial or semiaquatic adult with lungs or other respiratory structures to permit gas exchange with the atmosphere. However, even terrestrial adult amphibians have not completely left an aquatic existence. The skin of most amphibian species is highly permeable and unable to prevent desiccation, necessitating a life spent in wet or moist habitats. In addition, the vast majority of these species must return to the water or find other wet habitats to reproduce, because the amphibian egg is covered by a gelatinous envelope (as opposed to a hard shell) and will desiccate if deposited on dry land.

Amphibians have soft, glandular skin that contains mucus and poison glands but lacks scales. Their limbs lack nails or claws. The amphibian heart has three chambers, and adults of most species have lungs, although they are reduced or absent in some salamanders.

Amphibians are *ectothermic* vertebrates—they are unable to produce enough body heat through internal processes to maintain optimal body function. Instead, they obtain their body heat either directly or indirectly from the sun or another external heat source. In Delaware, many amphibians become inactive (hibernate) during the coldest months, burrowing underneath leaves, logs, rocks, or other debris, or into the substrate in freshwater habitats. Likewise, many amphibians become inactive (aestivate) during the hot, dry periods of summer, retreating to cool, moist microhabitats.

Living amphibians are divided into three distinct groups or orders: order *Caudata*—the salamanders; order *Anura*—the frogs; and order *Gymnophiona*—the caecilians. Only salamanders and frogs are found in North America. Amphibians known to occur in Delaware include eleven salamander and seventeen frog species.

SALAMANDERS (*ORDER CAUDATA*)

Salamanders are called the "tailed amphibians," as implied by the name *Caudata* (the Latin *cauda* means tail) because, unlike frogs, adult salamanders retain their tails after metamorphosis. Salamanders also differ in general appearance from frogs by having elongated bodies, and usually two pairs of relatively equal-sized limbs. They are similar in body shape to lizards, with which they are sometimes confused; lizards, however, have dry, scaly bodies and claws on their toes. Delaware salamander species are aquatic as larvae and terrestrial, semiaquatic, or aquatic as adults, with one exception: the Eastern red-backed salamander lacks an aquatic larval stage and is entirely terrestrial.

Like other amphibians, salamanders must live in wet or moist environments to avoid desiccation. Many terrestrial species hide under rocks or logs or in burrows during the day and emerge at night to feed. Others are *fossorial*, spending both day and night (and feeding) underground. All salamanders are carnivorous, feeding primarily on insects, spiders, worms, millipedes, and other invertebrates.

All of Delaware's salamanders reproduce by internal fertilization of their eggs without copulation. Courtship and mating may occur on the land or in the water, depending on the species. Elaborate rituals of "dancing," nudging, rubbing, grasping, and biting often are involved. If the female is receptive, the male deposits one or more spermatophores either directly on the substrate or on a fallen leaf, stick, rock, or other debris in the breeding habitat. The spermatophores are pyramid-shaped structures that are capped with sperm cells (termed the *sperm capsule*). Fertilization occurs when a female inserts the sperm capsule into her *cloaca*, bringing the sperm in contact with her eggs internally. The fertilized eggs are then deposited in aquatic habitats or moist microhabitats, and may be laid singly or in gelatinous masses or strands, depending on the species. Female salamanders of many of the species attend the eggs until hatching.

The aquatic larvae are typically dull in coloration without adult markings and therefore can be difficult to identify. Depending on the species, the duration of the larval period ranges from a few months to several years. Following metamorphosis, the subadults are similar to adults in appearance and behavior but are not sexually mature.

In Delaware, there are eleven salamander species representing three families:

- Mole Salamanders (*Family Ambystomatidae*)—mole salamanders are large and stout-bodied with relatively well-developed limbs and conspicuous parallel vertical grooves on their sides between their front and back limbs. Three species are found in Delaware: the spotted salamander, marbled salamander, and

Eastern tiger salamander. These species spend most of their lives underground or underneath logs or other debris, emerging during the breeding season to mate and lay eggs in dry or flooded vernal pools, depending on the species. The larvae develop in the vernal pools and have conspicuous external gills and a long dorsal fin that extends from just behind the gills to the tip of the tail. The subadults and adults have well-developed lungs.

- Newts (*Family Salamandridae*)—only one member of the Family Salamandridae is found in Delaware: the Eastern newt, which belongs to the genus *Notophthalmus*. This genus is unique in that instead of having the usual *two* postembryonic developmental stages found in most other amphibians (aquatic larva and terrestrial adult), its species often have *three* distinct developmental stages: aquatic larva, terrestrial subadult, and aquatic adult.
- Lungless Salamanders (*Family Plethodontidae*)—Plethodontidae is the largest family of salamanders in the world. As the common name for the family implies, adult plethodontids lack lungs, with oxygen exchange occurring instead through the skin and mouth lining. Sexually mature male plethodontids of most species have short, fleshy protuberances (*cirri*) on the upper lip and they also have mental glands under the chin that secrete chemicals used in courtship. For many plethodontids, elaborate courtship rituals culminate in a tail-straddle walk, in which the female straddles the male's tail and is led by the male over the top of his spermatophore. She then picks up the sperm capsule with her cloacal lips and inserts it into her cloaca, thus fertilizing her eggs. There are seven species in the Family Plethodontidae found in Delaware: the Northern dusky salamander, Northern two-lined salamander, long-tailed salamander, four-toed salamander, Eastern red-backed salamander, mud salamander, and red salamander.

FROGS (*ORDER ANURA*)

The order Anura and the term "frogs," as used here, refer to all frog species, including those species that are commonly referred to as "toads." Frogs are called the tailless amphibians because, unlike salamanders, all frogs lose their tails during metamorphosis to the adult stage. Frogs also differ in general appearance from salamanders by having a relatively short body, with a head that is not separated from the trunk by a discernible neck. The hind legs of frogs are longer than the front legs and have powerful muscles that give many frog species amazing jumping abilities.

All but a few species of frogs in North America are aquatic as larvae *tadpoles* and terrestrial or semiaquatic as adults. The majority of frogs live in moist or wet environments. Some live in or near wetlands and moist woodlands; others avoid desiccation by living under logs and debris or in underground burrows. Tadpoles are largely herbivorous, feeding primarily on algae and other aquatic vegetation (although fellow tadpoles are sometimes eaten). Adult frogs are carnivorous, feeding primarily on invertebrates such as earthworms, spiders, and insects. When they can, larger species such as the American bullfrog also take larger prey, including other amphibians and small reptiles, birds, and mammals.

All Delaware frog species are capable of producing sound. Vocal sacs connected to the throat are inflated during calling to modify and amplify the sound. A variety of vocalizations are made:

- *Advertisement call*—made by the males, it is the dominant sound heard during the breeding season and usually is quite loud. Distinctive for each species, this call presumably advertises the male's presence to other members of its species, defining calling sites and attracting females to the breeding area.
- *Rain call*—usually a weak version of the advertisement call, emitted for unknown reasons outside the breeding period, typically in association with rain events.
- *Release call*—often a chirp-like sound, usually accompanied by bodily vibrations, emitted mostly by males to discourage other males that are mistakenly attempting to mount them, or by females that are trying to thwart or terminate mating.
- *Alarm call*—often a loud squeak emitted as a frog attempts to escape a predator.
- *Aggressive call*—often a grunt, growl, or trill sounded by a male when he is defending a calling site.

Most frogs also have a well-developed sense of hearing, and many have a conspicuous *tympanum* (eardrum) on each side of the head.

Reproduction is accomplished in all Delaware frogs by external fertilization of the eggs. The males generally migrate first to the breeding sites, which are often fishless, non-flowing freshwater habitats such as freshwater marshes, wooded swamps, vernal pools, ditches, or ponds, but may also include the slow-moving portions of creeks and rivers. The males use the advertisement call to attract females to the breeding sites and, in some species, to the male's breeding territory. Once a male encounters a female, he attempts to clasp her in a mating embrace, called *amplexus*. In Delaware species, the male clasps the female just behind her front legs (*axillary* amplexus), with the exception of the Eastern spadefoot, the male of which clasps the female just in front of her hind legs (*inguinal* amplexus). The eggs are deposited in the water and are fertilized by the male as they are laid. Unlike some salamanders, Delaware frogs do not attend their eggs. The eggs hatch in a few days to a few weeks, depending on the species and the water temperature.

The tadpoles are typically dull in coloration and difficult to identify in the field. Tadpoles metamorphose from their aquatic form into tailless, carnivorous, terrestrial to semiaquatic juveniles (froglets or toadlets), which generally require one to three years to reach sexual maturity.

In Delaware, there are seventeen frog species, representing four families:

- Spadefoots (*Family Pelobatidae*)—spadefoots are similar in general body shape to true toads (see below), being broad-headed, short-legged, and stout. A broad, black, spade-like protrusion is found on the inside of each hind foot and is the source of their common name. This spade is used to dig backward into the ground. These toads are fossorial and typically spend most of their lives underground. Explosive breeders, they mate and deposit eggs in temporary pools on only a few very rainy days and nights each year. The egg and tadpole stages are aquatic and extremely short in duration. Only one species is found in Delaware: the Eastern spadefoot.
- True Toads (*Family Bufonidae*)—members of this family are stout-bodied and usually have thick, dry, glandular skin with well-defined warts. Most species have relatively short limbs and digits and therefore have reduced jumping ability (i.e., they hop instead of leap). Breeding males are often much smaller than breeding females; the males also have dark throats and dark pads on their thumbs and inner fingers. Some species, including those in Delaware, have prominent parotoid glands behind the eyes that produce defensive chemicals. Delaware species also have cranial crests—raised, bony ridges between and/or behind the eyes—although these are not visible in young toads. There are two species of true toads in Delaware, both in the genus *Anaxyrus*: the American toad and Fowler's toad.
- Treefrogs (*Family Hylidae*)—Hylidae is a very large and diverse frog family with representatives on all continents except Antarctica. The family is divided into four subfamilies, of which only the subfamily Hylinae is found in Delaware. Members of this subfamily are highly variable in size and external appearance. Most species are relatively smooth-skinned, and many are brightly colored. Most also have well-developed toe pads that enable them to climb shrubs and trees. There are seven species of treefrogs found in Delaware: the Eastern cricket frog, Cope's gray treefrog, gray treefrog, green treefrog, barking treefrog, spring peeper, and New Jersey chorus frog.
- True Frogs (*Family Ranidae*)—frogs in this family are variable in size, but many are relatively large. The external appearance of these frogs is usually what people think of as typical of a frog: a large head with a wide mouth; long, powerful hind legs with webbed feet; and a body color of some shade of green or brown. Breeding males have enlarged thumbs and front legs. Many species are semi-aquatic, spending much of their adult lives in or near water. There are seven species in Delaware, all in the genus *Lithobates*: the American bullfrog, green frog, pickerel frog, Southern leopard frog, mid-Atlantic coastal leopard frog, wood frog, and carpenter frog.

REPTILES (CLASS REPTILIA)

The fossil record suggests that the first true reptiles evolved in the early Pennsylvanian Period, about three hundred million years ago. Reptiles were the first group of vertebrates to live completely terrestrial lives. Although not all reptiles live on land, almost all species lay eggs or give birth on dry land. They can do this because of an evolutionary innovation: the *amniotic egg*. The amniotic egg is a fluid-filled envelope that surrounds the embryo and provides it with all the basic needs for its development. This mode of reproduction eliminates the aquatic, free-living larval stage found in most amphibians. In addition, *oviparous* (i.e., egg-laying) reptiles have amniotic eggs that are encased in a leathery or hard shell that protects the embryo from desiccation, damage from being buried in nests, and attack from small predators or infectious microorganisms.

Another adaptation that facilitates terrestrial existence is the relatively impermeable, scale-covered skin of reptiles. This scaly covering protects reptiles from abrasions and other injuries caused by contact with dry land and also greatly reduces the loss of body fluids in a terrestrial environment.

Reptiles, like amphibians, are ectothermic vertebrates. All but a few very large species obtain the vast majority of their body heat directly from the rays of the sun or indirectly from their surroundings. They can control their body temperature by moving from one area to another, such as moving from sun to shade. With the exception of sea turtles, all of Delaware's reptiles hibernate during the cold winter months, seeking shelter from freezing temperatures by crawling into crevices or by burrowing underground or into the substrate at the bottom of a wetland or an aquatic habitat. Unlike amphibians, reptiles have no larval stage. Both juvenile and adult reptiles have lungs, and the juveniles resemble the adults in body form although their coloration can be very different. In addition, except for the leatherback sea turtle, all reptiles with limbs have claws on their toes.

Living reptiles are divided into four distinct orders: order *Testudines*—turtles; order *Squamata*—lizards, snakes, and amphisbaenians; order *Crocodylia*—alligators, crocodiles, and gharials; and order *Rhynchocephalia*—tuatara (found only on small islands in New Zealand).

Regarding taxonomy within the order *Squamata*, herpetologists now believe that snakes should be considered a type of legless lizard. Nevertheless, in this chapter, snakes are treated as a group separate from lizards, and their main characteristics, unique or otherwise, are described separately.

Reptiles known to occur in Delaware include fifteen turtle, three lizard, and nineteen snake species.

TURTLES (*ORDER TESTUDINES*)

The order *Testudines* refers to all turtle species including those commonly referred to as tortoises, terrapins, cooters, and sliders. Turtles are not easily confused with any other type of animal. All have a protective covering or shell consisting of a *carapace* (upper shell) and a *plastron* (lower shell), which are connected by a bony or cartilaginous bridge. In some species, the plastron is hinged. The carapace is formed from numerous bones that are usually fused to each other and to the underlying vertebrae and ribs. In most species, the bones of the carapace and plastron are covered with large scales called *scutes*.

Another feature unique to turtles is that the shoulder and pelvic limb girdles are located inside the rib cage so the limbs can be retracted inside the shell to a greater or lesser extent depending on the species. The limbs of terrestrial turtles are modified for walking, whereas semiaquatic and aquatic turtles have varying degrees of webbing between their toes. Marine turtles have flattened limbs that serve as flippers. Most turtles have claws on their toes. Adult males typically have longer and thicker tails than females, and the female cloacal opening is located beyond the edge of the carapace.

Most turtles are semiaquatic or aquatic, spending their entire lives in, or basking near, water (except at nesting time, when the females seek drier land). Exceptions include the terrestrial box turtles in the genus *Terrapene* and tortoises in the Family Testudinidae. Most turtles are omnivorous, although with age some turtles' diets shift from omnivorous to herbivorous. Turtles do not have teeth but instead have a sharp, horny beak used for cutting or crushing.

Most of Delaware's turtle species hibernate underwater through much of the colder months, taking refuge in the soft bottom of their wetland or aquatic habitat. One exception is the Eastern box turtle, which typically burrows on dry land in loose soil or decaying vegetation or hibernates in animal burrows. The sea turtles are another exception. These aquatic turtles generally migrate south to warmer waters for the winter, although some apparently hibernate on the ocean bottom. Many turtles also burrow in the mud and become inactive during hot, dry summer weather, a process known as *aestivation*. Like other reptiles, turtles have lungs and obtain oxygen from the air. In addition, many turtles also obtain oxygen from the water through the skin, throat lining, and terminal end of the digestive system (sacks in the cloaca), especially during hibernation and aestivation.

Mating typically occurs in the spring, the fall, or both. It is preceded in many turtle species by courtship behavior that may include nudging, stroking, and/or biting of the female by the male. Internal fertilization of the eggs is accomplished when the male mounts the female and inserts his penis into her cloaca. All turtles lay their eggs on land. In Delaware, between late May and late July, females typically lay their eggs in cavities that they dig in the soil or sand with their hind legs. Some species lay more than one clutch during the nesting season. There is no maternal care of the eggs or the hatchlings. With the exception of an occasional loggerhead sea turtle, none of the various sea turtles (families Cheloniidae and Dermochelyidae) nest in Delaware.

Hatching typically occurs in August or September. Hatchlings are small but grow rapidly for the first few years. Turtles are very long-lived, with many species documented to live twenty to forty years, and some, like the Eastern box turtle, sometimes living more than seventy-five years. Sexual maturity is reached in two to thirty years, depending on the species.

In Delaware, there are fifteen turtle species representing five families:

- Snapping Turtles (*Family Chelydridae*)—snapping turtles are freshwater turtles with a large head, powerful jaws, and a long tail. They are primarily aquatic and feed on living and dead animals and aquatic vegetation. Although they can swim, they usually walk along the bottom of their aquatic habitats. The head, neck, and limbs of these turtles cannot be retracted fully into the shell. There is one representative of this family in Delaware: the snapping turtle.
- Musk and Mud Turtles (*Family Kinosternidae*)—these are weak-swimming, semi-aquatic turtles for which walking underwater or on land is the primary mode of locomotion. They have musk glands in the skin under their carapaces that release foul-smelling secretions. Two genera, *Kinosternon* and *Sternotherus*, are found in the United States and are each represented in Delaware by one species: the Eastern mud turtle and Eastern musk turtle.
- Pond, Marsh, and Box Turtles (*Family Emydidae*)—the family Emydidae is a diverse group of turtles. Most of the species in this family are aquatic or semiaquatic and have relatively flat carapaces. An exception in Delaware is the Eastern box turtle, which is terrestrial and has a domed carapace. All members of this family have a relatively large plastron, sometimes with a movable hinge. They typically also have small heads and tails, and many are habitual baskers. There are seven species of the family found in Delaware: the painted turtle, spotted turtle, bog turtle, diamond-backed terrapin, Northern red-bellied cooter, Eastern box turtle, and red-eared slider.
- Sea Turtles (*Family Cheloniidae*)—the family Cheloniidae is composed entirely of marine turtles, which come ashore only to nest. All species have low, streamlined carapaces that are covered with horny epidermal scutes. Their forelimbs are flipper-like with long digits and produce powerful swimming strokes, and their hind limbs are paddle-like and act as rudders. Their necks are short and cannot be retracted into the carapace. Most species are primarily tropical, but some forage and nest in temperate waters. Four members of this family may be seasonally found in the waters off Delaware: the loggerhead sea turtle, green sea turtle, Atlantic hawksbill sea turtle, and Kemp's ridley sea turtle.
- Leatherback Sea Turtles (*Family Dermochelyidae*)—the sole living member of this family, the leatherback sea turtle (*Dermochelys coriacea*), is the largest turtle in the world today. This turtle differs from other sea turtles in its lack of a bony, hard-shelled carapace. The carapace and limbs are covered with a rigid, leathery skin imbedded with a mosaic of small, bony plates. The limbs lack claws and are either flipper-like (forelimbs) or paddle-like (hind limbs). Leatherback sea turtles are primarily pelagic, but are sometimes observed close to shore, often following schools of jellyfish on which they feed.

LIZARDS (ORDER SQUAMATA)

Lizards are a diverse group of reptiles that inhabit a wide variety of habitats and geographical regions, although they are especially common in areas with warm, dry climates. Most species are terrestrial, arboreal, or fossorial; however, a few species are aquatic.

Most lizards have elongated bodies and four well-developed legs, each with five clawed toes. All are covered with scales, which may be smooth or keeled depending on the species. Unlike snakes, the majority of lizards have external ear openings and eyelids. Lizards hear well, and most can see well. They also have a good sense of smell. In addition to picking up odors through the nostrils, some lizards have a *Jacobson's organ*, a specialized chemosensory organ on the roof of the mouth (as do snakes), which facilitates sensing chemicals in the environment.

Many lizards have a breakaway tail that snaps off when grabbed by a predator. The discarded tail continues to wiggle, often distracting the predator long enough to allow the lizard to escape. The lizard eventually grows a new tail, although the new tail lacks vertebrae and usually is not as long or the same color as the original. Most lizards are fast-moving, active predators, scavengers, or herbivores. Most have small teeth, and the vast majority of lizards, including all Delaware species, are nonvenomous. Unlike snakes, the two halves of the lizard's lower jaw are fused together, limiting the size of prey it can swallow. All Delaware lizards are predators of terrestrial invertebrates, including a wide variety of insects, arachnids, millipedes, centipedes, and isopods. As they grow, lizards periodically grow a new layer of skin under the old one. The old skin typically is shed in patches, not in a single piece as with snakes. In colder regions, including Delaware, lizards spend the colder months hibernating underground or in rotting logs, tree stumps, woodpiles, or other crevices.

Lizard reproduction is accomplished by the internal fertilization of the eggs. Males have a pair of saclike copulatory organs called *hemipenes* that are stored inside the cloaca except during copulation. In many species, the males actively court the females (albeit briefly) before copulation. Courtship may include head bobbing, tongue flicking, chin rubbing, and other behaviors. After variable amounts of courtship, the male aligns his body lengthwise next to the female and wraps his tail around hers. Copulation occurs when the male everts a hemipenis and inserts it into the female cloaca. All Delaware lizards are oviparous, and lay their eggs under logs, stones, or other suitable cover or in soil or rotting vegetation.

There are three species of lizards known to occur in Delaware, representing two families:

- Iguanids (*Family Iguanidae*)—this family is very large and diverse, containing small- to medium-sized lizards that are found in a wide variety of habitats from Southern Canada through Central America. Most are terrestrial or semi-arboreal.

Some members of this subfamily have throat fans (*dewlaps*) or crests, and many use displays such as head-bobbing, nodding, and doing push-ups to declare territory, determine sex, or discriminate between species. Only one species, the Eastern fence lizard, is found in Delaware.

- Skinks (*Family Scincidae*)—members of this family are found on all continents except Antarctica and usually are small- to medium-sized lizards with smooth, often shiny scales. Most have cylindrical bodies, conical heads, and robust, medium-length tails. They are primarily terrestrial or semi-fossorial, although some are arboreal. Most are carnivorous, actively pursuing and feeding on invertebrates. There are two species of skinks known to occur in Delaware: the common five-lined skink and little brown skink. (A third species, the broad-headed skink, is found in Maryland very close to the Delaware border and may also occur in Delaware.)

SNAKES (*ORDER SQUAMATA*)

More than any other vertebrate group, snakes instill fear and disgust in many people. Whether these emotions are manifestations of some deep-seated fears that remain from times when humans lived in closer contact with the natural world, or are a result of myths and media hype passed down from generation to generation, the result is the same: many people do not like snakes, and many do what they can to get rid of them. Nevertheless, and paradoxically, people are also almost universally interested in these fascinating animals.

All snakes have limbless, elongate bodies that are covered with scales. The vertebral column is also elongated, and the number of vertebrae ranges from 141 to 435, each accompanied by a pair of ribs. Most snake species have only one functional lung, which may be almost as long as the body.

All snakes lack external ear openings and eyelids. They have long, forked, *protrusible* tongues, which they use to sample chemicals from the environment and deliver them to the Jacobson's organ. By using the tongue and Jacobson's organ, snakes can locate prey, detect danger, and identify other snakes within their species. Leglessness has not prevented snakes from exploiting a large variety of habitats. Some snakes spend most of their lives on the ground surface, whereas others are largely fossorial, arboreal, or aquatic. A few snakes are even marine. Snakes can be diurnal, nocturnal, crepuscular, or a combination of these.

Snakes cannot tolerate freezing temperatures, and therefore in colder regions (including Delaware) they spend the winter months inactive in a resting place, or *hibernaculum*. The hibernaculum may be located underground below the frost line or in a crevice or hole in a rotting log, tree stump, or debris pile out of reach of freezing temperatures. Some snakes hibernate solitarily, whereas others congregate in communal dens. Individuals of several different species sometimes use the same hibernaculum, and some snakes return to the same hibernaculum year after year.

Snakes are strict carnivores and are known to eat a wide variety of prey, including bird and reptile eggs. Small species tend to feed on invertebrates, whereas larger species often feed on fish, amphibians, reptiles, birds, or mammals. Some snakes eat their prey alive; others kill prey first by constricting them with their bodies. Still others have specialized teeth (*fangs*) that can deliver venom to help subdue their prey. Snake teeth are pointed and curved backward and cannot be used for chewing; therefore, snakes must swallow their prey whole. In fact, many snakes have jaw modifications that allow them to swallow relatively large prey whole. In these snakes, the jaw parts are less rigid than those of most other reptiles and are loosely connected to each other and to the cranium. An elastic ligament allows anterior separation of the lower jaw halves, permitting an extra-wide gape.

Snakes reproduce through internal fertilization of eggs. Like other squamates, male snakes have a pair of saclike copulatory organs (*hemipenes*) that are kept inside the cloaca except during copulation. Only one hemipenis is used at a time during mating. Snakes are either *viviparous* (live-bearing) or *oviparous* (egg-laying), depending on the species. Parental care of offspring is limited, although some species do attend eggs after their deposition, offering protection from predators.

As snakes grow, they shed their skin periodically, usually in one piece from head to tail. Many species, including those in Delaware, reach sexual maturity between one and four years of age. Contrary to popular belief, most snakes are not aggressive, and will flee when approached by a human. If cornered, however, many snakes will resort to defensive striking or biting. When captured, some snakes thrash about and secrete a foul-smelling fluid, known as musk, from the anal scent glands near the base of the tail.

Delaware has nineteen species of snakes, representing two families:

- Colubrid Snakes (*Family Colubridae*)—this is by far the largest family of snakes in the world, and has been divided into various subfamilies, three of which are represented in Delaware as described here.

    Harmless Live-bearing Snakes (Subfamily *Natricinae*)—this subfamily contains small to moderately large snakes that may be aquatic, terrestrial, or semifossorial. Those in Delaware feed primarily on cold-blooded prey such as frogs, salamanders, fish, and invertebrates, and are not constrictors. Many are noted for discharging foul-smelling musk when handled. There are eight species known to occur in Delaware: the red-bellied watersnake, common watersnake, queen snake, Dekay's Brownsnake, red-bellied snake, Eastern ribbonsnake, common garter snake, and smooth earthsnake.

    Harmless Rear-fanged Snakes (Subfamily *Xenodontinae*)—the term "rear-fanged" is used for this subfamily because most members have two enlarged, faintly grooved rear teeth, and some, like the Eastern hog-nosed snake, can deliver mildly venomous saliva. Members of this subfamily in Delaware are terrestrial or fossorial, and they prey mostly on cold-blooded vertebrates or invertebrates. Some species apparently use constriction to subdue their larger prey. All in Delaware are oviparous. There are three species known to occur in Delaware: the common wormsnake, ring-necked snake, and Eastern hog-nosed snake.

Harmless Egg-laying Snakes (Subfamily *Colubrinae*)—this subfamily represents the largest and most diverse group of colubrids. All species are oviparous. Those in Delaware may be terrestrial, semi-fossorial, or largely arboreal. Some species are constrictors, and many feed on small mammals, although some eat a variety of cold-blooded vertebrates and invertebrates. There are seven species known to occur in Delaware: the Northern scarlet snake, Northern black racer, red cornsnake, Eastern ratsnake, Eastern kingsnake, Eastern milksnake, and rough green snake.

- Vipers (*Family Viperidae*)—most vipers are heavy-bodied, venomous snakes with a broad, triangular head and vertical pupils. All have a pair of hollow fangs on the upper jaw that are used to deliver venom to prey or to ward off predators. The fangs are located near the front of the mouth and are hinged at the base so that they fold back along the roof of the mouth when the snake is at rest and move forward as the mouth opens to strike. All New World species have a pair of heat-sensing pits between the eyes and nostrils. These pits contain cells that can detect differences in temperature of 0.001° Celsius, an adaption that allows these pit vipers to locate and track warm-blooded prey in total darkness, such as in a rodent burrow. This family is represented in Delaware by only one species, the Eastern copperhead.

HERPING: FINDING, OBSERVING, AND DOCUMENTING AMPHIBIANS AND REPTILES

HOW TO FIND AND OBSERVE AMPHIBIANS AND REPTILES

Most herptiles spend the majority of their lives under cover or in other ways out of sight of humans. However, by knowing the preferred habitats and the behavior of herptiles, a patient herper can locate many Delaware species with relative ease. Suggestions shown with each amphibian and reptile group will aid the searcher.

GENERAL GUIDELINES

- Minimize disturbance to animal habitats while searching.
- If you turn over a log, rock, board, or other debris, make sure to return it to its original position; underneath most objects is a fragile, small ecosystem that may have taken many years to develop.
- Avoid destroying any logs or stumps.
- Be careful not to disturb mating pairs or eggs.

SALAMANDERS

- To find terrestrial or semiaquatic salamanders, look under logs, rocks, or debris in woodlands and in or near small wooded streams, spring seeps, freshwater marshes, wet meadows, or bogs.
- To find aquatic salamanders, breeding adults, or larval salamanders, search various freshwater habitats, especially streams, vernal pools, and ponds, at night with a bright flashlight.
- Searching for salamanders on rainy nights with a bright flashlight can be especially productive.

FROGS

- Look for terrestrial species on the ground while walking through woodlands or other terrestrial habitats.
- Look for semiaquatic species along the shoreline or at the surface of freshwater habitats.
- To find breeding frogs of most species, visit freshwater habitats (especially vernal pools and marshes) on a warm night in late winter, spring, or early summer, particularly during or just after rainfall. Listen for frogs calling and search with a headlamp or bright flashlight.
- Use triangulation to locate an elusive, calling frog whereby two or more searchers stand five to twenty feet apart and point toward the calling frog with their flashlights. Look for the frog where the light beams intersect.

TURTLES

- To find aquatic or semiaquatic turtles, visit a freshwater habitat on a warm morning or early afternoon. Look for turtles basking on logs, rocks, or stumps in the water or on the shoreline.
- Look for terrestrial turtles (e.g., Eastern box turtles) while walking through woodlands, scrub-shrub areas, fields, or wet meadows in spring to early summer, especially after rainfall.
- Look for sea turtle tracks in the sand on Atlantic Ocean beaches.

LIZARDS AND SNAKES

- Search for basking lizards and snakes on warm, sunny mornings on debris piles, fallen trees, rocks, or stone foundations. Abandoned house sites and roadside dumps are often good places to search.
- Look for terrestrial snakes and lizards by walking quietly through woodlands or fields, watching and listening for movement. Turn over logs, rocks, boards, and other debris. Roadside debris piles can be especially productive.
- Look for semiaquatic snakes basking at the edges of freshwater habitats.

ROAD CRUISING

Another way to search for herpetiles is by car. When "road cruising," consider the following:

- Choose a seldom-traveled road and drive slowly.
- Pick a warm night to observe snakes warming on the roadway.
- Pick a rainy night to see many frogs or salamanders (and sometimes snakes) moving across the wet road.
- Search in pairs, if possible—the driver can watch the road, and an observer with a flashlight can jump out to check sightings.

HELPFUL EQUIPMENT

- Binoculars and spotting scopes can be useful in identifying distant herpetiles, especially basking turtles, lizards, or snakes. Close-focusing binoculars are especially useful in allowing the observer to approach an amphibian or a reptile within six to eight feet and get excellent close-up views of the animal without disturbing it.
- A flashlight is essential for nighttime herping. Invest in a high-quality, bright flashlight.
- A headlamp is also good to use because it keeps the hands free; however, it may not be bright enough to serve as the primary searchlight.
- Hip boots or chest waders are helpful for exploring many aquatic and wetland habitats.
- Knee-high rubber boots are adequate for exploring shallow-water habitats and are also useful in drier habitats for protection against chiggers and ticks.

HOW TO CATCH AND RELEASE AMPHIBIANS AND REPTILES

It usually is not necessary to catch an amphibian or a reptile to identify it. Field guides provide descriptions of species that focus on distinctive, readily observable features or field marks that generally can be seen from a distance. In addition, an observer can gain much enjoyment by watching the undisturbed animal. However, capturing is sometimes necessary to make a positive identification (particularly for beginners) and to measure or examine an animal more closely. Adherence to the important general guidelines below and the suggestions under each amphibian and reptile group will help keep the animal, its habitat, and you safe.

GENERAL GUIDELINES

- Never disturb mating or nesting animals.
- Take care not to harm the animal during or after capture.

- Never handle amphibians or reptiles with hands sprayed with insect repellent.
- Minimize the length of time an amphibian or a reptile is handled.
- Always return an animal to the exact place of capture.
- After handling an amphibian or a reptile found under an object (e.g., log or rock), always replace the object first, and then release the animal next to the object.
- Always wash your hands thoroughly after handling any amphibian or reptile.

SALAMANDERS AND FROGS

- Wet your hands before handling any amphibian to minimize damage to the animal's protective layer of slime.
- Never grab a salamander by the tail. The tail may break off.
- Aquatic salamanders, salamander larvae, and tadpoles can be captured with a small plastic container or dip net.
- Most adult frogs can be caught by hand, although a long-handled net may be helpful.

TURTLES

- Most turtles can and will bite in self-defense. Keep your fingers away from the turtle's mouth.
- Never grab a turtle by the tail (except snapping turtles, see below), because this can injure the turtle.
- Aquatic turtles are likely to quickly slip into and underneath the water's surface when approached. A long-handled dip net and a great deal of patience often are required to catch one.
- When handled, the Eastern snapping turtle can be dangerous because of its powerful bite. Leave this turtle alone whenever possible. If you must move a snapping turtle, the safest method is to pick it up by the tail with a gloved hand, taking care to hold the turtle's body far away from your own. A large plastic garbage can or other container may be helpful in transporting the turtle to a safer location.

LIZARDS

- Lizards are highly mobile and difficult to catch.
- Capturing lizards is easiest in the morning, before the sun warms their bodies.
- Never grab a lizard by the tail, as the tail can break off.
- A lizard noose can be a good way to capture lizards.

SNAKES

- Before capturing a snake, be sure that it is not venomous (the copperhead is Delaware's only venomous snake).

- Wear gloves and long sleeves while capturing snakes. Many snakes will bite in self-defense, and this bite can be painful.
- Many snakes will remain motionless when approached, relying on camouflage and immobility to render them invisible. In such circumstances, an observer may be able to approach slowly and grab the snake with one hand, just behind the head, keeping a firm hold but not squeezing. The other hand should be used to support the body of the snake.
- Special tools (e.g., snake sticks and snake tongs) are helpful for capturing and handling snakes, but should be used with care to avoid injuring the snake.
- If bitten by a nonvenomous snake, clean the wound with soap and water and apply an antiseptic.
- If bitten by a venomous snake, remain as calm as possible and seek medical attention immediately. With proper medical treatment, the possibility of fatality is extremely low.

WHAT TO DO IF YOU FIND A TURTLE CROSSING A ROAD

Do not endanger yourself or others just because you want to save the turtle! If it is safe to do so, you can pull off the road and attempt a rescue. If the habitat is adequate, place the turtle on the side of the road in which it was heading. If the habitat is highly disturbed, note the location where the turtle was found, and take the turtle to a knowledgeable person (e.g., staff at a state park or an environmental center) to decide where to release the turtle.

PRECAUTIONS REGARDING COLLECTION

REGULATORY GUIDELINES

- It is illegal to collect or possess any of Delaware's state-designated endangered or threatened amphibians or reptiles without appropriate state and local permits. If the species is federally listed, a federal permit is required.
- Under current regulations, the sale of native amphibians and reptiles collected from the wild is prohibited in Delaware.
- Any collection of common species should be for scientific, classroom, or home study purposes only, according to specific provisions in current laws.

ETHICAL GUIDELINES

- Proper containers should be used for transporting these animals. Plastic containers with adequate ventilation are ideal for most herpetiles. Because snakes are escape artists, containers used to hold them must be secure. Plastic containers with screw-on lids and pillowcases that can be tied at the top work well.

- Captured animals must not be subjected to high temperatures or dry conditions. They easily can die in a hot car. A cooler can be used for a limited period of time to help keep animals in a suitable temperature range during transportation.
- Amphibians and reptiles brought into the classroom or into the home should be returned to the site of collection as soon as possible.
- If you are going to keep an amphibian or a reptile, research how to take care of it. There are many excellent websites with animal care information.
- Consider purchasing a captive-bred animal instead of collecting one. A captive-bred animal usually does better in captivity than an animal born in the wild, and you will not be depleting the natural population.
- Never release into the wild a captive-bred herp, or even a wild-caught herp that has been raised as a pet.

HOW TO DOCUMENT OBSERVATIONS

Whenever possible, observations of herpetiles (especially particularly rare or uncommon species) should be documented. Documentation must include the date and location of the observation in order to be of any scientific use. The best way to document an observation is to take a good photograph. In addition, it is helpful to provide field notes that include a description of the habitat, the weather conditions, a detailed description of the animal (including size and coloration), and a description of the animal's behavior.

An excellent way to document herpetile observations is to use *Herpmapper*, an online database for worldwide amphibian and reptile records. A photograph (and/or sound recording for frog species) is required to document an occurrence. Records can be made in the field using the *Herpmapper* mobile app or by entering the required information on the website Herpmapper.org. Data from Delaware that is recorded in the *Herpmapper* database is currently being used by the Delaware Division of Fish and Wildlife to guide conservation efforts.

Photographs for documentation or identification purposes can be taken with any camera as long as the resulting image of the amphibian or reptile provides an adequate rendering of the animal. Of course, the quality of the photograph will increase with the skill of the photographer and the quality of the photographic equipment, but basic point-and-shoot and cell phone cameras generally are adequate for documenting all herpetiles. It is best to take several photographs at different angles and distances to ensure that a positive identification can be made.

HOW TO CONFIRM AN IDENTIFICATION

If you find an amphibian or a reptile for which you would like an identification confirmed, you may contact the Delaware Nature Society or Delaware's State Herpetologist at the Delaware Division of Fish and Wildlife.

CONSERVATION OF DELAWARE'S HERPETOFAUNA

CAUSES OF AMPHIBIAN AND REPTILE DECLINE

By far the major cause of amphibian and reptile decline in Delaware is habitat loss and degradation. The clearing of woodlands, the filling and draining of wetlands, and the spreading of invasive, non-native plants have had devastating effects. In addition, various types of pollution, such as sediment runoff, fertilizers, and toxic chemicals (including pesticides and herbicides) continue to degrade habitats.

Other causes of decline include: collection of herpetiles from the wild; predation by non-native animals (particularly cats); and death on roadways by motor vehicles. Particularly vulnerable to motor vehicle strikes are amphibians that migrate across roadways to breeding areas on warm, rainy nights, snakes that warm themselves on road surfaces, and the Northern diamond-backed terrapin, which frequently crosses coastal roadways to nest.

CONSERVATION AND MANAGEMENT OF AMPHIBIANS AND REPTILES

Actions needed to protect and increase the remaining herpetile populations in Delaware include the following:

- Protect as much acreage of woodlands, stream corridors, and wetlands as possible, on both private and public lands.
- Manage natural areas to maximize biodiversity, providing quality habitats for the complete spectrum of native species, including herpetiles. Especially important is the retention of microhabitats, including unmowed areas, logs, stumps, rocks, and leaf litter.
- Enact and strictly enforce laws to reduce habitat loss and degradation. Many freshwater wetland habitats have inadequate or no legal protection and consequently are at high risk for degradation or destruction for development. Upland forests are also insufficiently protected. Unless stricter laws are enacted, only habitats that lie within protected areas such as parks, wildlife areas, refuges, or preserves will sustain healthy amphibian and reptile populations.
- Strictly enforce laws to prevent commercial collecting of all nongame herpetofauna.
- Support and strengthen existing federal, state, and local laws that protect rare or endangered species and fully protect their habitats.
- Conduct research and maintain databases on species distribution, life history, and causes of decline, and apply the findings to land management practices.
- Promote environmental and science education programs that include the importance of protecting herpetiles and their critical habitats.

- Publicize the values and needs of Delaware's amphibians and reptiles to raise consciousness and gain public support for their protection.
- Support agencies and organizations (both public and private) that are working to complete and advocate for these various and necessary actions.

LEARNING RESOURCES

Books helpful with guidance and identification include:
- J. F. White, Jr., and A. W. White, *Amphibians and Reptiles of Delmarva*, rev. ed. (Centreville, MD: Tidewater Publishers, 2007)
- R. Powell, R. Conant, and J. T. Collins, *Peterson Field Guide to Reptiles and Amphibians of Eastern and Central North America*, 4th ed. (New York: Houghton Mifflin Harcourt, 2016)
- H. R. Cunningham and N. H. Nazdrowicz, *The Maryland Amphibian and Reptile Atlas* (Baltimore: Johns Hopkins University Press, 2018).

CHAPTER NINE

Birds

IAN STEWART

INTRODUCTION

It is hard to imagine a world without birds, and many people who have lived their entire lives in Delaware do not realize what an exceptional state it is for birds and birding. Delaware is a genuine "four-season state" that offers very different birding experiences at different times of year. Despite its small size, Delaware hosts about 175 species of breeding birds, with another 150 or so either passing through the state during migration or spending the winter here.[1] Even people who profess to have no interest in birds could probably identify at least a half-dozen species they have casually noticed around their yard or workplace or—at the very least—through their association with sports teams or their presence on greeting cards.

Birds are attractive for many reasons and have stimulated an interest in nature in many people. Most of them are active during the day and make little attempt to hide from view. Many, even common backyard birds like the American robin (*Turdus migratorius*), blue jay (*Cyanocitta cristatus*), and Northern cardinal (*Cardinalis cardinalis*), are colorful or have attractive songs. Birds are found in all habits including urban locations sparsely occupied by other forms of wildlife. They are present all year round, including in the depths of winter, and the fact that different species come and go with the seasons adds a fascinating dimension to our daily routines. Birds can also be attracted to backyards that provide food and water, and some may also breed in these yards if there are suitable native species of shrubs or trees, or artificial nest boxes.

Beyond its familiar species, Delaware also has had more than its share of rare vagrants, so that the official list of species ever recorded in the state currently stands at 420. This remarkable total is due to three main factors. First there is an unusual diversity of habitats contained in the second smallest state. Delaware has deciduous forests, farmland, marshes, sandy bays, pine forests, freshwater lagoons, mudflats, ponds, rivers, estuaries, and rocky seashores. Second, all of these sites are at least partially accessible to birdwatchers, and many are preserved as state parks and wildlife refuges. Third, Delaware has a large and active network of birdwatchers and, because the state is so small, their coverage is thorough.

The state's *avifauna* has been well-documented thanks to a Breeding Bird Atlas, a survey that took place between 1983 and 1987. During this five-year project, the state was divided into 222 blocks, which were searched by an incredible total of 271 volunteers to determine which birds were definitely or probably breeding in each one. The survey data were published in a magnificent book that featured a biographic review of each breeding species, and of each migratory and wintering species, and contained several excellent essays about Delaware's birds. The Breeding Bird Atlas was repeated a quarter-century later (from 2008–2012) to test for changes in the breeding status of each species, and the results are currently being prepared for publication. The birds of Delaware also feature strongly in *Birds of Maryland, Delaware, and the District of Columbia*, a 2019 book about mid-Atlantic birds.[2]

Delaware benefits from having a lively birding organization, the Delaware Ornithological Society (www.dosbirds.org), which hosts regular indoor and outdoor meetings and publishes articles on local birds. Delaware birders also have a strong presence on social media with the Delaware Birding Facebook group currently boasting over four thousand members.

A PRIMER OF BIRD BIOLOGY

Birds are warm-blooded animals, meaning that they can maintain a constant body temperature regardless of external environmental conditions and weather, and were traditionally thought to have evolved from reptiles.[3] Indeed, most authorities now consider them to be a subcategory of reptiles ("living dinosaurs") because they share several reptilian characteristics, including nucleated red blood cells, scaly legs, and the production of an amniotic egg (soft-shelled in reptiles, hard-shelled in birds). Unlike reptiles, however, birds do not have teeth, and they are the only animals to possess feathers. Remarkably, a fossil was discovered in a Bavarian quarry in the 1850s of an *Archaeopteryx*, thought to have lived approximately 150 million years ago, which is considered a "transitional form" between reptiles and birds that shows features of both groups.

Birds are thought to have undergone a large diversification about sixty-five million years ago, following the extinction of the dinosaurs. Modern birds are divided into two broad categories: *passerines* (perching birds with three thin toes pointing forward and one backward); and *non-passerines* (birds with thicker toes or webbed feet).

Passerines are much more numerous and are thought to have evolved more recently than non-passerines. Like mammals, birds' warm-blooded metabolism has profound consequences for their physiology. The high metabolic

FIGURE 9.1. Wintering gulls and seaducks at the Indian River Inlet. (Photo courtesy of Ian Stewart.)

requirements of maintaining a high body temperature (40°C/104°F), coupled with the high energetic demands of flight, mean that small birds spend most of their lives looking for food. Birds also have a series of air sacs throughout their chest and abdomen that are interconnected to the lungs in such a way that air flows in one direction. This enables them to extract as much oxygen as possible from each breath.

Birds also need to retain as much metabolic water as possible, and so they excrete *nitrogenous waste* in the form of uric acid. This is the white precipitate many people assume is bird feces. It is not. Bird feces is the dark brown lumps mixed in with the uric acid splash.[4]

Feathers are made of the protein keratin, and are thought to have initially evolved as elongated, frayed scales that helped insulate the body. Most of a bird's feathers still help it minimize heat loss by trapping a layer of warm air near the body, although the feathers of the wings and tail became much longer and broader to enable it to fly. The wing feathers are mostly used to power flight while the tail is mostly used for directional changes. As they grow, feathers receive continuous blood from the body, but once they are fully grown, their base becomes sealed, so that feathers are "dead" structures that gradually become abraded and worn. Consequently, most birds replace all of their feathers once a year during late summer or early fall through a process known as *molt*.[5]

Feather coloration is highly variable between birds, and is usually caused by the presence of pigments. *Carotenoid pigments* are responsible for the various shades of red, orange, and yellow, while *melanin pigments* are responsible for the various shades of browns, grays, and blacks. Blue feathers do not contain pigments, but get their appearance because of the way light waves reflect from their microstructure.[6]

In many birds, the males are much more colorful than the females. This is thought to have arisen through sexual selection, where the color or brightness of a male's plumage indicates his desirability as a partner, perhaps because of his age, disease resistance, or health. Female preference for these males, in other words, has caused these evolutionary changes. Natural selection has probably simultaneously favored females that are less colorful or even dull, since in most species the female spends much more time than the male *brooding*, sitting motionless in the nest either incubating its eggs or keeping its young nestlings warm, and they are less likely to be detected by a predator if they blend into their surroundings.

Over 90 percent of birds are *socially monogamous*, meaning that a male and a female form a pair and rear their young together.[7] Most passerine birds build some form of grassy nest in which the female lays the eggs, then incubates them for several weeks. Incubating females usually lose the feathers on their belly and breast to produce an area of bare, well-vascularized skin, known as a *brood patch*, which helps transfer body heat to the eggs. In passerines, the nestlings are *altricial*, meaning that they hatch blind and naked and are fed by both parents for up to three weeks before they can leave the nest. In non-passerines, such as ducks and shorebirds, the young are *precocial*, meaning they hatch at an advanced stage of development and can walk as soon as they hatch. Nest predation is a major source of loss, especially for small birds. Common nest predators of both eggs and nestlings are: snakes; predatory mammals such as weasels, raccoons, and skunks; and larger birds like blue jays, crows, and hawks.

About half of the 175 species that breed in Delaware migrate south for the winter, either to the Gulf Coast or across the water to Central or South America. Migration is strongly linked to diet, since almost of all these migrants are insect- or worm-eating birds that cannot find enough food to survive a typically cold Delaware winter. Migration has always been a well-studied branch of ornithology, but has recently gained even more prominence thanks to technological advances that allow birds to be tracked remotely. One example of this is the Motus network (*motus* being the Latin word for movement), in which automated receiving towers (including four in Delaware) detect the presence of animals up to fifteen kilometers away that have been fitted with electronic tags. Thankfully, much of the Motus data can be viewed by anyone at www.motus.org, where one can find the origin and sometimes the destination of birds detected by Delaware's towers with just a few clicks.

HOW TO BE A BIRDER

The most important piece of equipment for birding by far is a decent pair of binoculars. Just as a poor-quality pair will frustrate you, a good pair will heighten your enjoyment of birding and can now be bought for $200 or less. Although this is not an insignificant amount, think of it is an investment that will last years if not decades. The most common have 8× or 10× magnification, and it is best to try them out in a store rather than buying them online, since the same pair of binoculars can feel very different to different people. If you become interested in shorebirds or waterfowl, a telescope becomes almost essential because its greater magnification (often 20–45×) makes it easier to view birds that are often far off in the distance. Telescopes require the use of a freestanding tripod (the sturdier the better), although they can be attached to a window mount to allow birding from the comfort of your car. This can be an attractive proposition on a cold and blustery winter day.

As Jon Cox has written elsewhere in this volume, photography provides an extra dimension to birding, and advances in digital technology have made it much easier to get good quality images of the birds you see, even for a relative beginner. Many point-and-shoot cameras now have ultra-zoom lenses offering 40× magnification or more, although for better results it is best to purchase a digital single-lens reflex (SLR) camera with a telephoto lens, which can be found for less than $500 during sale periods. Once you are safely back home, photographs also make it possible to examine and identify birds you were unsure of in the field, and good quality pictures are a pleasant souvenir of a particularly memorable sighting.

An easy and enjoyable way to start learning about birds is to attract them to your yard with a variety of food types, either by providing abundant native plants (as Doug Tallamy outlines elsewhere in this volume), or through the use of different styles of feeders. The best all-around food is black oil sunflower, as this is readily consumed by most common backyard birds, including finches, cardinals, chickadees, and titmice. Feeders also provide a spot that makes birds relatively easy to photograph from your window.

As with any other hobby or pastime, the key to improvement is practice. There is a wealth of field guides and bird books to help you become familiar with each bird species, but there is no substitute for time spent actually watching birds. The best thing to do is to learn the common or most visible birds first, and then slowly expand your repertoire to those that are much less common or more difficult to find. Learning the songs of each species can be a huge help—most experts identify the majority of birds not by sight but by their vocalizations, especially for birds singing high in the canopy. Learning birdsongs is not easy, and can be overwhelming in the field, so again it is best to start by learning those of the birds you encounter everyday, either in situ or though recordings.[8]

Another way to become a better birder is to take part in some of the organized walks held throughout the state, as these are a great opportunity to learn from other birders and to share your own identification tips. Many people find that birding in a group is much more enjoyable than birding alone. Plus, the greater number of eyes and ears means that groups usually find more birds. The Delaware Nature Society runs four free bird walks per week during spring and fall, and the Delaware Ornithological Society runs regular trips all year round, each one to a different location. Visiting a bird banding station is also an excellent opportunity for seeing differences in color and body proportions between each species.

Another way to become more proficient at birding is to learn the most common birds associated with each habitat, so you know what to expect from each site. This eases the identification of any mystery birds because it narrows down the list of likely suspects in a location.

As you become more experienced, you will start to take a holistic approach to seeing each bird in its environment and noticing subtleties in their distribution. Birds exhibit microhabitat preferences, so even though multiple species may be present in a single patch of deciduous woodland, they will feed or nest at different heights or in different trees or bushes. Many ornithologists have spent their careers studying these differences.[9]

CITIZEN SCIENCE AND INTERNET BIRDING

Until the end of the twentieth century, most research on birds and other animals and plants was performed by university or government scientists using data they collected themselves. However, some of these scientists realized that there are an enormous number of citizens interested in nature who are willing to help collect data on their behalf. As one branch of so-called citizen science (which Tara Trammell explains elsewhere in this volume), this has led to the accumulation of a massive amount of data from a much broader network of locations than the scientists themselves could have possibly collected.

Arguably the greatest advance in birding in the twentieth century has been *eBird* (www.eBird.org), a user-friendly, interactive website developed by the Cornell Lab of Ornithology that allows anyone to set up an account and start entering their bird sightings, which in turn can be analyzed by scientists. These sightings can be from anywhere: a backyard, a local park, wildlife refuge, or a beach vacation. All of the data (except for reports of sensitive species) can be viewed by the public, so users can see which species other birders have been seeing and can thus plan their birding trips to the most productive "hot spots" or sites where a target species (such as a rare bird) has recently been seen.

Sightings from across the continent have advanced our knowledge of species distributions during both summer and winter, as well as of the timing of their migration. Codes can be added for the sex and age (if known) of any birds seen, as well as for any evidence of breeding, which has helped scientists test for sex-specific differences in winter distribution and migration timing. This has helped refine our knowledge of each species breeding range. Although eBird started in the United States, it is now truly global, with sightings from 99 percent of countries, representing over 99 percent of the world's approximately 10,500 bird species. The eBird site also allows users to submit photographs and sound recordings, which the Lab of Ornithology use to improve *Merlin*, a free mobile app that helps people identify birds of North America. Merlin can be used to identify unknown birds based on a series of questions and suggested possibilities. During migration periods, Cornell also publishes *BirdCast* (https://birdcast.info), which predicts which species will start appearing at certain times in different regions of the country based on the direction and strength of the prevailing winds. For Delaware, winds from the south or southwest are most favorable for spring migration, while winds from the northeast are best for fall migration.

Nest Watch (www.nestwatch.org) is another citizen science scheme developed at Cornell. Participants submit data from nest boxes checked once or twice per week, and these data are then used by scientists to examine continent-wide patterns in key life history components such as laying dates, clutch size (number of eggs laid per attempt), hatching success, fledging success, and the number of clutches per season. Erecting "trails" of wooden nest boxes began in earnest in the 1960s as a way of counteracting the decline in Eastern bluebirds. Thanks to the expanding network of trails in public parks, golf courses, and large suburban yards, Eastern bluebirds are now common throughout most of the East, and several other species are also taking advantage of the nest boxes. A great way to gain insights into the breeding biology of several bird species—and to contribute data to science—is to put up nest boxes in your backyard, or to help check boxes at one of several existing trails across Delaware, and submit these data to Nest Watch. There are many resources available for those who wish to build or maintain boxes for multiple species (see www.sialis.org, www.nabluebirdsociety.org).

Another way to become involved in citizen science is to take part in one of the long-running bird counts that take place each year, such as the seven Christmas Bird Counts in Delaware, and the Global Big Day (the first Saturday in May).

BIRD CONSERVATION: A CALL TO ACTION

Many North American birds are in serious trouble. The North American Bird Conservation Initiative's recent *State of North American Birds* report summarized a conservation vulnerability assessment made for all 1,154 bird species that occur in Canada, the continental United States, and Mexico. Each species was given a "Concern Score" from one to twenty based on a combination of factors including their population size, range, and population trend. Species were placed on a watch list if they had a Concern Score of fourteen or higher, or a Concern Score of thirteen plus a population decline index of five (the steepest). Fully 37 percent of the species (432 of them) were placed on the watch list. Most of these birds were birds of the ocean or tropical forests but forty-two of them were species that occur in Delaware, sixteen of which are regular breeders.[10]

The main threats to birds are usually *habitat loss* and *habitat fragmentation*. Habitat loss tends to be more dramatic, especially when a whole forest or field is ploughed under to build a new development, since this leads to the loss of almost all of the breeding and wintering species there. But habitats can also be lost gradually, as when a suburban development encroaches into a forest. The effects of habitat fragmentation can be more subtle. Forests can be chopped up by logging, agriculture, or the construction of new roads or isolated houses. This may appear less problematic, as it can leave large patches of woodland remaining, but unfortunately, many bird species are "area-sensitive" and their breeding success declines as their forest habitat shrinks.[11] Smaller patches also have a greater proportion of *edge habitat*, and birds nesting nearer the edges of patches suffer more losses from predators and parasitic species like Brown-headed cowbirds (*Molothrus ater*) (see species description below). Moreover, most forest birds require a minimum area of habitat for breeding, and once a forest is trimmed below this threshold, many birds will abandon it.

Once habitat is lost to development, it is usually lost forever. Old developments or shopping malls are much more likely to be replaced with updated buildings than to be torn down and have their land returned to forests or farms. Delaware has the sixth-highest population density in the nation and the fourteenth-fastest rate of growth. The population increased from 900,000 in the 2010 census to 980,000 in the spring of 2019, a growth of almost 10 percent in less than a decade.[12] More people means the construction of more housing and infrastructure like roads, as well as restaurants and stores, which in turn means a continued loss of wildlife habitat. Hundreds of years of diminishing forestland makes it all the more imperative that private landowners landscape with the native trees, shrubs, and flowers that birds require as sources of food and habitat.

Some of Delaware's birds are also compromised by habitat losses that occur thousands of miles away. About fifty of our breeding species spend the winter in the lush forests of Central and South America, which are steadily being cleared to create open areas for coffee plantations and grazing land for beef cattle. Migrant songbirds face particular challenges, since their habitats and breeding grounds are being curtailed at both ends of their journeys. Our birds may fly to countries with practically no laws against shooting or trapping them. All American birds and their eggs and nests are protected, except for house sparrows, starlings, and feral pigeons.

For other birds, the problem may not just be habitat loss, it may be the disappearance of a food source, or outright persecution. The bobolink (*Dolichonyx oryzivorous*) is a small, seed-eating grassland blackbird that breeds in the upper half of the Eastern United States. Because it is undergoing a steep population decline, it is now on the watch list, and is actively conserved in the United States. However, bobolinks are reportedly shot on their South American wintering grounds because local farmers view them as pests because they feed on the crops in large flocks. The threatened red knot (*Calidris canutus*) has been in serious decline in the United States because of a decline in their primary food, the eggs of horseshoe crabs, since local fishermen and crabbers collect the adults to use as bait.[13] A state moratorium was passed on collecting horseshoe crabs, but this was soon overturned after fishermen argued that this infringed on their livelihood.

As Doug Tallamy points out in his chapter, probably the most overarching threat to birds is the collapse of many insect populations, both in the United States and beyond. Insects and other arthropods are a keystone of many avian food webs. The adults of many large bird families, such as warblers and flycatchers, only eat insects, and insects are also sometimes eaten by species, like sparrows and finches, that mostly eat seeds. Importantly, however, almost all passerines feed their nestlings on insects (and especially on the caterpillar larvae of insects), with insects providing some or all of the nestling diet in eighty of our 175 breeding species. Insect numbers are in steep decline

FIGURE 9.2. A red knot surrounded by semipalmated sandpipers at Port Mahon. Note the metal leg band on the red knot. (Photo courtesy of Joe Sebastiani.)

nationwide due largely to the increased use of pesticides in agriculture and the three hundred million acres we have now planted in monoculture field crops such as corn, soybeans, and wheat, which only support a tiny number of insect species.[14]

Another factor in insect decline is the explosion in non-native and invasive plants, since these do not support the same diversity or biomass of arthropods as our native plants do.[15] Many of our insects (especially caterpillars) are narrowly host-specific and have not evolved to subsist on alien plants. Some of these non-native plants, like Bradford pear (*Pyrus calleryana*, 'Bradford'), English ivy (*Hedera helix*), and Japanese barberry (*Berberis vulgaris*), are commonly sold to homeowners by garden stores and national chains that are either unaware or unconcerned that these plants have such a deleterious impact on ecological food webs.

Seabirds and waterbirds are especially threatened by many sources of contamination and pollution, including the constant residual presence of lead shot, industrial waste in rivers, and catastrophic events such as oil spills. In recent years, evocative images of seabirds killed by choking on plastics have highlighted the importance of reducing the amount of our trash that winds up in our oceans. Beyond this are the mind-boggling impacts of domestic and feral cats. One study projected that as many as 3.7 billion birds may be killed each year in the United States alone by outdoor cats.[16]

Another primary underlying problem is that landscapes that birds find attractive are often not the same as those that many humans find attractive. Many suburbanites like to maintain large, close-cropped, treeless yards, and yet these are virtually useless for birds, since they contain few plants that can host the insects that birds require to feed, and the openness of lawns makes birds an easy target for predators.

Although this prognosis may sound grim, there are several short-term and long-term ways in which people can help birds. A better option for backyards would be to let the grass grow long, perhaps with the addition of some native meadow wildflowers such as goldenrods (*Solidago* sp.) and coneflowers (*Echinacea* sp.) that would attract insects during the summer and provide seeds during the winter. One trick that is often underappreciated: if lawns are left unmowed until spring, they provide winter cover and roosting sites for birds like sparrows. Adding several native shrubs and trees, especially high-value species like oaks (*Quercus* sp.) and native cherries (*Prunus* sp.), would provide birds with nesting sites, food sources, and refuges if predators appear.

One powerful short-term measure is to keep cats indoors and attempt to persuade others to do likewise.[17] Other excellent ideas: persuade the owners of high-rise buildings in our cities to turn off most of their lights at night, since bright lights are a distractive hazard to migrant birds;[18] and remove invasive alien plants from backyards and wildlife areas and replant them with native equivalents, which are now readily available from several local suppliers. A particularly rewarding short-term measure is to become involved in conservation-related projects such as trash cleanups and tree-plantings, since these have a direct, tangible impact, which may motivate more people to become involved in larger projects. This is related to another idea, which is working to stimulate an interest in birds (and healthy bird habitats) among families, and particularly their children, so that birds continue to have a voice once the current generation has passed.

Finally, a common-sense measure to protect rare breeding or wintering birds is to restrict access to them, as at Cape Henlopen State Park, where barriers prevent humans from walking or driving into the colonies of birds like the beach-nesting piping plovers (*Charadrius melodus*). Long-term conservation measures include pressuring elected representatives to enact bird-friendly legislation. For example, an executive order was recently signed in New Castle County to direct that only native plants be used to landscape county-owned parks. Another measure would be to persuade stores or seed suppliers to provide customers with a greater selection of native plants, or to ban the sale of the most noxious and aggressive alien plants.

Perhaps the most important long-term step would be to urge legislation at the federal level to address climate change. Global warming has already impacted the lives of many birds; tree swallows, for example, now lay eggs approximately nine days earlier than they did fifty years ago.[19] Sea level rise is a genuine threat to some of Delaware's most emblematic birds, including rails, which breed in brackish marshes. If these marshes become flooded with seawater, the rising salinity levels will alter the types of plants supported, which could compromise traditional habitat and nesting areas for rails.

Ultimately, it is incumbent upon us to fight for birds, since human impacts on birds and their habitats cannot be solved by the birds themselves. If we want birds to be there for future generations to enjoy, the time to act is now.

THE MAIN BIRD FAMILIES OF DELAWARE

WATERFOWL (SWANS, DUCKS, AND GEESE)

Waterfowl are a morphologically and ecologically diverse set of birds, but as their group name suggests, all are associated with watery habitats in which they swim or paddle with their webbed feet. Swans and geese are large and long-necked and mostly graze on grass or submerged vegetation, whereas ducks tend to be smaller and more omnivorous. Ducks are further divided into "dabblers," which dip their broad, flattened bill under the surface of the water in search of floating invertebrates and soft vegetation, and "divers," which swim underwater to pursue fish, crustaceans, and mollusks. Swans and geese are sexually *monomorphic*, with the sexes looking alike, whereas most ducks are sexually *dimorphic*, with the males being more colorful and patterned than the females.

FIGURE 9.3. Snow geese gathering in a stubble field in winter. (Photo courtesy of Ian Stewart.)

Ducks provide an excellent introduction to birding, as their bright colors and distinctive patterning make the males relatively simple to identify, and they are usually easy to study as they float slowly along in open water. Even small lakes in urban parks can be surprisingly rewarding sites for watching ducks at close range. Sea ducks are usually harder to watch because they are further away and frequently disappear behind the waves, plus the sea fret and cold wind hampers observation. However, a relatively sheltered place to watch wintering sea ducks at close range is the Indian River Inlet, which regularly hosts dozens of the spectacular long-tailed duck (*Clangula hyemalis*) and all three species of scoters (*Melanitta* sp.), sturdy black ducks that dive for crustaceans and mollusks.

Delaware is a nationally recognized center for migrating and wintering waterfowl because of its extensive wildlife refuges, coupled with its inlets, marshes, and miles of seashore. On a good day in late winter it is possible to see over twenty species of waterfowl. Most waterfowl breed far to our north in the tundra and only a handful remain in Delaware during the summer, with the most common by far being the mallard (*Anas platyrhynchos*) and Canada goose (*Branta canadensis*). The wood duck (*Aix sponsa*) is our only cavity-nesting duck, and now breeds throughout the state in artificial nest boxes designed to replicate tree holes. These nest boxes are best placed on poles near quiet ponds or marshes, as wood ducks are skittish and easily disturbed by humans.

Canada geese were formerly only present during the winter, but—thanks in part to a year-round supply of food found on the region's farms—many now stay here year-round and breed throughout the state near any standing body of water. This brings them into conflict with humans as their copious green droppings can be slippery to pedestrians and are unpleasant to see and smell. Flocks of geese are also hazardous to drivers because small "gaggles" walk fearlessly across roads. However, because they are a native species, they are protected by law, so they cannot be controlled unless a special permit is obtained.

The snow goose (*Chen caerulescens*) is a striking black-and-white bird that breeds in the Canadian Arctic but winters in the Midwest and along the East Coast, with coastal Delaware being a popular location because of its flat stubble fields. Snow geese can be seen in flocks of over a thousand birds grazing in the farm fields along Route 9 in Kent County. Witnessing one of these flocks taking off is one of the great experiences in Delaware birding.

BIRDS OF PREY (HAWKS, FALCONS, AND VULTURES)

These species deploy a range of hunting styles to capture small birds or mammals, but can also catch and consume large insects or snakes, or feed on carrion. Distinctive features include a hooked bill for tearing flesh, and strong talons with which to grasp or kill prey. Many birds of prey display *reversed sexual size dimorphism*, in which the females are significantly larger than males. The reason for this is unknown.

A familiar bird of prey for many Delawareans is the red-tailed hawk (*Buteo jamaicensis*). This hawk is common year-round throughout the state and any large soaring hawk is likely to be a red-tail. Two other common birds of prey that look very similar are the Cooper's hawk (*Accipiter cooperii*) and the sharp-shinned hawk (*Accipiter striatus*). They are both primarily woodland hawks that regularly hunt small- to medium-sized birds in backyards, although the Cooper's hawk is the only one of the pair found year-round (the sharp-shinned is only present during winter and migration). These hawks are usually sit-and-wait predators, standing unnoticed in a tree, then swooping down on birds that are distracted while feeding. Their wings are broad and rounded and adapted for maneuverability rather than speed.

In contrast, falcons are fast-flying birds of open habitats with narrow, pointed wings adapted for speed. The peregrine falcon (*Falco peregrinus*) is a species that almost went extinct in the United States because of the insecticide DDT, but since that chemical was banned in 1972, the peregrine has rebounded and is now relatively common in cities (including Wilmington), where they prey upon feral pigeons as well as other birds. The American kestrel (*Falco sparverius*) is a small, cavity-nesting relative of the peregrine that is in serious decline in the mid-Atlantic region, where the number of breeding pairs has dropped by 88 percent since the 1970s. The reason for the decline is unclear, but suggestions include a lack of cavity-nesting sites, heightened predation by the increasingly common Cooper's hawk, or a lack of large insect prey because of greater pesticide use. Kestrels are birds of open areas that can sometimes be seen hovering in midair as they search the ground for their prey, which includes small rodents, grasshoppers, small birds, and snakes.

The bald eagle (*Haliaeetus leucocephalus*) is the national symbol of the United States, and this status—combined with its distinctive appearance—has helped it become a high-profile conservation success story. As was the case in many states, the bald eagle was almost extirpated as a breeding bird in Delaware primarily because of DDT. Numbers gradually recovered following the 1972 DDT ban, and now over fifty pairs breed in Delaware each year. Ospreys (*Pandion haliaetus*) are another charismatic fish-eating species that have had a successful recovery following the ban on DDT and the construction of wooden nesting platforms placed in marshes.

Vultures are large birds of prey that are often maligned by the general public, perhaps because of their habit of feeding on dead or rotting animals, but also because of their sinister black-feathered, bald-headed appearance. However, they play an important ecological role as scavengers cleaning up carrion that would otherwise accumulate along roadsides and around human habitation. The two species of American vultures—the turkey vulture (*Cathartes aura*) and black vulture (*Coragyps atratus*)—are both common throughout Delaware. They look superficially similar, but with practice can be distinguished with the naked eye, since turkey vultures glide on long, two-tone wings held in a shallow dihedral angle, while black vultures flap more frequently on wings that are shorter and have white tips. At close range, adult turkey vultures have red heads, whereas black vultures have gray heads.

GALLINACEOUS BIRDS (GAMEBIRDS)

Gamebirds are ground-nesting, round-bodied, omnivorous birds that have long been hunted for their meat. Delaware has few gamebirds left, and those that are present may descend from birds released here for hunting. Bobwhite quail (*Colinus virginianus*) are small, streaky brown birds more likely to be detected by their "bob-white!" call than seen. They were once widely distributed over the state but are now localized to a handful of sites including Bombay Hook National Wildlife Refuge. The wild turkey (*Meleagris gallopavo*) is a large gamebird that also formerly occurred throughout Delaware but was hunted, and its habitat contracted, to the point that it almost went extinct. Wild turkeys were reintroduced in the 1970s and are now fairly common in the lower two-thirds of the state, mostly in woodlands surrounded by open fields. The declines in quail and turkeys were probably attributable to a combination of hunting and the predation of eggs, young, and incubating adults by red foxes (*Vulpes vulpes*) and other carnivores.

HERONS AND EGRETS

Herons are large, long-legged birds native to watery areas, and since Delaware is blessed with many creeks, lakes, and marshes, we are also blessed with a dozen of these breeding species. Herons are renowned for eating fish, some of which are remarkably large, but they are actually omnivores that will eat amphibians, mollusks, reptiles, and even small mammals. Almost all herons are colonial and occupy traditional nesting areas (*heronries*) in trees for decades. Most of Delaware's breeding herons are concentrated at Bombay Hook and at Pea Patch Island off the coast of Delaware City. The Pea Patch Island heronry contains thousands of breeding herons and is the largest in the United States north of Florida. Consequently, it was designated a state nature preserve in 1988.[20]

SHOREBIRDS

Shorebirds (also known as "waders") are a diverse group of small- to medium-sized birds native to sandy shores, mudflats, ponds, and grasslands that usually feed by probing the soil or sand for invertebrates such as worms. Shorebirds vary strikingly in the length of their legs and bills, such that different species probe for prey buried at different depths, or in the case of some short-billed species, simply pick it off the surface. Only a handful of shorebird species spend the summer or winter in Delaware, but during both northward and southward migration, Delaware is a magnet for shorebirds, which can be seen in the thousands at sprawling coastal wildlife refuges like Bombay Hook and Prime Hook. The Delaware Bay is a world-renowned stopover point for migratory shorebirds that winter in South America but breed in the North American tundra. Since this is a journey of several thousand miles, many shorebirds stop here to refuel on invertebrate-rich mudflats and shores. A species inextricably linked with the Delaware Bay is the red knot, a medium-sized brick-red shorebird whose diet (as discussed above) is largely comprised of the eggs of horseshoe crabs (*Limulus polyphemus*), a distinctive arthropod more closely related to arachnids than to crabs.

GULLS AND TERNS

Gulls are large, powerful birds that are essentially omnivorous, while terns are more angular and forage for fish by plunging into the water from above. Several species of these mostly white birds are localized breeders on the sands and marshes of lower Delaware, though many thousands more winter here. The ring-billed gull (*Larus delawarensis*) is an especially abundant species during winter and is often found in flocks in parking lots.

OWLS

Owls are more common throughout Delaware than the average person realizes. Several species are year-round residents but are difficult to see during the day because they roost quietly deep inside the branches of trees where they are camouflaged by their streaky brown or gray plumage. Most are only active at night when it is too dark to see them, but they can be detected by their penetrative, eerie calls. The small screech owl (*Megascops asio*) emits a peculiar series of high-pitched descending whinnies and wails, the crow-sized barred owl (*Strix varia*) drawls its memorable "who cooks for you" call, and the large great horned owl (*Bubo virginianus*) gives the low, rhythmic, repeated "who who" hoot familiar to any fan of scary movies. Great horned owls start nesting in January, and hearing a pair of these dueting in a stand of conifers is a Christmastime treat. Most owls are nocturnal predators of small mammals like voles and mice, but larger owls will also take hawks and smaller owls. Owls swallow their prey whole, then later regurgitate indigestible material like fur and bones in the form of gray pellets, which can be picked apart to identify their prey, usually by the skull and (specifically) the teeth.

HUMMINGBIRDS

Like all of the other states in the Eastern United States, Delaware has only one species of hummingbird, the ruby-throated (*Archilochus colubris*). Hummingbirds are actually fairly common breeders in Delaware, but their tiny nests are difficult to find as they are high up in trees and camouflaged by lichen and moss. Interestingly, in recent years several Western species of hummingbird have appeared each fall in Delaware and other Eastern states, some of which remain through the winter, and so it is best to leave feeders up until Thanksgiving.

FLYCATCHERS

Flycatchers are relatively dull olive or brown birds that spend most of their lives sitting quietly on the edges of deciduous trees before sallying forth to catch insects in midair. Their plain colors and reclusive behavior render them fairly cryptic, so they are most easily detected by their distinctive vocalizations. The most widely distributed and easily seen flycatcher in Delaware is the Eastern phoebe (*Sayornis phoebe*), which nests below bridges and roof overhangs and is known for its constant tail wagging. The explosive "pit-za" call of the Acadian flycatcher (*Empidonax virescens*) and whistled "pee-a-whee" of the Eastern wood-peewee (*Contopus virens*) reveal the commonality of both species in Delaware.

SWALLOWS

Swallows are long-winged aerodynamic birds that catch insects in their bills in mid-flight. The barn swallow (*Hirundo rustica*) is a beautiful iridescent blue bird well known throughout Delaware because it builds its mud nests inside barns and on porches, sometimes in surprisingly urban locations. Several other swallows are associated with humans: the tree swallow (*Tachycineta bicolor*) commonly occupies nest boxes and the purple martin (*Progne subis*), our largest swallow, only breeds in artificial martin houses or gourds. Many swallows are in decline in certain parts of North America, probably (again) due to the presence of fewer aerial insects, which in turn likely results from the widespread use of pesticides and a tendency toward agricultural monocultures.

MIMIC THRUSHES

This trio of long-legged and long-tailed birds are common throughout Delaware, although only the Northern mockingbird (*Mimus polyglottus*) is familiar to non-birders because it is often found in backyards and urban parks where it sings loudly and aggressively chases humans who venture too close to its nest. The gray catbird (*Dumetella carolinensis*) and brown thrasher (*Toxostoma rufum*) are both common countryside birds, and the former is easily identified by its cat-like mewing call.

THRUSHES

These medium-sized insect and worm-eating birds are mostly restricted to large deciduous forests, with the exception of the ubiquitous American robin (*Turdus migratorius*), which is also found in parks and backyards, and the Eastern bluebird (*Sialia sialis*), which lives in open, lightly wooded areas where there is short grass. Our two other breeding thrushes, the Veery (*Catharus fuscescens*) and wood thrush (*Hylocichla mustelina*) are both fabled songsters; the ethereal song of the latter is regarded as one of the most beautiful of all. Unfortunately, both of these species are in decline, with the wood thrush being listed as near-threatened, having declined by over 50 percent since 1966.[21]

WARBLERS

Warblers are a diverse group of small colorful birds that feed on insects gleaned from leaves or picked from the soil surface. Probably no other group of birds makes people anticipate the arrival of spring more than these jeweled beauties. Over a dozen species breed in Delaware, and as many as twenty-five can be seen in a single day during peak migration. During migration most warblers feed high in trees, and the strain of hours spent watching them flitting above induces a common condition known as "warbler neck." The breeding species tend to be easier to see because they nest and feed low down or on the ground.

The most widely distributed breeding warbler across Delaware and the entirety of North America is the common yellowthroat (*Geothlypis trichas*). These breed in open habitats in dense thickets, especially of thorny plants like blackberries (*Rubus* sp.), and are identified by their bright yellow coloration and the black "bandit mask" of the male. Their constant lisping "wichity wichity" call is a common summer refrain. Another common and distinctive warbler is the ovenbird (*Seiurus aurocapilla*), a large, streaky, thrush-like warbler of deciduous woods usually found walking along the ground where it also builds its oven-shaped nests. Its loud ascending "teacher teacher" song, given from high in the canopy well into August, is a classic sound of Delaware's forests. The distribution of breeding warblers in Delaware is closely tied to habitat differences. A handful of species only breed in wooded swamps or pinewoods, while several of the rarer species are only found in the large tracts of hilly deciduous woodlands of the Piedmont. This includes the cerulean warbler (*Setophaga cerulea*), which is showing the fastest population decline of any warbler.

Warbler-watching is also popular during fall migration, which is more protracted because there is less pressure to arrive on the wintering grounds quickly. Since warblers are almost exclusively insectivores, few are found in Delaware during the winter, although the yellow-rumped warbler (*Setophaga coronate*) overwinters in coastal areas in significant numbers where it switches to a diet of berries.

BLACKBIRDS

Blackbirds are streamlined black or dark-colored birds renowned for the large flocks they form in winter. Even non-birders are astounded by the spectacle of long streams of blackbirds, which leave their marshland roosting sites each morning and head inland to farm fields where they use their long-pointed bills to probe the soil for insects, larvae, and seed before streaming back to their roosts later in the afternoon. These streams are mostly made up of red-winged blackbirds (*Agelaius phoeniceus*), but usually also contain many common grackles (*Quiscalus quiscula*) and brown-headed cowbirds. The cowbird is a notorious *brood parasite* that lays its pale, spotted eggs in the nests of other birds, especially open-cup-nesters such as cardinals, robins, and song sparrows (their nests are actually cup-shaped). The host parents either do not recognize the egg to be parasitic or are unable or reluctant to evict it and end up rearing the cowbird nestling themselves.

Since the cowbird nestling is usually much larger than their own nestlings, it requires a disproportionately large amount of food, and the hosts' own nestlings often die of starvation or are significantly undersized. However, because cowbirds are a native species, it is unlawful to remove cowbird eggs found in the nest of another species.

FINCHES, CARDINALS, AND BUNTINGS

These are colorful seed-eating birds, several of which are common throughout Delaware and easily attracted to feeders. The Northern cardinal is a stunning red bird recognized by schoolchildren and their parents alike, and holds the record for being the official state bird of the most states (seven). The American goldfinch (*Spinus tristis*) is another common backyard bird that is unmissable in the summer after it molts into its bright yellow and black breeding plumage. An unusual subcategory of these birds are the "winter finches," such as the pine siskin (*Spinus pinus*) and the two species of crossbills (*Loxia* spp.), which only appear in Delaware in certain winters. This appears to reflect a failure of the cone crop in the boreal forests of the Northern United States and Southern Canada, where they breed and would otherwise spend the winter. In years when the cone crop is particularly sparse, large numbers of winter finches irrupt south in search of food.

SPARROWS

Sparrows are a species-rich group of mostly small, brown, streaky birds found in many habitats throughout Delaware. Most are seed-eaters, although they feed their nestlings on invertebrates. Ten species breed across the state, although some of these are rare and localized, and several more winter here in large numbers. The most common and widespread breeder is the song sparrow (*Melospiza melodia*), the reference species from which all other sparrows are distinguished. Sparrows rely upon the deep cover of bushes and grasses for nesting and feeding, so learning their calls is the first clue to their identity. Two songs heard throughout the state are the "maids-maids-maids, teakettle-ettle-ettle" of the song sparrow and the "Oh, sweet Canada Canada Canada" of wintering white-throated sparrows (*Zonotrichia albicollis*).

10 PLACES TO WATCH BIRDS IN DELAWARE

The list below features the three most species-rich sites in the state (current eBird total of present species given in parentheses), as well as seven other sites that contain a different composition of habitats and therefore of birds. Most have an entrance fee (indicated by an asterisk), although state parks are free from December through February. The Delaware Birding Trail

FIGURE 9.4. A great blue heron surrounded by shorebirds at Bombay Hook. (Photo courtesy of Ian Stewart.)

(http://www.delawarebirdingtrail.com) is a superb resource with directions to these and other sites and information on their specialty birds.

1. *Bombay Hook National Wildlife Refuge*, Smyrna (323 species). Contains an extensive system of lagoons with periodically exposed mudflats, marshes, and woodlands, most of which can be viewed from a driving trail. An excellent site to see rails, shorebirds, waterfowl, and winter raptors, as well as a rare-bird hotspot.*
2. *Cape Henlopen State Park*, Lewes (298 species). A great site for fall hawk-watching and winter seabird-watching. A roped-off section of the dunes is home to several rare nesting species.*
3. *Prime Hook National Wildlife Refuge*, Milton (298 species). A similar set of habitats to Bombay Hook, though not quite as accessible.*
4. *Ashton Tract*, Port Penn (243 species). An area of Augustine Wildlife Area purchased in part with funds raised by the Delaware Bird-a-Thon and viewable from an ADA-compliant observation platform. Also a good habitat for wintering sparrows.*
5. *Indian River Inlet*, Rehoboth Beach (233 species). The prime site in Delaware for wintering sea birds, including occasional rarities such as auks.
6. *Middle Run Natural Area*, Newark (220 species). An area of thickets and woodlands that has been managed for biodiversity by the Delaware Nature Society, which also created an informative "birding trail" there. Best during spring and fall migration, but also has some unusual breeding species.
7. *Ashland Nature Center*, Hockessin (214 species). The headquarters of the Delaware Nature Society has a series of walking trails especially popular with birders during spring and fall migration, but also hosts a fall hawk watch and a bird blind.*
8. *Mispillion Harbor*, Milford (196 species). The DuPont Nature Center here is an outstanding place to view the flocks of feeding shorebirds, as is nearby Slaughter Beach.
9. *White Clay Creek State Park*, Newark (192 species). A great site to spot woodland passerines year-round and an excellent location for migrants, especially warblers.*
10. *Trap Pond State Park*, Laurel (157 species). A beautiful Southern-style wooded swamp surrounded by cypress trees that is home to some unusual breeding species.*

NOTES

1. Gene Hess, Richard L. West, Maurice V. Barnhill, and Lorraine M. Fleming, *Birds of Delaware* (Pittsburgh: University of Pittsburgh Press, 2000).
2. Bruce Beehler and Middleton Evans, *Birds of Maryland, Delaware, and the District of Columbia* (Baltimore: Johns Hopkins University Press, 2019).
3. Frank B. Gill, *Ornithology*, 3rd edition (London: W. H. Freeman, Ltd., 1996).
4. Ibid.
5. Ibid.

6. Geoffrey E. Hill and Kevin J. McGraw, *Bird Coloration, Volume 1: Mechanisms and Measurement* (Cambridge, MA: Harvard University Press, 2006).
7. Gill, *Ornithology*.
8. Lang Elliot and Marie Read, *Common Birds and Their Songs* (Boston: Houghton Mifflin, 1996).
9. Robert H. MacArthur, "Population Ecology of Some Warblers of Northeastern Coniferous Forests," *Ecology* 39 (1958): 599–619.
10. North American Bird Conservation Initiative, *The State of North America's Birds 2016* (Ottawa, ON: Environment and Climate Change Canada, 2016), www.stateofthebirds.org.
11. Thierry Boulinier, James D. Nichols, James E. Hines, John R. Sauer, Curtis H. Flather, and Kenneth H. Pollock, "Forest Fragmentation and Bird Community Dynamics: Inference at Regional Scales," *Ecology* 82 (2001): 1159–69.
12. See http://worldpopulationreview.com/states/delaware-population/.
13. Allan Baker, Patricia Gonzalez, R. I. Guy Morrison, and Brian A. Harrington, "Red Knot (*Calidris canutus*)," in *The Birds of North America*, ed. A. Poole (Ithaca, NY: Cornell Lab of Ornithology, 2013).
14. Francisco Sáncho-Bayez, and Kris A. G. Wyckhuys, "Worldwide Decline of the Entomofauna: A Review of its Drivers," *Biological Conservation* 232 (2019): 8–27.
15. Doug Tallamy, *Bringing Nature Home* (Portland, OR: Timber Press, 2007).
16. Peter P. Marra and Chris Santella, *Cat Wars: The Devastating Consequences of a Cuddly Killer* (Princeton, NJ: Princeton University Press, 2016).
17. Ibid.
18. See http://www.lightsoutwilm.com/.
19. Peter O. Dunn, and David W. Winkler, "Climate Change Has Affected the Breeding Date of Tree Swallows throughout North America," *Proceedings of the Royal Society of London, Series B*. 266 (1999): 2487–90.
20. Hess, West, Barnhill, and Fleming, *Birds of Delaware*.
21. North American Bird Conservation Initiative, *The State of North America's Birds 2016*.

CHAPTER TEN

Citizen Science and the Scientific Method

TARA TRAMMELL

INTRODUCTION

As our world becomes increasingly dominated and influenced by human activities, it is vital that we broaden the public's understanding of human impacts on the environment. We are entering a new era that some scientists call the *Anthropocene*, a proposed epoch that defines the current global geologic time period as largely determined by human activities.[1] Scientists debate the official designation of this new epoch: some argue about the difficulty of finding clear evidence for a globally identifiable marker of human-caused global change;[2] others contend that the overwhelming variety and magnitude of human impacts are clear evidence that we are living in a distinct, human-influenced period.[3]

Beyond arguments over definitions, there is little question that the rapid global changes to our natural world require that we take proactive roles in shaping and restoring our ecosystems toward balance and sustainability.[4] In this unprecedented time, citizen awareness of, engagement with, and restoration of our natural world at a global scale is perhaps more important than at any other time in human history.

One of the most effective approaches for generating enthusiasm for environmental science is through *citizen science* programs, in which people not formally trained in the sciences play important roles in collecting data and/or samples for scientific research. Part of this effort requires taking the mystery—and even the fear—out of the acquisition of scientific knowledge. Even the words "science" or "scientists" can evoke many different reactions, from excitement to boredom to anxiety, depending on how we are introduced to the scientific disciplines. For many of us, our first exposure to science occurs in school, and this experience can shape our attitude toward and curiosity about science for many years.

While formal school curriculum advances are working to increase childhood excitement and enthusiasm toward science, traditional pedagogy clearly has not enticed enough students toward scientific fields of study, as reflected in raw overall numbers and in the gender and racial diversity of professional scientists. We still have a lot of work to do to make all data equally and equitably available and to create a diverse scientific community. Beyond the training of professional scientists, it is increasingly clear that we must also increase citizen participation in science, to help us all understand and work to solve complex global environmental challenges.

SCIENCE AND THE SCIENTIFIC METHOD

What is science? Posing this question to a citizen group, academic class, or scientific audience can generate a variety of answers depending on the perspective, experience, and knowledge of the person responding. But in the most basic sense, science is simply a field of inquiry built on credible, repeatable studies that yield credible, repeatable evidence, both within and across research groups. Scientists may seek, for example, to identify consistent observable patterns in the natural world, such as the number of native species in a particular habitat (to quantify its *biodiversity*) or concentrations of carbon dioxide in our atmosphere. As reliable patterns are discovered, scientists seek to discover the underlying processes or mechanisms that cause those patterns. Thus, scientific discovery is inherently a collective endeavor, as each new study builds upon the last.

As conclusions across research studies converge toward consistent and reliable evidence, scientists construct theories about natural phenomena. A *scientific theory* does not have the same meaning as the use of the word "theory" in everyday language. Many times, someone will say, "I have a theory" to mean they have a guess or hunch (and sometimes scientists will use this language in casual conversation as well). However, it is important to understand that a scientific theory is distinctly different from an individual having a "theory." A scientific theory is built upon repeatable verification that has withstood the test of time. In some instances, a scientific theory is discovered independently, such as the theory of evolution by natural selection, which was simultaneously proposed by Charles Darwin and Alfred Wallace. But because Darwin had conducted years of meticulous study to provide evidence to support his theory prior to publication, he is more recognized for the development of the theory of natural selection.

RESEARCH AND THE SCIENTIFIC METHOD

The pursuit of knowledge happens in various ways, such as through philosophical discourse, democratic processes, or academic scholarship. In the scientific disciplines, the cornerstone of discovery is *research*, a systematic investigation designed to establish facts and generate new conclusions.

The scientific process starts with having a firm understanding of what we currently know and what we do not know in order to focus our efforts on the most pressing problems of our time. This requires significant *secondary research*, the gathering of information from previous research and scholarly activities. This process is similar to the process students use to write research papers, or journalists or research analysts conduct to gather data and evidence for new ideas, concepts, or conclusions about a particular subject. This step is vital for having the greatest impact on new research directions.

Having a firm understanding of the current state of knowledge leads scientists to study new phenomena, or to conduct new scientific research in order to answer new questions. This process of gathering new data is *primary research*, the foundation of science. Scientists conduct primary research through the "scientific method." The essential steps in the scientific method are as follows:

Observation: The scientific method begins with an individual making an observation and posing a question about the observed phenomena. For example, you may walk into a local forest for a hike and observe a thick understory plant community, or—alternatively—the absence of plants growing in the understory. These observations can lead to many questions, such as: Why is there a thick understory in some forest locations and not in others? Which plants are thriving in the forest with a thick understory? And why are native tree species not regenerating (i.e., reproducing) in the forest with an undeveloped understory? At this stage, many scientists gather information and existing data to determine whether their question on the observed phenomena has been studied previously. This information-gathering step provides a basis for understanding the current state of knowledge, that is, to understand the current "knowns and unknowns" associated with the question at hand.

Hypothesis formulation: Once a scientist has made an observation, they put forth a *hypothesis*. A hypothesis is a proposed explanation for an observation, a speculation or conjecture as a starting point for study. Hypotheses must be based on specific, testable expectations that explain the observed phenomena. Previous research provides a basis for expectations and predictions. For example, based on previous studies of forest plant communities, we might hypothesize that deer browse and/or a non-native plant invasion are negatively affecting native tree regeneration. Carefully constructed predictions are necessary in order to test our hypotheses. To continue with the previous example:
1. If deer browse reduced native tree regeneration, then native tree seedlings and saplings should increase in abundance in deer exclosures (fenced areas that exclude deer).
2. If non-native invasive plants reduce native tree regeneration, then native seedlings and saplings should increase in forest locations with the repeated removal or eradication of non-native invasive plants.

These predictions provide a basis for interpreting whether the data support or disprove a hypothesis, and provide a basis for developing study designs.

Investigation: Scientists next conduct *experiments* to test whether the results of their study support or disprove their formulated hypotheses and predictions. To accurately test predictions, experiments or studies must reduce potential confounding factors that could affect their results and isolate the factors predicted to control the observed phenomena. The goal of rigorous experimental/study design is to diminish any possible errors. Study, investigation, and experimentation represent the data-gathering stage of the scientific process, and there are various potential methods and approaches to employ during that stage depending on the field of study and the questions posed by the researcher. For example, to test our predictions posed above about deer and invasive plant pressures on native trees, a *manipulative field study* would be the best approach. Researchers could build a deer exclosure and count native tree seedlings within and outside the exclosure. To test for the effect of invasive plants on native seedlings, invasive plants could be selectively removed in some *treatment plots* and allowed to remain in other *control plots*, and then the researcher would count the number of native seedlings present in both the control and removal plots.

Analysis: The final step of the scientific method is the analysis of collected data. *Statistics*, a branch of mathematics, helps with the analysis of large quantities of data and supports data interpretation. Scientists use statistical methods to determine whether hypotheses are supported or rejected, and this provides credibility to the methodology employed and the conclusions made by the researchers. This analysis stage includes analyzing evidence and interpreting results in the context of previous research in hopes of making new discoveries.

THE SCIENTIFIC PROCESS

The scientific method is often described as a linear, pragmatic process, but in reality, science is an iterative cycle: the final stage of analysis and interpretation typically leads to refined hypotheses and predictions or entirely new questions. Most often, the actual practice of science does not follow the steps of the scientific method in a unidirectional path. Throughout the process of conducting scientific research, scientists strive to work with objectivity, or with at least as much objectivity as possible. As humans, it is difficult to eliminate all subjectivity during any intellectual endeavor, but the aim of scientific discovery requires that we strive for objective interpretation of research results.

One of the cornerstones of science, and of all scholarship, is the process of professional critique and *peer review*. All published research must withstand a rigorous review process by multiple experts in the field. While this process can be difficult, especially for students and early career professionals, it is necessary for strict interpretation and the maintenance of standards of objectivity. Reviewers spend a great deal of (typically unpaid) time reading and examining work by other scholars, and in many cases, this scrutiny enhances the presentation and credibility of the research project.

Science improves as more minds analyze and interpret research discoveries. Thus, an advantageous characteristic for scientists (and scholars from all disciplines) is the ability to constructively receive and incorporate criticism into our own work, or—when necessary—to respectfully refute criticism. The scientific process is inherently collaborative, whether during the inception of research ideas, throughout research activities, or at the end via the peer review process. As a rule, the best science combines perspectives and intellect from many individuals.

RESEARCH APPROACHES

An important part of understanding research findings that are reported in many popular outlets, such as the news media, is recognizing the type of research conducted. Depending on the research, experiments and studies can take several different forms, such as the classical laboratory experiment, a manipulative field study, a natural experiment, or an observational study. There are three elements of a well-designed traditional experiment:

1. *Control* (a standard for comparison that does not include the manipulated treatment or variable);
2. *Replication* (multiple sets of the treatments and controls); and
3. *Randomization* (randomly selecting individuals, locations, or objects for treatments).

Both the classical lab and manipulative field experiments include these elements, yet there are benefits and tradeoffs to lab experiments versus field investigations. In the laboratory setting, it is easier to isolate the treatment or variable of interest by eliminating or reducing any potential confounding factors. However, lab experiments can never precisely replicate conditions in the field. An experiment conducted in a manipulated field setting has the benefit of realism, yet a trade-off of field studies is the inability to completely isolate the treatments or variables of interest. Therefore, it depends on whether realism or control is more important in deciding which type of experiment to conduct in order to answer the study questions.

Another experimental approach to understanding natural phenomena are *natural experiments,* which are designed specifically to study such phenomena, such as a prescribed fire (designed, for example, so ecologists or firefighters can better understand how to manage landscapes and blazes). Alternatively, some natural experiments are serendipitous, as when scientists study ecosystems that suddenly experience a natural disturbance, such as a tornado, wildfire, or hurricane. Finally, some researchers use a natural experimental approach to study human impacts on ecosystems. Global changes, such as increases in temperatures, carbon dioxide (CO_2) concentrations in the atmosphere, and invasive species spread, and altered biogeochemical cycles (e.g., nitrogen), are difficult to study in situ, and many experimental approaches focus on one or a combination of global changes (e.g., increased CO_2 and nitrogen or increased CO_2 and temperature).

Cities, or urbanized landscapes, already experience higher temperatures, increased CO_2 concentrations, more invasive species, and altered biogeochemical cycles simultaneously, and therefore can act as a natural experiment to study global environmental changes.[5] In order to use urban ecosystems as a proxy for global change, all other environmental or site conditions need to be held constant, or as similar as possible. For example, research utilizing urban forest patches to examine how combined influences of increased temperatures, CO_2, nitrogen, and other pollutants alter ecosystem processes, like primary production (i.e., plant growth), requires that the forests have similar physical (e.g., bulk density, texture) and chemical (e.g., pH) soil properties as well as similar vegetation structure (e.g., canopy and understory density) and composition (e.g., dominant tree species). This study design allows for a comparison of factors surrounding the forest (urbanization) to be the experimental focus of the research. Natural experiments can incorporate realism while isolating factors of interest.

For many scientific disciplines and new areas of research, observational studies are the first stage in the process of discovery. Observation is the basis for building knowledge about patterns in the natural world. Many research projects that incorporate citizen scientists are conducting important Earth observation research, which is vital in this era of rapid global change. These types of research projects are observational studies or long-term monitoring projects. Thus, it is important to explain how the resulting types of data are most useful in understanding our natural world, and explain a common misconception with observational research: that observed patterns or relationships can always be attributed to the predicted cause of the observed phenomena. For example, in monitoring plant biodiversity in forests that experience non-native plant invasion it would seem obvious that such an invasion affects native plants, but this is not necessarily true, and this is why.

"Correlation does not equal causation" is a saying that my PhD advisor taught me many years ago, and I teach this to my students in almost every class every semester. I use a ridiculous example to make this point very clear to my students. Over the last couple hundred years, as global temperatures have increased, the number of active pirates in the oceans has decreased. Does this suggest that if we added more pirates to the oceans, then global temperatures would decrease? Most students laugh. Yes, there is a correlation between these two variables in time, but that does not mean that either variable is causing the change in the other variable (i.e., pirates do not affect global temperatures, and global temperatures are not the cause of reduced piracy).

If we return to our example of non-native plant invasion in forests, then this dynamic is less obvious for many individuals, including budding scientists. Here is an example of what might happen when scientists study the abundance of different plants in forest ecosystems: in a study of multiple forests, scientists find that the number of non-native invasive plants increases as the number of native plants decreases; some might prematurely conclude that the invasive plants are causing the decrease in native plants. While this may be true, it is also possible that there are other factors causing the decline in native plants, such as:

- A change in environmental conditions that favors non-native invasive plants, such as greater nitrogen deposition in the forest leading to high-nitrogen soils that native plants are not accustomed to or evolved to withstand; or
- Land use legacies, such as the abandonment of former farms in the Eastern United States. Once forests regrew, they remained infested with multiflora rose, which had once been used as living fence line. In this case, non-native invasive plants were poised to establish in regenerating forests before native plants arrived to the newly forming forest.

These two examples show that unexpected (or unpredicted) factors could be the cause of reduced native plant abundance even as non-native invasive plants have increased. However, it is also possible that non-native invasive plants alter soil conditions and processes creating a positive feedback loop to promote their own dominance over native plants.[6] The observational study establishes the patterns observed in nature: forests with more invasive plants have fewer native plants.[7] Further studies, such as manipulated field or lab experiments are then necessary to establish how invasive plants directly influence native plants.

The study or experimental design of a research project determines how translational or transferable the research findings are for other systems or situations. The primary goal of scientific research is to contribute to building our understanding of the physical, chemical, and biological processes in the world and create approaches to maximize human innovation. Citizen scientists are a vital component of and contributor to the scientific discovery of our natural world as we develop widespread understanding of global change.

THE CREATIVE PROCESS

The scientific process, then, follows scientific principles, pursues objective conclusions, and lets evidence speak for itself. However, science also embraces a creative process. Scientists are passionate about their fields (like biology or chemistry), organisms (like trees or bacteria), systems (like marine or forest), approaches to research (like DNA sequencing), and the development of new technology (like remote sensing of the Earth). This passion can be highly personal, and is often based on individual experiences and preferences. While the direction of their research is subjective (that is, determined by their passions and interests), the interpretation of their research findings must remain objective. The reporting of methods and results in scientific writing can be very dull, but the presentation and interpretation of research outcomes nonetheless requires a creative process in scientific manuscripts. Especially in a world in which politics has entered the scientific conversation, telling a convincing story—in oral presentations to scientific and public audiences as well as in written manuscripts—is an important process for scientists.

Traditionally, graduate programs in the STEM fields do not train students in the art of communicating their research to public audiences. Yet, this is extremely important in getting citizen scientists—to say nothing of journalists and policymakers—excited about or engaged in science, and many scientists spend a lot of time learning to articulate their research to broader audiences. Crafting a good story can capture an audience's attention, and the greatest impact for science is in communicating research findings to *everyone*.

CITIZEN SCIENCE

Across the U.S. and around the world, millions of citizen scientists are participating in data collection for research. A *citizen scientist* can be defined as "a volunteer who collects and/or processes data as part of a scientific enquiry."[8] This definition can be expanded to include citizens who contribute to research by participating in data collection on their property[9] and/or doing interviews and surveys about their preferences, actions, and behaviors regarding the natural world.[10] Additionally, citizens who get actively involved in community-driven questions[11] can also establish new data collection or research projects that are of particular concern to local residents, such as investigating the expansion of invasive species or urban air quality issues.[12]

The variety of research projects citizens are getting involved in is endless. For the purposes of this book, this chapter focuses on projects associated with the natural world, and I provide examples of projects wherein citizens collect data on plants, birds, and other organisms. In addition, the different levels of participation (from granting permission to access private property to engaging in active data collection) and scales of projects (from global-scale projects to those of local organizations) will be discussed in order to introduce the varying levels of citizen involvement in research science projects.

RESEARCH PARTICIPATION ON PRIVATE PROPERTY

As our understanding of global environmental change expands, professional scientists are increasingly interacting and collaborating with social scientists to study coupled *human-natural systems*, such as urban or agricultural landscapes. Urban environments are considered the most human-dominated ecosystems, which is appropriate considering the infrastructural and technological advances within our cities. While cities do not cover a large proportion of total land area (for example, in the U.S., the current estimate of urban land area is just 6 percent of the total conterminous land area[13]), the magnitude of their environmental impact is disproportionately high, such as with fossil fuel emissions. Our suburbs also create ecological challenges: over a decade ago, it was estimated that the largest irrigated crop in the United States was turf grass (or lawns), surpassing even the vast corn crops that cover much of the middle third of the United States.[14] Thus, the magnitude of the influence of lawn management on water, energy, and nutrient cycles is quite high—to say nothing of the vast acres of former forestland and meadows that have been replaced by monoculture turf grass. A socio-ecological approach (that is, taking a coupled human-natural approach) is necessary to create a comprehensive understanding of lawn management impacts on global change, such as biodiversity loss and effects on global carbon cycles.

In recent years, residential landscapes have received increasing attention from the scientific community due to the spread of lawn ecosystems across the U.S. and the tight linkages of human preferences, attitudes, and behaviors with ecosystem response to human actions, such as both biodiversity and nutrient loss.[15] Planting preferences (e.g., shrubs and flowers) and yard management (e.g., weeding and the use of chemicals) can control plant composition in residential landscapes,[16] which in turn influences other organisms, such as insect populations.[17]

Conducting research in these areas requires the permission of homeowners to access their private property, to participate in surveys or in-person interviews, and/or to act as stewards of scientific equipment on their property.[18] This type of research is not possible without the countless homeowners who have shared their personal landscapes and management practices with researchers. The lessons learned from these research projects have been vital for improving lawn and yard management practices that have the potential to increase biodiversity and other ecosystem services across developed landscapes. With the projected rise in urban and suburban populations and managed land area, promoting improved private land care practices is imperative for informing citizens on the importance of sustainability and the power of individual and community-based actions.

In 2012, a large, multi-investigator project began studying residential yards across the United States and included both social scientists and ecologists. The purpose of this research was to study whether similar yard management practices lead to homogenization in ecological structure (e.g., plant composition) and function (e.g., soil carbon sequestration) at continental scales, with implications for impacts on air and water quality and on human well-being.[19] The initial stages of this research included over nine thousand respondents to a telephone survey[20] and over 140 homeowners who gave permission to access their private yards for research. The yard research involved daylong site visits for plant species identification, in-person interviews, one-meter-deep soil collections, and long-term sensor deployment for microclimate data collection. These research efforts resulted in over twenty peer-reviewed publications, countless presentations at national scientific meetings, and the continuation of research on yard management practices. The success of the initial project, which would not have been possible without participation by homeowners (i.e., citizens), led to the second and current stage of the research project, called *Alternative Ecological Futures for the American Residential Macrosystem*. The project team is researching multiple yard management practices, including those of yards with high- and low-intensity lawn care, yards designated as biodiverse, and yards that incorporate a hydrologic feature, such as *xeriscaping* (landscaping that reduces or eliminates the need for supplemental water from irrigation) or rain gardens. The purpose of this project is to evaluate the ecological implications of alternative yard management practices (e.g., plant, insect, and bird biodiversity, soil carbon and nitrogen cycling, nutrient leaching, plant water use) and to determine factors that drive change and/or maintenance in residential land use and yard management practices. It is the team's goal to be able to provide recommendations for best practices to increase yard biodiversity as well as enhance carbon storage and reduce nutrient losses in runoff.

These research efforts, it is hoped, will contribute to a complex understanding of the human and natural controls on residential landscapes, which could help homeowners understand how their own properties could become part of an expansive ecosystem across the United States. Research that occurs in locations where citizens consistently interact with each other, such as urban landscapes, provides many opportunities for scientists to engage with citizens, hopefully enhancing citizen perception of and comfort with the scientific process.

ACTIVE PARTICIPANTS AND DATA COLLECTORS

Citizen scientists can also participate actively in research in a variety of ways: through school projects; volunteering during implementation of research projects; data collection events for research; or individual data collection. Monitoring and observation of biodiversity across the world is popular among avid naturalists and many citizens, and is—considering the troubled current state of global biodiversity—a crucial area of research.[21] Therefore, many of the current citizen science efforts across the nation and around the world discussed here focus on biodiversity observations.

BioBlitz is a targeted event focusing on the identification and inventory of as many plant and animal species as possible in a specified location and length of time (typically twenty-four hours). A BioBlitz can occur in any location (urban, suburban, or rural) at any scale (schoolyard or country). Various groups of people (e.g., scientists, teachers, students, volunteers) and organizations can conduct and/or organize BioBlitz events. BioBlitz has grown to include related biodiversity events like *Blogger Blitz*, where participants pledge to conduct individual species inventories that are mapped jointly with the goal of raising awareness of biodiversity. Since the inception of BioBlitz, organizations such as the National Geographic Society and the U.S. National Park Service have organized annual events to create consistent records of biodiversity. In North America, Europe, Asia, and Australia, countries organize BioBlitz events regularly to engage citizens and scientists in taking biodiversity inventories.

Such events are an excellent way to raise awareness about biodiversity and potential threats to biodiversity. To standardize identification and inventory approaches, an online platform, *iNaturalist*, was developed to assist with consistency in species records and expertise in species identification. A joint initiative by the California Academy of Sciences and the National Geographic Society, iNaturalist (www.inaturalist.org) now has over 750,000 members. This community of scientists and naturalists is a social network in which members share photographs of organisms observed in nature, providing a record of species in a specific location at a specific time. The shared biodiversity information provides potential data for scientific study and/or educational purposes. iNaturalist is a "species identification system and an organism occurrence recording tool" with two primary goals: to connect people with nature and produce biodiversity data that is scientifically valuable. Individual users can record and monitor their own biodiversity observations, and receive professional feedback on species identification by scientists and taxonomic experts. Once species observations receive the status of "research grade," the data is uploaded and indexed with the Global Biodiversity Information Facility, an international network funded by governments around the world to provide global, open access biodiversity data (https://www.gbif.org).

The iNaturalist platform includes the capacity to record observations of any species around the world, providing naturalists access to information on species occurrences worldwide, whereas other online platforms are focused on the characteristics of species (e.g., invasive species) or on taxonomic groups (e.g., birds). *EDDMaps* (Early Detection and Distribution Mapping System) developed by the University of Georgia's Center for Invasive Species and Ecosystem Health, for instance, is a web-based mapping system for documenting the distribution of invasive species across the United States (www.eddmaps.org). In addition to coalescing databases on invasive species, volunteer observations of invasive species provide real-time tracking. The goal of the program is to facilitate Early Detection and Rapid Response programs (EDRR) for invasive species and share data freely with scientists, researchers, educators, land managers and owners, and more.

EDDMaps provides local and national distribution maps through their website. For example, a quick search for a common non-native invasive plant in Delaware, the multiflora rose (*Rosa multiflora*), yields several records across the state. While this mapping system has the potential to identify invasive species at early detection, its effectiveness is dependent on the extent of volunteer observations. The Delaware Invasive Species Council (https://delawareinvasives.net) is currently working on developing web-based apps for Delaware citizens to assist with identifying the distribution of common invasive species threats to our state. The goal is to create maps that demonstrate the spread of species that threaten local ecosystems. The potential for citizens to aid in detection of invasive plant spread is substantial, as the accuracy of volunteer identification and counts of invasive plants is comparable to that of botanical experts or others professionally trained in invasive plant identification.[22] Expanded public engagement in the detection and distribution of invasive plants could greatly benefit land management and restoration efforts.[23]

Conservation research programs can also capitalize on the expertise of naturalists that focus on specific taxa, such as birds. The birding community includes avid birdwatchers that keep meticulous checklists of personal observations. As you learned in Ian Stewart's chapter on birds in this volume, the website *eBird* embraces the idea "that every birdwatcher has unique knowledge and experience" (https://ebird.org/about). The free online platform and mobile app encourages equitable access to one of the largest citizen science projects across the world focused on biodiversity: eBird observations have recorded more than one hundred million birds each year worldwide. Platform users enter data that systematically documents bird distribution, abundance, and habitat, and regional experts review any unusual bird records. This rich data collection is freely available to anyone, and an important source of conservation efforts and for scientific peer-reviewed journal articles informing decision-makers and research worldwide.

There are many rigorous efforts to observe and monitor not only biodiversity distribution nationally and globally but also changes in the function of species across the world. Over ten years ago, the USA National Phenology Network was established to create a national phenology program (https://www.usanpn.org). *Phenology* is the study of plant and animal life cycles and the timing of the birth, growth, reproduction, and death of organisms. Phenology is "nature's calendar": the timing of the "bud bursts" of spring flowers, of fall leaf color, and of nest building by birds. The purpose of the National Phenology Network is to collect and share data on plant and animal phenology across the United States, and the network consists of thousands of individual volunteers, including citizen scientists, educators, resource managers, and research scientists. Large-scale efforts in monitoring phenology over time have shown that spring events are now occurring earlier and fall events later for many species. However, not all species respond in similar ways or at similar paces to such changes, leading to what is called a "phenological mismatch," which can alter feeding habits and trophic nutritional dynamics. Therefore, monitoring changes in phenology over time is an important indicator of climate change. The National Phenology Network, while focused on the United States, includes a list of the phenology monitoring networks from around the world on their website. Incorporating citizen scientists into efforts such as monitoring phenology provides citizens the opportunity to experience firsthand how global climate change can affect plants and animals.

Beyond purely participating in research, it is also essential that citizens better understand the benefits that nature provides people, known as *ecosystem services*. The U.S. Forest Service has developed an online toolkit, *i-Tree* (https://www.itreetools.org), that estimates the benefits (such as carbon sequestration) that urban trees provide to people. While initially developed in the United States, i-Tree is now used throughout the world, and one of the i-Tree tools, Eco, has been translated into Spanish (i-Tree Mexico) for widespread use south of our border. Thousands of communities, organizations, municipalities, and individuals use i-Tree to document ecosystem services provided by urban trees. The i-Tree website provides in-depth protocols for data collection and support for online tools that calculate urban tree benefits, which helps support advocacy for trees in our built landscapes.

Municipalities, organizations, and research programs have engaged citizens in collecting urban tree information not only for the evaluation of ecosystem service provision by trees, but also for an estimation of urban tree canopy and conditions. Municipal tree management decisions, such as tree treatment or removal of ash trees in cities where the invasive emerald ash borer threatens the ash trees, are supported by citizen scientists that collect urban tree data, enabling science-based management decisions.[24] Citizen science involvement in inventorying and monitoring urban trees is growing, and in several cities across the U.S., volunteer-collected urban tree identification and size data were within 90–93 percent accuracy of data collected by professionally trained individuals, demonstrating the value of volunteers in aiding data collection.[25] Public engagement with urban forest assessment encourages resident participation in the protection and management of urban greenspaces.[26]

There are many examples of national and global networks that support participation by citizens in scientific projects, and these programs are essential for studying large-scale changes in biodiversity and species functions. Local-scale efforts are equally important for encouraging citizens to engage in science and feel the rewards of their efforts. While The Nature Conservancy (TNC) is a national organization, the state chapter in Delaware is working to become a center for citizen science locally. The Delaware TNC chapter has several successful citizen science programs, such as the Stream Stewards Program (established in 2016 to train citizens in collecting water quality data), and the chapter is continuing to expand their citizen science programs (e.g., urban tree monitoring in Wilmington). Whether at local, regional, national, or global scales, citizen science is vital for engaging everyone in scientific discovery and connecting everyone to nature.

CITIZEN SCIENCE GOALS

Citizen science has the potential to make significant contributions to science. Information on large-scale species distribution and abundance[27] would not be possible without the vast number of citizen observations unobtainable by scientists alone. Many scientific fields are enhanced by citizen participation, such as conservation biology, resource and land management, and environmental protection, and citizens also contribute to policymaking through public engagement.[28] While some scientists may question the validity of data collection by non-scientists, efforts by citizen scientists can be rigorous when the participants are properly trained and the studies are properly designed and evaluated.[29] A group of individuals and organizations established the international Citizen Science Association (https://www.citizenscience.org) to support the development and organization of citizen science projects across the world, by uniting experts and citizens toward working on scientific research together. Such organizations aid in the implementation of rigorous citizen science projects.

Citizen science not only enriches the current understanding of our natural world, but also has the potential to cultivate science literacy and public action in support of science and environmental initiatives. For example, as part of the National Institute of Invasive Species Science, citizen scientists have been trained on invasive species monitoring techniques, which has resulted in increases in citizens' expressed intention to participate in activities that help improve the forest environment.[30] One way to increase the effectiveness of citizen science programs is to involve participants in the entire research process, from study/experiment design to the interpretation and implementation of results.[31] While citizen science is not a new concept or endeavor (indeed, historically speaking, science was typically conducted mostly by volunteers), the inclusion of citizens in modern scientific discovery has blossomed over the last ten years and—again, considering the need for Earth observation studies, restoration efforts, and many other types of scientific discovery—should continue to be a high priority.

The most successful citizen science projects will emerge from mutual understanding and respect wherein citizens learn the scientific process and precisely follow methods for data and/or sample collection, and scientists communicate the infinite value of citizen participation and openly share results. This reciprocal esteem will enable high-quality data collection while contributing to scientific discovery and creating solutions for environmental challenges.

Solving global environmental challenges, such as climate change or the spread of invasive species, will require both a robust understanding of the scientific evidence and a concerted effort among all of us. Citizen engagement with, participation in, and understanding of the scientific process is vital as we work toward improving human impacts on our environment. Thank you to all who participate and work to learn about and understand our world!

NOTES

1. F. C. Finney, "The 'Anthropocene' as a Ratified Unit in the ICS International Chronostratigraphic Chart: Fundamental Issues that Must Be Addressed by the Task Group," *Geol. Soc. Lond. Spec. Publ* 395 (2014): 23–28.
2. P. L. Gibbard, and M. J. C. Walker, "The Term 'Anthropocene' in the Context of Formal Geological Classification," *Geol. Soc. Lond. Spec. Publ* 395 (2014): 29–37.
3. Richard T. Corlett, "The Anthropocene Concept in Ecology and Conservation," *Trends in Ecology & Evolution* 30 (2015): 36–41.
4. F. Stuart Chapin III, and Erica Fernandez, "Proactive Ecology for the Anthropocene," *Elem Sci ANth* 1 (2013).
5. Nancy B. Grimm, Stanley H. Faeth, Nancy E. Golubiewski, Charles L. Redman, Jianguo Wu, Xuemei Bai, and John M. Briggs, "Global Change and the Ecology of Cities," *Science* 319, no 5864 (2008): 756–60.
6. Tara L. E. Trammell, Haylee A. Ralston, Shannon A. Scroggins, and Margaret M. Carreiro, "Foliar Production and Decomposition Rates in Urban Forests Invaded by the Exotic Invasive Shrub, *Lonicera maackii*," *Biological Invasions* 14, no. 3 (2012a): 529–45.
7. Tara L. E. Trammell, and Margaret M. Carreiro, "Vegetation Composition and Structure of Woody Plant Communities Along Urban Interstate Corridors in Louisville, KY, USA," *Urban Ecosystems* 14, no. 4 (2011): 501–24; Jonathan Silvertown, "A New Dawn for Citizen Science," *Trends in Ecology & Evolution* 24, no. 9 (2009): 467–71.
8. Silvertown, "A New Dawn for Citizen Science"; Tara L. E. Trammell, et al., "Plant Nitrogen Concentration and Isotopic Composition in Residential Lawns across Seven US Cities," *Oecologia* 181, no. 1 (2016): 271–85.
9. E.g., Trammell, et al., "Plant Nitrogen Concentration."
10. Meghan L. Avolio, Diane E. Pataki, Tara L. E. Trammell, and Joanna Endter-Wada, "Biodiverse Cities: The Nursery Industry, Homeowners, and Neighborhood Differences Drive Urban Tree Composition," *Ecological Monographs* 88, no. 2 (2018): 259–76.
11. Rick Bonney, Jennifer L. Shirk, Tina B. Phillips, Andrea Wiggins, Heidi L. Ballard, Abraham J. Miller-Rushing, and Julia K. Parrish, "Next Steps for Citizen Science," *Science* 343 (2014): 1436–37.
12. Air Quality, West Oakland Environmental Indicators Project (2013), www.woeip.org/air-quality/.
13. Collin Homer, et al., "Completion of the 2011 National Land Cover Database for the Conterminous United States—Representing a Decade of Land Cover Change Information," *Photogrammetric Engineering & Remote Sensing* 81, no. 5 (2015): 345–54.
14. Christina Milesi, Steven W. Running, Christopher D. Elvidge, John B. Dietz, Benjamin T. Tuttle, and Ramakrishna R. Nemani, "Mapping and Modeling the Biogeochemical Cycling of Turf Grasses in the United States," *Environmental Management* 36, no. 3 (2005): 426–38.
15. Peter M. Groffman, et al., "Ecological Homogenization of Residential Macrosystems," *Nature Ecology and Evolution* 1, no. 7 (2017): 0191.
16. Tara L. E. Trammell, et al., "Climate and Lawn Management Interact to Control C_4 Plant Distribution in Residential Lawns across Seven U.S. Cities," *Ecological Applications* (2019): e01884. See also Avolio, Pataki, Trammell, and Endter-Wada, "Biodiverse Cities."
17. Desiree L. Narango, Douglas W. Tallamy, and Peter P. Marra, "Nonnative Plants Reduce Population Growth of an Insectivorous Bird," *Proceedings of the National Academy of Sciences* 115, no. 45 (2018): 11549–54.
18. E.g., Sharon J. Hall, et al., "Convergence of Microclimate in Residential Landscapes across Diverse Cities in the United States," *Landscape Ecology* 31, no. 1 (2016): 101–17.

19. Peter M. Groffman, et al., "Ecological Homogenization of Urban USA," *Frontiers in Ecology and the Environment* 12, no. 1 (2014): 74–81.

20. Colin Polsky, et al., "Assessing the Homogenization of Urban Land Management with an Application to US Residential Lawn Care," *Proceedings of the National Academy of Sciences* 111, no. 12 (2014): 4432–37.

21. Intergovernmental Science-Policy Platform on Biodiversity and Ecosystem Services (IPBES). United Nations, Science and Policy for People and Nature, https://www.ipbes.net.

22. Alycia W. Crall, Gregory J. Newman, Thomas J. Stohlgren, Kirstin A. Holfelder, Jim Graham, and Donald M. Waller, "Assessing Citizen Science Data Quality: An Invasive Species Case Study," *Conservation Letters* 4, no. 6 (2011): 433–42.

23. Rebecca C. Jordan, Wesley R. Brooks, David V. Howe, and Joan G. Ehrenfeld, "Evaluating the Performance of Volunteers in Mapping Invasive Plants in Public Conservation Lands," *Environmental Management* 49, no. 2 (2012): 425–34.

24. Richard Hallett, and Tanner Hallett, "Citizen Science and Tree Health Assessment: How Useful Are the Data?" *Arboriculture & Urban Forestry* 44, no. 6 (2018): 236–47.

25. Nick Bancks, Eric A. North, and Gary R. Johnson, "An Analysis of Agreement between Volunteer- and Researcher-Collected Urban Tree Inventory Data," *Arboriculture & Urban Forestry* 44, no. 2 (2018): 73–86; Lara A. Roman, Bryant C. Scharenbroch, Johan P. A. Ostberg, Lee S. Mueller, Jason G. Henning, Andrew K. Koeser, Jessica R. Sanders, Daniel R. Betz, and Rebecca C. Jordan, "Data Quality in Citizen Science Urban Tree Inventories," *Urban Forestry & Urban Greening* 22 (2017): 124–35.

26. Lara A. Roman, Lindsay K. Campbell, and Rebecca C. Jordan, "Civic Science in Urban Forestry: An Introduction," *Arboriculture & Urban Forestry* 44, no. 2 (2018): 41–48.

27. Mark Chandler, et al., "Contribution of Citizen Science toward International Biodiversity Monitoring," *Biological Conservation* 213 (2017): 280–94.

28. Duncan C. McKinley, et al., "Citizen Science Can Improve Conservation Science, Natural Resource Management, and Environmental Protection," *Biological Conservation* 208 (2017): 15–28.

29. Alycia W. Crall, Rebecca Jordan, Kristin Holfelder, Gregory J. Newman, Jim Graham, and Donald M. Waller, "The Impacts of an Invasive Species Citizen Science Training Program on Participant Attitudes, Behavior, and Science Literacy," *Public Understanding of Science* 22, no. 6 (2012): 745–64.

30. Tabea Turrini, Daniel Dorler, Anett Richter, Florian Heigl, and Aletta Bonn, "The Threefold Potential of Environmental Citizen Science—Generating Knowledge, Creating Learning Opportunities and Enabling Civic Participation," *Biological Conservation* 225 (2018): 176–86.

CHAPTER ELEVEN

Weather and Climate Change

JENN VOLK

Our natural systems are dependent on weather and climate conditions. This chapter will provide an overview of weather and climate elements as well as past climate trends and future climate projections for Delaware. It will then further examine how climate change may impact species, habitats, and ecosystems, and review available adaptation and mitigation strategies for natural systems.

WEATHER VERSUS CLIMATE

It is not uncommon for people to use the terms "weather" and "climate" interchangeably. In fact, the two terms mean very different things; the difference comes down to spatial and temporal scales. *Weather* refers to the instantaneous atmospheric conditions for a specific location; it is basically what is happening outside where you are today. Weather is measured over short time periods (minutes to months) for localized areas, and (as all picnic planners know) it can change rapidly. *Climate,* on the other hand, refers to the average atmospheric conditions for an entire region over a much longer period of time—it essentially refers to a region's long-term weather pattern. When discussing climate, the minimum period of time considered is typically twenty years, and changes in climate are typically observed much more slowly. To remember the difference between weather and climate, remember this phrase, "Climate is what you expect, weather is what you get!"[1]

UNDERSTANDING THE MAIN ELEMENTS OF WEATHER

Weather is caused by the Earth's atmosphere responding to variable heating from the Sun. Since the Earth is a sphere, only half of the planet faces the Sun (and is thus warmed) at a time. Earth's axial tilt of about 23.4 degrees also causes different parts of the planet to be closer to (or further from) the Sun, and thus warmer (or cooler) as our planet rotates around it throughout the year. This variation is why we have seasons. And, since different surfaces on Earth (like oceans, or forests, or pavement) absorb different amounts of the

Sun's radiating heat, the warming that does take place can be highly (spatially) variable. As the Sun's heat reflects back off these surfaces, it warms the air in the lowest layer of our atmosphere, called the *troposphere*, where most of our weather occurs.[2] *Air temperature* is the measure of how hot or cold the atmosphere is at the Earth's surface; it is usually measured with a thermometer in degrees Fahrenheit (F) or Celsius (C).[3]

Our atmosphere, which is made up of gases, has weight. As the Earth rotates, gravity pulls those gases closer to the surface. The pressure exerted by the atmosphere on a particular location is known as *atmospheric pressure*, or *air pressure*.[4] Atmospheric pressure is measured with a *barometer*, and is reported in units of millibars or "inches of mercury." Air pressure is greatest at low altitudes, where there is more air overhead exerting pressure downward, and it decreases with elevation. The pressure at a given location will also change as the air temperature changes. When air warms, the space occupied by gas molecules expands, so that there are fewer molecules per unit of gas, and hence there is less weight and pressure exerted on that location. Thus, warm air generally brings low pressure systems, and cool air brings high pressure systems.[5]

Because warm air is lighter, it rises, allowing cooler, heavier air to move in to replace it. These air currents, known as *wind*, move along a gradient from high to low pressure areas.[6] Because the Earth is rotating, winds tend to curve to the right in the Northern Hemisphere and to the left in the Southern Hemisphere, a phenomenon called the *Coriolis effect*. In a high-pressure system, wind direction moves clockwise and away from the center of the system. Conversely, in a low-pressure system, wind direction runs counterclockwise, toward the center of the system, creating an upwelling effect at the center.[7]

Local wind direction can be determined with a wind vane or wind sock, and is reported as the direction from which the wind originates (i.e., a *Nor'easter* is a storm with wind coming from the northeast). Wind speeds can be measured with an *anemometer* (named after the Greek word for wind, *anemos*), which often consists of spinning cups or blades that move with the wind and calculate the speed of the rotations. In the United States, wind speeds are often reported in miles per hour (mph), though other units are commonly used in the meteorological community. Winds are labeled as ranging from *calm* (one mph or less) to *hurricane* (seventy-four mph or greater).[8] They occur at all levels of the atmosphere, and upper-atmospheric winds, like the *jet stream*, push weather all around the planet.[9]

Humidity is the term that describes the amount of water vapor—the gaseous form of water—present in the air.[10] *Absolute humidity* is calculated as the mass of water vapor in a volume of air, in units such as grams per cubic meter. *Relative humidity*, measured with an instrument called a *hygrometer*, is expressed as a percentage of the maximum amount of water that air can hold at a given temperature.[11] Warm air contains more water vapor than cold air for one main reason: the warmer the air, the faster molecules will move, and the greater the number of water molecules that will move from liquid into vapor. Consider the feel of a hot, sticky, late summer afternoon: if the relative humidity is 100 percent, the air is fully saturated with water vapor and precipitation—perhaps a late afternoon thunderstorm is possible.

Clouds form when a parcel of warm moist air rises and cools to below the *dew point*, the temperature at which water vapor molecules slow down and begin to condense.[12] Those molecules will aggregate around tiny micrometer-sized particles (like dirt, smoke particles, or ocean spray, for example) called *condensation nuclei* to form a cloud droplet. Many droplets of water or ice together form clouds. Clouds, in turn, affect how much light reaches the Earth's surface during the day, so clear skies result in warmer days, while cloudy skies result in cooler days; and they affect the amount of radiation leaving the Earth at night, so clear skies cause cool nights, while cloudy skies make for warmer nights.

There are three main types of clouds, though there are many modifications and combinations of those types. *Cirrus* clouds occur high in the atmosphere, are composed of ice crystals, and appear thin and wispy. *Cumulus* clouds generally have a flat bottom and a puffy or fluffy top and occur at mid- and low levels. *Stratus* clouds appear as broad and wide layers like a blanket and also occur at mid- and low levels. Finally, *nimbo-form* clouds, which can combine with any of the primary three cloud types, produce precipitation and can have great vertical height into the atmosphere.[13]

Precipitation is the process through which water vapor further condenses in the atmosphere to form droplets large enough (and with enough weight) that they fall back to the earth as *drizzle* (raindrops with a diameter less than 0.5 mm), rain, snow, sleet, freezing rain, or hail. *Snow* forms when the temperature in the atmosphere and on the ground is at or below freezing (32°F/0°C). *Sleet* occurs when there is a wedge of warm air between two cold air masses, resulting in the refreezing of rain or snow into ice pellets as it falls. *Rain* that falls as a liquid but freezes upon contact with the cold ground is known as *freezing rain*. *Hail* forms during a thunderstorm when updrafting air lifts water droplets high into the atmosphere, where it freezes and, as it falls to the ground, hailstones can continue to grow. Precipitation, especially rain, is highly variable over space.[14] The absence of precipitation is known as *drought*, of which there are five levels: abnormally dry, moderate, severe, extreme, and exceptional.[15] A *rain gauge* can be used to measure precipitation in person whereas *radar* can be used to remotely observe and measure precipitation.

So temperature, atmospheric pressure, wind, humidity, cloudiness, and precipitation are the six main elements that describe the weather we experience. When one steps back and looks at weather patterns with a wider view, we can speak more broadly about *air masses*. Air masses are large bodies of air with relatively similar temperature and humidity throughout.[16] They are characterized by where they originate, such as over land (continental), over water (maritime), and by latitude (arctic, polar, or tropic). When air masses converge, they form a *front*. In a cold front, cool dense air moves in to replace warm air; in a warm front, the opposite is true. *Stationary fronts* occur when two masses meet, but neither moves, and an *occluded front* is a hybrid mass where a cold front typically overtakes a warm front. Frontal movements result in changes to our weather conditions, sometimes in extreme ways.[17]

EARTH'S CLIMATE SYSTEM

If we step further back to look at the long-term weather conditions of an area, we begin to study climate. To understand climate, one should fully appreciate Earth's *climate system*. There are five main components: the atmosphere, cryosphere, hydrosphere, lithosphere, and biosphere. The *atmosphere*, as we've already discussed, contains the gases covering Earth: 78 percent of it is made up of nitrogen; 21 percent is oxygen; 0.9 percent is argon; and 0.1 percent is composed of trace gases. The *cryosphere* includes all the Earth's surface areas containing frozen water in the form of ice or snow, including ice sheets, ice caps, glaciers, snow-covered areas, permafrost, and frozen waters. The *hydrosphere*, includes all the Earth's fresh and saline waters, both on the surface and below ground, including oceans, seas, rivers, lakes, aquifers, etc. The *lithosphere* is the solid outer part of the Earth, including the crust and upper solid mantle. The *biosphere* includes all of the parts of Earth containing life, whether on land, in water, or in the air.[18]

Each component of the climate system interacts with the others through physical, chemical, and biological processes to exchange water, energy (in the form of heat), and gases. Climate is stable when incoming solar radiation equals the outgoing radiation emitted by all of the components of the climate system. A number of constant factors impact a region's climate, including its latitude, altitude, proximity to mountains and oceans, and local topography. The amount of solar radiation (and warmth) reaching a particular location depends on its latitude, with high latitudes (near the poles) receiving less solar radiation on an annual basis than low latitudes (near the equator). This is similar for gains in elevation: as you go up in elevation, air cools, and higher-altitude locations have climates that resemble higher-latitude locations. Additionally, by interfering with the movement of air masses, mountains have the ability to produce clouds and precipitation, and can impact the winds and weather systems of surrounding areas. Oceans and other large bodies of water moderate the temperatures of coastal areas, and the temperature differences between land and sea can produce winds that result in increased storms. Finally, the physical topographic features of an area, such as the slope of the land, can impact the amount of sunlight that the land can absorb and the size of the land area exposed to winds.[19]

Climate variations occur when there is a change to any one (or more) of the climate system components, to how they interact with each other, or to an external force that results in a change to the amount of radiation entering or leaving Earth's system. Some of the interactions between the components create positive or negative *feedback loops*. Positive loops amplify the net climate change, and negative loops dampen climate changes. Many climate variations are natural, while others (as we are becoming all too aware) are human-induced.

Variable factors impacting climate could include global wind and pressure regimes and ocean currents, the nature of the Earth's surface, and *anthropogenic* (human-caused) effects. Large-scale wind patterns, such as the trade winds in the tropics and the jet stream in the Northern hemisphere, influence regional climates. When atmospheric and oceanic circulation are coupled (the El Niño Southern Oscillation, for example), worldwide weather patterns can be changed for years. Natural (or human-caused) changes to the surface of the Earth can have a number of impacts. A highly reflective surface (such as snow cover) decreases the amount of solar radiation that can be absorbed (this is known as the *albedo effect*). The moisture of a surface impacts whether solar energy is used to dry or warm the surface. A rough surface (created by trees or buildings) creates more turbulence in the air, which increases mixing and decreases extreme conditions at the surface. Finally, human impact on climate ranges from the clearing of forests to the increased levels of carbon and other gases in our atmosphere generated through the combustion of fossil fuels and other industrial practices.[20]

THE GREENHOUSE EFFECT

While the major atmospheric gases (nitrogen, oxygen, and argon) do not absorb or emit *solar* (short-wave) or *infrared* (long-wave) radiation, a number of the trace gases (carbon dioxide, methane, nitrous oxide, and ozone) and water vapor do. Of the solar, short-wave radiation that enters Earth's system, 22.5 percent is reflected back to space by clouds, *aerosols* (suspended fine solid or liquid particles), and the atmosphere; 19.5 percent is absorbed by the atmosphere; 9 percent is reflected back by the Earth's surface; and 49 percent is absorbed by the surface. The heat that is absorbed by the surface returns to the atmosphere as heat (5 percent), through *evapotranspiration* (the sum of evaporation, the conversion of liquid water to water vapor, and transpiration, the movement of water through a plant including evaporation off of stems, leaves, and flowers) (16 percent), and as thermal infrared, long-wave radiation (79 percent). A small portion (10 percent) of this long-wave radiation passes directly back out to space, while the rest (90 percent) is absorbed by the gases in the atmosphere. Some (38 percent) of the radiation absorbed by the atmosphere is re-emitted back to space and the rest (62 percent) is radiated back to the surface, where it is absorbed and further warms the Earth. This effect, where the atmosphere captures and reradiates heat back to Earth, is known as the *greenhouse effect* and the gases that are doing this work have been dubbed "greenhouse gases" (GHGs).[21]

The greenhouse effect is a natural climatic process, but the concentrations of GHGs in the atmosphere has significantly increased since the Industrial Revolution, which began around 1750. Since this time, the insulation effect has been amplified and global temperatures have increased.[22] The Intergovernmental Panel on Climate Change (IPCC), a large global group of independent scientific experts convened by the United Nations, has concluded that atmospheric carbon dioxide levels in recent centuries have increased from 280 to 400 parts per million, and that there is more than a 95-percent probability that the temperature warming of our planet is due to human-produced GHGs.[23] The National Oceanic and Atmospheric Administration's Annual Greenhouse Gas Index measures atmospheric gases at 168 sites across forty-six countries. In 2019, the index had a value of 1.43, which indicates that the heat-trapping capacity of our atmosphere has increased by 43 percent since 1990, which was the baseline year.[24]

Globally, carbon dioxide (CO_2) emissions are greatest among the influential GHGs, making up 76 percent of total GHG emissions. About 65 percent of CO_2 emissions are produced through the burning of fossil fuels for transportation, electricity, and other industrial processes, while 11 percent are from the burning of forests and other land uses (like agriculture).

Methane (CH_4) makes up 16 percent of global GHG emissions. The industrial production of natural gas, petroleum products, and coal all result in CH_4 emission, as does livestock agriculture and the decomposition of waste in landfills and wastewater treatment processes. Nitrous oxide (N_2O) makes up 6 percent of global GHG emissions and is primarily generated through agriculture soil management activities, like the application of fertilizers, with fuel, industry, and waste management serving as additional, smaller sources. Finally, fluorinated gases (f-gases) make up the final 2 percent of global GHG emissions. These human-generated compounds are primarily used as refrigerants and are also produced as a byproduct of industrial processes.[25]

While present at lower levels, methane, nitrous oxide, and f-gases are more effective at trapping radiation than carbon dioxide is. The *global warming potential* (GWP) is a measure to allow comparison of the impacts of the emission of one ton of each greenhouse gas over a century to the impact of the emission of one ton of carbon dioxide, which has a GWP of 1.0. Methane has a GWP of 28–36, nitrous oxide has a GWP of 265–298, and the GWP of f-gases range from thousands to tens of thousands. So while carbon dioxide is the most abundant greenhouse gas, these other compounds are also very influential in the warming process.[26]

CLIMATE AND CLIMATE TRENDS

GLOBAL PATTERNS

There are a number of systems for classifying climate, with the Köppen-Geiger system being the most common. The Köppen-Geiger system broadly categorizes the Earth's climate into various zones, which generally follow ecological regions: Zone A corresponds to tropical climates; Zone B consists of dry climates; Zone C includes moist subtropical mid-latitude climates; Zone D is assigned to moist continental mid-latitude climates; and Zone E is assigned to polar climates. Each zone is further broken down into additional subzones. The first subdivision is based on precipitation patterns, including: f for zones that are wet year-round; m for those featuring monsoons; s for zones with a dry summer season; and w for those with a dry winter season. The final subdivision is based on temperature patterns, including: a for zones with a hot summer; b for those with a warm summer; c for those with a cool summer; and d for zones featuring very cold winters. Figure 11.1 shows the present-day Köppen-Geiger-classified climatic zones on Earth.[27]

As a result of increased levels of GHGs in the atmosphere, global temperatures have risen. According to an October 2018 special report by the IPCC, "Human activities are estimated to have caused approximately 1.0°C [1.8°F] of global warming above pre-industrial levels, with a likely range of 0.8°C [1.4°F] to 1.2°C [2.1°F]."[28] The atmospheric warming has resulted in a number of other global changes all further impacting our climate. Much of the heat in the atmosphere is absorbed by the oceans, causing ocean temperatures to warm as well.

Warmer air and water temperatures have caused ice sheets at the poles and sea ice in the Arctic to melt. Similarly, glaciers in all major mountain systems worldwide have retreated, and snow-covered areas have decreased. With snow and ice melt across the globe entering our oceans and water expanding because it is warmer, sea levels are rising at increasing rates. Changes in the reflectivity of the surface are also playing into feedback loops, further exacerbating surface warming. The severity and frequency of extreme events, like heat waves and storms, also appear to be on the rise. Finally, our oceans are also absorbing the increased CO_2 in the atmosphere, which is resulting in our oceans becoming more acidic, with dire implications for marine life, notably the world's endangered coral reefs.[29]

DELAWARE PATTERNS

Delaware, along with most of the Central and Southern parts of the United States, is situated within the Köppen-Geiger Cfa zone, which is characterized as humid subtropical (C), wet year-round (f), with a hot summer (a) (see Figure 11.1).[30] Constant factors that impact Delaware's climate include our latitude, altitude, and proximity to large bodies of water. As the second smallest state in the nation, Delaware stretches from 38.4° to 39.8° north latitude, which is not a large range, but is large enough to produce some climatic differences across the state. With a mean elevation of only sixty feet and altitudes ranging from sea level (zero feet) at the coast to 450 feet above sea level at a high point near the Pennsylvania border, Delaware is also the lowest-lying state in the nation, and this too impacts our climate. Finally, because Delaware is part of the Delmarva Peninsula, our climate is influenced by the Atlantic Ocean and Delaware Bay to the east and the Chesapeake Bay to the west; together, they have a tendency to moderate temperatures compared to locations further inland at this same latitude and altitude.[31]

FIGURE 11.1. Earth's climatic zones characterized using the Köppen-Geiger system. (Based on "JetStream Max: Addition Köppen-Geiger Climate Subdivisions," National Weather Service, https://www.weather.gov/jetstream/climate_max.)

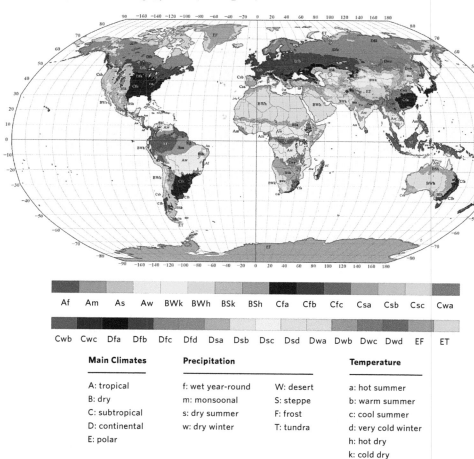

Main Climates

A: tropical
B: dry
C: subtropical
D: continental
E: polar

Precipitation

f: wet year-round
m: monsoonal
s: dry summer
w: dry winter

W: desert
S: steppe
F: frost
T: tundra

Temperature

a: hot summer
b: warm summer
c: cool summer
d: very cold winter
h: hot dry
k: cold dry

According to the Office of the Delaware State Climatologist, the mean annual temperature in the state is recorded as 54.0°F in Northern New Castle County and 58.0°F along the coast of Sussex County.[32] For the 1981–2010 period, at the Dover monitoring station, January was on average the coldest month (35.2°F mean) and July the warmest month (77.7°F mean) of the year (Figure 11.2). The maximum temperature recorded in Delaware was 110°F on July 21, 1930, in Millsboro. Millsboro is also the site of the coldest recorded temperature in the state, which was −17°F on January 17, 1893.[33]

Precipitation in Delaware is highly variable from year to year, but annually averages about forty-five inches. At the Dover location, rainfall averaged three to a little over four inches each month between 1981 and 2010 (see Figure 11.2). The greatest amount of rainfall recorded in the state within a twenty-four-hour period is 8.5 inches, which was recorded at the Dover station on July 13, 1975. Dover also holds the record for greatest twenty-four-hour snowfall, which was set on February 19, 1979, at twenty-five inches.[34]

Though small in size, Delaware is rich in data, including weather data available through the Delaware Climate Projections Portal (climate.udel.edu/declimateprojections/). In 2014, as part of the Delaware Climate Change Impact Assessment report, Delaware's State Climatologist, Dr. Daniel Leathers, analyzed historical (1895–2012) climate data for the state to identify statistically significant trends in climate variables.

Temperatures in Delaware have increased from an annual statewide mean temperature of 54°F in 1895 to 56°F in 2012, a 2°F increase over 100+ years, or 0.2°F increase per decade (Figure 11.3). This increase is consistent with and at the upper edge of the range of temperature increase reported by the IPCC in 2018. The increasing temperature trend of 0.2°F per decade in Delaware was also found for each season, except autumn, where the increase was 0.1°F per decade. Minimum temperatures have also increased, which is evidenced by fewer nights with minimums below 32°F and more nights with minimums above 75°F. And the length of the growing season has increased across the state with earlier dates of last freezes in the spring and later dates of first freezes in the fall.[35]

With respect to precipitation, there were no statistically significant statewide trends for the period of 1895–2012. The only season that did have a statistically significant trend was autumn, where statewide precipitation increased at a rate of 0.27 inches per decade. During this period of time, precipitation was highly variable both within single years and from year to year.[36]

Additionally, *tide stations* at a number of locations along Delaware's coast have also been recording sea level trends since the early 1900s. At the tide gauge in Lewes, Delaware, the sea level has risen thirteen inches over the last century, or by 0.13 inches per year. This rate is nearly twice the global rate of sea level rise, which is 0.07 inches per year. Delaware's rate of sea level rise is faster because our land surface is also sinking due to the diminishing weight of ice that once overlaid our continent.[37]

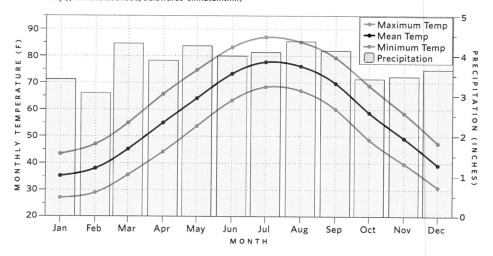

Figure 11.2. Delaware Climate Normals for the 1981–2010 period in Dover. ("Delaware Climate Information," Office of the Delaware State Climatologist, http://climate.udel.edu/delawares-climate.html.)

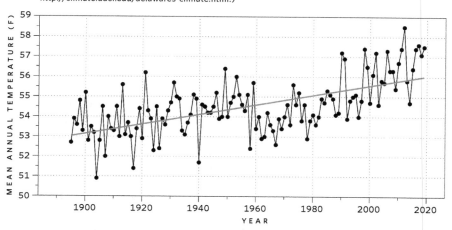

Figure 11.3. Delaware statewide mean annual temperature for the period of 1895–2019. ("Delaware Climate Information," Office of the Delaware State Climatologist, http://climate.udel.edu/delawares-climate.html.)

CLIMATE PROJECTIONS

GLOBAL

Predicting how our climate will change in the future is an exercise in complex modeling. Climate models, often called *general circulation models* (GCMs), are based on physical processes that mathematically move energy and matter between the components of our climate system. Models are calibrated and validated with observed data to ensure that they can reproduce past conditions accurately. Often, the results of multiple models will be considered together to get a better perspective of the range of possible future outcomes and of the level of certainty that a prediction will occur. The IPCC relies on this multiple model approach and has examined how climate will change under a number of carbon concentration scenarios.

The IPCC 2018 special report shares that, "Global warming is likely to reach 1.5°C [2.7°F] between 2030 and 2052 if it continues to increase at the current rate." The report further goes on to explain that, even if all current emissions ceased today, the greenhouse gases that have already been emitted by humans from the pre-industrial period up to the present will continue to result in global warming, up to as much as another 0.5°C [0.9°F], for hundreds to thousands of years. Other projections identified in the 2018 report include increases in mean temperature in most land and ocean regions, a rise in extreme heat in most inhabited regions, heavy precipitation in several regions, and higher probabilities of drought in some regions. And, notably (and ominously) for places like Delaware, sea levels will continue to rise beyond 2100, though the magnitude of rise will depend on future emission pathways.

DELAWARE

In 2014, as part of the Delaware Climate Change Impact Assessment report, the Delaware Division of Energy and Climate worked with climate modeling consultants to localize global climate models to predict future scenarios for Delaware. Thirteen global climate models were used in this exercise to project climate indicators through 2100. Model-simulated trends were compared with observed data from Delaware for the 1960–2011 period to assess the models' ability to accurately predict the recent past. A "low" scenario was run to simulate a future where clean energy sources are used in the near future, resulting in a gradual decrease in atmospheric CO_2 concentrations, and a "high" scenario simulated a continued reliance on fossil fuels and growing GHG emissions into the future. More than 160 climate indicators related to temperature and precipitation were assessed.

The results of this modeling exercise concluded that average and seasonal temperatures in Delaware are expected to increase. By 2100, annual average temperature could increase by as much as 3.5°F under the low scenario to as much as 9.5°F under the high scenario. The growing season will continue to lengthen, and temperature extremes are projected to change. For example, the number of very hot days is projected to increase from less than one a year (the historical average) to three days per year (in the low scenario) to as many as thirty days per year (in the high scenario) by 2100. Conversely, there will likely be fewer very cold days. Heat waves are also expected to become longer and more frequent, especially under the high scenario. And the growing season will increase by thirty days (low scenario) to as much as forty-eight days (high scenario).

Consistent with precipitation observations in Delaware, precipitation projections show a high degree of variability too. But, overall, the models predict an increase in precipitation by as much as 10 percent by 2100 with an increased probability of rainfall extremes as well. Little to no change was predicted in the number of dry days, which indicates that precipitation intensity will increase if overall rainfall increases on the same number of wet days. Most of the precipitation increase is expected to occur in fall and winter, as opposed to spring and summer. There was also model agreement on increases in the frequency of heavy rain events, broadly defined as events producing a half-inch to eight inches over a one-day to two-week period, by 2100.

Finally, with regard to sea level rise, the Delaware Sea-Level Rise Technical Committee prepared in 2017 a series of recommendations for state officials to consider when conducting future sea level rise planning scenarios and activities. After reviewing published reports and assessments on past and projected sea level rise rates for Delaware and the world, the committee recommended that three levels of sea level rise be used in future planning scenarios and activities. The low scenario projects the sea level to rise by 1.71 feet (0.52 meters), the intermediate scenario projects a rise of 3.25 feet (0.99 meters), and the high scenario projects seas to rise by 5.02 feet (1.53 meters). The report further explains there is a 98 percent probability that sea levels will rise an additional foot (0.3 meters) by 2100 and an 87 percent probability they will rise two feet (0.61 meters) by 2100.[38]

POTENTIAL CLIMATE CHANGE IMPACTS TO ECOSYSTEMS AND WILDLIFE

Past climate trends are projected to continue into the future, resulting in warmer temperatures and wetter overall conditions with more temperature and precipitation extremes and loss of land to sea level rise. It is anticipated that such climate change will have direct and indirect impacts on the distribution of plant and animal species, on their habitats, and on the function of ecosystems within Delaware. These impacts may be further exacerbated by external stressors that are already challenging our natural systems. Such stressors include population growth, pollution, invasive species, and land use changes that result in the loss of natural areas and habitat fragmentation. The 2018 IPCC special report acknowledges that human and natural systems have already been impacted by global warming and that "land and ocean ecosystems and some of the services they provide" have been altered. The report further shares that, with another 1°C [1.75°F] increase in temperature, approximately 4 percent of the terrestrial land area worldwide will transform from one type of ecosystem to another. And, "of 105,000 species studied, 6 percent of insects, 8 percent of plants and 4 percent of vertebrates are projected to lose over half of their climatically determined geographic range for global warming of 1.5°C [1.75°F]."

CARBON DIOXIDE IMPACTS

Interestingly, increasing CO_2 in the atmosphere—the very reason we are experiencing warmer temperatures—also has a direct impact on natural systems, as plant growth is stimulated by the extra CO_2. Some studies show that weedy and invasive plants like poison ivy, honeysuckle, and kudzu thrive under the elevated CO_2 levels and respond with faster growth rates than native plants.[39] Poison ivy reportedly becomes more toxic.[40] And leaf chemistry is altered by the higher CO_2, which could impact herbivore food sources.[41]

TEMPERATURE IMPACTS

Temperature increases alone have a number of potential impacts as temperature is a primary factor in plant growth and development. Each plant species has an optimum temperature for both vegetative development and reproductive development, so temperature impacts on plants are species-dependent.[42] Certain perennial plants like fruit trees and flowering bulbs require a chilling period as part of their life cycle to initiate flowering and seed production. If winters are warmer, these plants may not experience adequate chilling, and flowering could be reduced or buds could drop off. Frost and freeze damage is possible if temperatures warm in the late winter or early spring for long enough to initiate vegetative growth but then drop again, a phenomenon dubbed "false spring." As springs more consistently warm earlier, leaf-out and bloom times will also change.

Perhaps the greatest impacts temperature can have on plant development though occur during the reproductive stage, as extreme high temperatures can impact pollen viability and hence the fertilization and formation of fruit or grain. Additionally, warmer evening temperatures during the summer growing season can increase plant respiration rates, and raise the potential for plants to reach maturation faster at a smaller size, with lower yields of fruit or grain. This can be especially pronounced in non-perennial plants, where growth can be significantly impacted by the weather of a single growing season. One trade-off to warmer temperatures, though, is a longer growing season for farmers, which allows for more time for plants to grow and the possibility of producing multiple crops in a single season.

These temperature impacts to plant growth may also result in changes to the composition of plant communities, as the range of plants change in both terrestrial and aquatic environments. Some species will see their range increase while that of others will decrease, and a general migration of species northward is expected. The ability of plants to migrate often depends on seed dispersion and the mode of transport for that dispersion. The level of habitat fragmentation is also an external factor that can impact the ability of a species to migrate. Under changing and new conditions, native plants may be stressed and become vulnerable to the expansion of invasive or undesirable plants like weeds, as well as to insects and diseases. Invasive species could take advantage of weakened ecosystems and out-compete native species. In some cases, as species migrate into new areas that are now climatically suitable for them, they may become invasive or harmful to the natural system.

Shifts in the life cycle and spatial distributions of plants have the potential to disrupt ecosystems as asynchronies develop between pollinators and plants, herbivores and plants, and predators and prey. The study of plant and animal life cycle interactions in relation to variations in climate is known as phenology: "Advancement in spring phenology of 2.8 ± 0.35 days per decade has been observed in plants and animals in recent decades in most Northern Hemisphere ecosystems (between 30°N and 72°N), and these shifts have been attributed to changes in climate."[43] Phenological shifts may result in some species-to-species interactions becoming decoupled and loss of ecosystem functionality.

If bloom times shift, then nectar and pollen will be available at different times, which may not sync with the presence of pollinating insects. In the Northeast, native bees have advanced spring bloom arrival by an average of ten days between 1880 and 2010 due to increased warming; plants have also shown earlier bloom times, so synchrony thus far has been maintained. Animals with high-energy needs could be especially impacted if their plant food sources are not available even over short periods of time. Temperatures at wintering grounds have been changing more slowly than temperatures at spring breeding grounds, so the availability of food for migrating animals, birds especially, could become a concern. The predator-prey relationship could also be disrupted if herbivore prey shift with their food source. Some shifts may occur synchronously, while others may result in the loss of historical interactions and potentially create novel species interactions as depicted by Figure 11.4.[44]

Insects are especially sensitive to warming temperatures, and populations for some species could increase since longer warm periods will allow for more reproduction, and warmer winters may not keep populations in check. The Centers for Disease Control reports that ticks are one example of an insect population that is increasing in size and geographic range due to climate change.[45] However, bats, which feed on insects, may face reduced

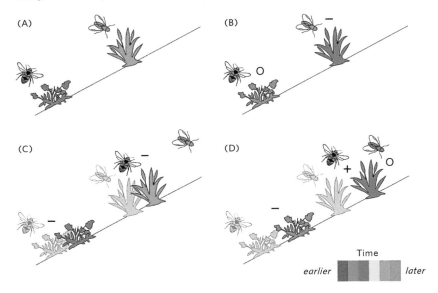

FIGURE 11.4. Potential results of shifts in timing and spatial distribution of plants and pollinators. (Morton and Rafferty, "Plant-Pollinator Interactions under Climate Change: The Use of Spatial and Temporal Transplants.")

hibernation periods in Northern regions since insects, their primary food source, could be present longer.

In the aquatic environment, warmer air temperatures result in warmer water temperatures. The amount of oxygen dissolved in water could fluctuate beyond normal thresholds as biological and chemical processes respond to the increase in temperature. Elevated levels of nutrients from human activities already impact dissolved oxygen levels and have been documented to stress many aquatic systems in Delaware,[46] and water quality could be further impacted by rising temperatures. Warmer temperatures could also reduce water quantity in aquatic environments, which will clearly impact aquatic plants and animals. Natural processes like evapotranspiration will increase with warmer temperatures, but humans will also compete with natural systems for water. Water demand for domestic use, irrigation, and power generation increases at least 3 percent for every 1 percent increase in maximum air temperature.[47] Many temperature impacts are then often further exacerbated by changes to water regimes.

PRECIPITATION IMPACTS

Plant productivity generally increases under wetter conditions, but only up to a point when there is too much water. Changes to the timing, amount, or intensity of precipitation will impact natural systems as the hydrologic cycle will change in response. If the timing of precipitation is altered, the seasonal flow of water to streams, rivers, and downstream estuaries, and the plants and animals that inhabit and rely on these environments, could also be impacted. And with warmer temperatures, there is the potential for more precipitation to occur as rain than as snow, which could decrease meltwater, limiting spring flows, which then has impacts on fish spawning. Groundwater recharges during rainy periods, so precipitation timing changes could also impact the amount of water stored underground at various times of year. This could have negative consequences if groundwater supplies are not sufficient to match water demands by both humans and ecosystems during warm periods.

Changes in the amount of precipitation on an annual average basis—and especially seasonally—could impact entire natural systems. Too much water, particularly over a short period of time, results in flooding, erosion, and runoff. Some plants could get washed away or drown under standing water, while fungi (both beneficial and harmful forms) could spread since spore production and distribution increases under wet conditions. Sediment and nutrient transport increase with erosion and runoff. Excess nutrients fuel algae growth on the water's surface, which blocks sunlight from reaching beneficial submerged aquatic vegetation. Dissolved oxygen levels can dramatically fluctuate and, at very low levels, aquatic life that can relocate will do so while those species that cannot move away fast enough may succumb to the adverse conditions.

Conversely, too little precipitation produces dry conditions that stress plants and animals in both terrestrial and aquatic environments. Availability of water is critical for all life, but some animal species depend on water at specific times during their reproductive cycle. For example, many amphibians reproduce in water and will be negatively impacted if seasonal water levels drop and the presence of breeding areas like vernal pools decrease. Lower water levels also mean warmer water temperatures, which could impact water quality and aquatic life. Plants stressed by too little water are more susceptible to pests and disease. If drought conditions occur along with higher temperatures, then evapothen transpiration will further reduce water levels and soil moisture content and stresses to plants and animals will worsen. The National Climate Assessment reports that short droughts in high temperatures can kill trees that would have survived longer droughts at low temperatures. Finally, as we have seen in dramatic fashion in California in recent years, forest fires become an increased risk under dry conditions.

The increased frequency of extreme precipitation events will also increase the risk for habitat damage. As already mentioned, extreme rain events could produce flooding that washes out and drowns vegetation. High winds during storms could down trees, while ice and hail could also damage plants. Some of the precipitation impacts in coastal environments are also intensified by sea level rise.

SEA LEVEL RISE IMPACTS

Coastal areas rely on the accretion of sand and soil at the rate of sea level rise or beaches, flats, dunes, and wetlands could be inundated. Our planet's warming has resulted in sea level rise outpacing accretion rates in most places around the globe. As sea levels continue to rise, the most low-lying areas will gradually disappear, but impacts are also experienced further inland. Coastal areas episodically experience flooding and overwash during tidal cycles and storm events. The salty water salinizes soils of coastal forests, farms, and yards, making them unfit to grow certain species of plants. Rising sea levels also result in saltwater intrusion into groundwater supplies and push salty water further up fresh rivers and streams.

Coastal areas are critical habitats for many species of birds, reptiles, amphibians, mammals, fish, and shellfish. As discussed elsewhere in this volume, species-to-species interactions are important, and in the coastal zone, many of these animal species are dependent not only on their natural habitat, but also on the presence of each other. Horseshoe crabs lay eggs that migratory shore birds feed on, for example. If the horseshoe crab no longer has a suitable habitat, shorebirds would also be negatively impacted. Many coastal plants and animals also have specific ranges for the salinity, temperature, and inundation frequency they can tolerate. As these conditions change due to sea level rise, the wildlife will have to evolve accordingly or its existence could be threatened. For example, "ghost forests," or the dead trees ringing estuaries, are becoming more prevalent in the Northeast portion of the United States. Matthew Kirwan and Keryn Gedan explain that forest conversion to marsh occurs over three steps, beginning with sapling death, followed by opening of the canopy and invasion of shrubs and Phragmites (large perennial wetland grasses), and finishing with the death of adult trees, which give ghost forests their spooky appearance.[48]

The ability of our coastal ecosystems to migrate inland depends on a few factors. First, there must be space available for the plant and animal communities to move into without being blocked by human infrastructure like roads or development. There also need to be certain physical geologic features and plant assemblages present to allow the transformation to become established—most notably, for an adequate sediment supply to keep up with the sea level rise. If these conditions cannot be met, then wetlands and other coastal features will be replaced by open water.

CLIMATE ADAPTATION AND MITIGATION STRATEGIES FOR NATURAL RESOURCES

Given that plant and animal species, habitats, and ecosystems are impacted by the changes occurring in our climate, humans can help these natural systems through implementing climate adaptation and mitigation strategies.

These two terms, *adaptation* and *mitigation*, like weather and climate, are often confused and used interchangeably. Mitigations are actions to slow, stop, or reverse the magnitude of climate change. Adaptations, however, are efforts that accept that climate change is occurring and try to deal with the change by making adjustments to limit vulnerability. And, sometimes these strategies can function simultaneously.

MITIGATION STRATEGIES

The most obvious mitigation strategies revolve around reducing GHG emissions, with carbon as the primary focus, through improved energy efficiency and use of alternative energy sources like wind and solar power. Another pathway that can be simultaneously pursued is to capture and store carbon from the atmosphere to effectively offset emissions and work toward a net zero or even negative balance. This can be done through using natural processes like planting trees and managing soils, which are less costly and easier to deploy, or highly technological strategies that require more research and development and will be more costly to deploy. According to the National Climate Assessment report, "forests and wood products stored about 16 percent (833 teragrams, or 918.2 million short tons, of CO_2 equivalent in 2011) of all the CO_2 emitted annually by fossil fuel burning in the United States." The IPCC sees land use and land use change as critical mitigation pathways that can help keep global temperature rise from exceeding 1.5°C.

The U.S. Climate Alliance, a bipartisan coalition of U.S. governors committed to the GHG reduction goals articulated in the 2015 Paris Agreement, is leading a Natural and Working Lands (NWL) Initiative. The member states in the alliance all have agreed to reduce GHG emissions by 26–28 percent below 2005 levels by 2025. Delaware's Governor John Carney has signed on to the alliance and to the NWL initiative, which seeks to protect and enhance the carbon sequestration and storage occurring on natural and working lands. These lands include forests, grasslands/rangelands, wetlands, croplands, and urban greenspaces.[49] The Delaware Forest Service estimates there are about nineteen million tons of carbon stored in forests across the state, approximately two thirds of which are in the aboveground live biomass.[50] Efforts are underway in Delaware not only to calculate current carbon storage on natural and working lands within the state, but also to identify best practices for land conservation, restoration, and management to decrease emissions, reduce losses of carbon already stored, and increase storage capacity on these lands.

A number of strategies to achieve these goals are under consideration across the nation and in Delaware. The protection of existing forest stands is of utmost priority in order to maintain the carbon stores already in place. *Afforestation*, the planting of trees where there had not previously been a forest, and *reforestation*, the replanting of trees in a previously forested area, will help to increase current storage levels. Land use planning and conservation and preservation programs help to preserve and promote these natural areas.

But maintaining and restoring forests does not mean people can no longer harvest trees. Indeed, having established and thriving wood and forest product markets is also vitally important to keep forests *as* forests. The carbon emissions from harvesting trees are lower than those produced to make other structural materials like concrete and steel, so there is now a renewed emphasis on promoting the use of wood. Besides producing wood, forests can also provide bioenergy resources, recreation opportunities, and ecosystem service markets. And the shade provided by trees in developed areas also helps to lower energy costs, thus providing multiple mitigation benefits.

In agricultural areas, carbon sequestration and storage can be increased through better soil management practices. Reducing tillage to low- or no-till levels not only reduces carbon emissions through limiting tractor passes on fields, but also reduces erosion in these areas. Using cover crops, which cover fields after their cash crops have been harvested, also protects soils from erosion and helps keep carbon in the soil. Other conservation practices, like the use of forested buffers, or strips of trees along waterways, not only reduce erosion but also help capture and store carbon from the atmosphere.

ADAPTATION STRATEGIES

A number of adaptation strategies are available to help ecosystems cope with warming temperatures, wetter conditions, and more frequent extreme events. The National Climate Assessment advises that it is more effective to focus on whole system management rather than on one species at a time since populations with high genetic diversity are likely to be more flexible. In order to do this, some planning work must be done to identify what actions will protect and promote biodiversity and be resilient through future changes and impacts. Often, this means recognizing where there are weak links in life histories and ecosystem assemblages.

As you have learned in several other chapters in this volume, Delaware (and much of the United States) has experienced a significant drop in biodiversity. But Master Naturalists (among others) are in an excellent position to restore lands to promote biodiversity. Forests can be managed to have a healthy structure, with a range of species, ages, and sizes represented. They can also be managed to increase growth through responsible fertilization and irrigation; control of weeds, pests, and diseases; and use of fast-growing stocks of trees with short rotations. Harvesting can be done sustainably where it is done with species and density in mind while minimizing disturbances to the ecosystem. When replanting, species or ecotypes suitable for future climates can be used, which will help with the migration process. Since extreme temperatures during pollination are especially dangerous to plant success, plants that shed pollen during cooler periods, like the early morning, or flower over longer periods, may be more successful in future climates. Finally, seed banking (the storage of seeds to preserve genetic diversity), biobanking (the storage of plant and animal samples for research), and captive breeding can be done to conserve and reintroduce species to new zones.

With respect to sea level rise, the Delaware Sea Level Rise Advisory Committee has recommended four strategies for adaptation:[51] we can *avoid* building in the coastal zone by limiting new development in these areas through easements, setbacks, and transfers of development rights; we can *accommodate* for sea level rise by raising homes and structures along coasts and manage drainage with green infrastructure like rain gardens and bioswales, which are linear grassed or vegetated channels for conveying stormwater; we can try to *protect* the coasts through beach renourishment, and building bulkheads and dikes; or we can *retreat* and strategically over time relocate homes and businesses out of the areas to be inundated. As we have seen, coastal ecosystems will migrate inland if allowed the space to do so.

Public outreach is a final strategy to help with implementing adaptation and mitigation strategies. Yale and George Mason University have conducted eleven nationwide surveys on American views regarding climate change over the last five years. The researchers found that there were six unique audience segments: alarmed, concerned, cautious, disengaged, doubtful, and dismissive. Over this period of time, the surveys indicate the proportion of Americans that believe global warming is happening has increased by 11 percent, and more now believe that it is human-caused (an increase of 15 percent). Additionally, the percentage of Americans who are worried about global warming and fear it will harm Americans has also increased over this five-year period (increases of 16 percent and 12 percent, respectively).[52]

Interestingly, still in 2018, only 57 percent of Americans believed that most *scientists* agree global warming is happening. In reality, conservative estimates are that 97 percent of climate scientists are convinced climate change is happening and caused by human activity. Researchers associated with this study have concluded that "Public misunderstanding of the scientific consensus—which has been found in each of our surveys since 2008—has significant consequences."[53] Thus, public education and outreach on climate topics is essential to expose people to the facts of climate science so everyone better understands climate change's current (and potential) impacts and becomes willing to participate in adaptation and mitigation strategies.

RESOURCES

DELAWARE RESOURCES

- *Office of the Delaware State Climatologist*—the principal scientific extension service for weather and climate information in Delaware, situated within the Department of Geography in the College of Earth, Ocean, and Environment at the University of Delaware, http://climate.udel.edu/.

- *Delaware Climate Data*—the Office of the Delaware State Climatologist curates climate normals (three-decade averages of climate variables), precipitation frequency estimates, and standardized precipitation indices for the state, http://climate.udel.edu/data.html.
- *Delaware Environmental Observing System (DEOS)*—a product of the Delaware State Climatologist office, DEOS is a real-time environmental data service that provides access to the latest conditions, summaries, and management applications, http://deos.udel.edu/.
- *Delaware Environmental Monitoring & Analysis Center (DEMAC)*—DEMAC promotes and coordinates environmental monitoring and provides an effective and user-friendly interface to share information, http://demac.udel.edu/.
- *Delaware Climate Projections Portal*—model projection data from the Delaware Climate Impact Assessment Report are available for general information, download, and visualization, http://climate.udel.edu/declimateprojections/.
- *Delaware Sea Grant*—an education and outreach arm of the UD College of Earth, Ocean, and Environment, the Sea Grant helps communities develop sustainable coastal economies based on a thriving natural environment, https://www.deseagrant.org/.
- *Delaware Cooperative Extension*—an education and outreach arm of the UD College of Agriculture and Natural Resources, the extension connects the public with university knowledge, research, and resources to address, youth, family, community, and agricultural needs, http://extension.udel.edu/.
- *DNREC Division of Climate, Coastal, & Energy*—a division of the Delaware Department of Natural Resources and Environmental Control that integrates applied science, education, policy, and incentives to address Delaware's climate, energy, and coastal changes, https://dnrec.alpha.delaware.gov/climate-coastal-energy/.

REGIONAL RESOURCES

- *Northeast Climate Hub*—regional climate hubs led by U.S. Department of Agriculture personnel from the Agriculture Research Service and Forest Service bringing together research and services to ensure robust and healthy agricultural production and natural resources under increasing climate variability and change, https://www.climatehubs.oce.usda.gov/hubs/northeast.
- *As If You Were There*—a project of the Northeast Climate Hub, these immersive 360-degree tours feature demonstrations of climate adaptation and mitigation practices on agricultural and forest lands, https://www.climatehubs.oce.usda.gov/hubs/northeast/project/360.
- *Chesapeake Bay Program Climate Resiliency Workgroup*—this group coordinates climate-related efforts to address climate resiliency across the Chesapeake Bay Watershed, https://www.chesapeakebay.net/who/group/climate_change_workgroup.

NATIONAL RESOURCES

- *National Oceanic and Atmospheric Administration (NOAA)*—NOAAs mission is: to understand and predict changes in climate, weather, oceans, and coasts; to share that knowledge with others; and to conserve and manage coastal and marine ecosystems and resources, www.noaa.gov.
- *National Weather Service (NWS)*—a program under NOAA, the NWS provides weather, water, and climate data, forecasts, and warnings to protect life and property and enhance the national economy, www.noaa.gov.

- *SKYWARN*—an NWS volunteer program to obtain critical weather information from trained severe weather spotters, https://www.weather.gov/SKYWARN.
- *Community Collaborative Rain Hail Snow Network (CoCoRaHS)*—a nonprofit, community-based network of volunteers working together to measure precipitation across the nation, https://www.cocorahs.org/.
- *Meteorological Phenomena Identification Near the Ground (mPING)*—a free app that allows the NOAA National Severe Storms Laboratory to collect public weather reports through a free app on smartphones and mobile devices to fine-tune forecasts, https://mping.nssl.noaa.gov/.
- *Students' Cloud Observations On-Line (S'COOL)*—a National Aeronautics and Space Administration (NASA) platform to collect citizen science data on clouds, https://scool.larc.nasa.gov/rover.html.
- *U.S. Drought Monitor*—a partnership between the National Drought Mitigation Center, NOAA, and the U.S. Department of Agriculture release a map each Thursday showing which parts of the U.S. are currently experiencing drought, https://droughtmonitor.unl.edu/.

NOTES

1. "What's the Difference Between Weather and Climate?" *National Aeronautics and Space Administration*, https://www.nasa.gov/mission_pages/noaa-n/climate/climate_weather.html.
2. Eric Danielson, James Levin, and Elliot Abrams, *Meteorology* (Boston: WCB McGraw-Hill, 1998).
3. "Glossary," *National Weather Service*, https://w1.weather.gov/glossary/.
4. Ibid.
5. Danielson, Levin, and Abrams, *Meteorology*.
6. "Glossary," *National Weather Service*.
7. "Origin of Wind," *National Weather Service*, https://www.weather.gov/jetstream/wind.
8. Danielson, Levin, and Abrams, *Meteorology*.
9. "Origin of Wind," *National Weather Service*.
10. "Glossary," *National Weather Service*.
11. Danielson, Levin, and Abrams, *Meteorology*.
12. Ibid.
13. "The Four Core Types of Clouds," *National Weather Service*, https://www.weather.gov/jetstream/corefour.
14. Danielson, Levin, and Abrams, *Meteorology*.
15. "United States Drought Monitor," *National Drought Mitigation Center*, https://droughtmonitor.unl.edu/.
16. Danielson, Levin, and Abrams, *Meteorology*.
17. "Air Masses," *National Weather Service*, https://www.weather.gov/jetstream/airmass.
18. A. P. M. Baede, E. Ahlonsou, Y. Ding, and D. Schimel, "2001: The Climate System: An Overview," in *Climate Change 2001: The Scientific Basis. Contribution of Working Group I to the Third Assessment Report of the Intergovernmental Panel on Climate Change*, eds. J. T. Houghton, Y. Ding, D. J. Griggs, M. Noguer, P. J. van der Linden, X. Dai, K. Maskell, and C. A. Johnson (New York: Cambridge University Press, 2001).
19. Danielson, Levin, and Abrams, *Meteorology*.
20. Ibid.
21. Baede, Ahlonsou, Ding, and Schimel, *Climate Change 2001*.
22. Ibid.

23. "Summary for Policymakers" in *Climate Change 2014: Mitigation of Climate Change. Contribution of Working Group III to the Fifth Assessment Report of the Intergovernmental Panel on Climate Change*, eds. O. Edenhofer, et al. (New York: Cambridge University Press, 2014).
24. Theo Stein, "Rising Emissions Drive Greenhouse Gas Index Increase," *National Oceanic and Atmospheric Administration Research News*, last modified May 21, 2019, https://research.noaa.gov/article/ArtMID/587/ArticleID/2455.
25. "Global Greenhouse Gas Emissions Data," *United States Environmental Protection Agency*, last modified April 13, 2017, https://www.epa.gov/ghgemissions/global-greenhouse-gas-emissions-data.
26. Stein, "Rising Emissions."
27. "JetStream Max: Addition Köppen-Geiger Climate Subdivisions," *National Weather Service*, https://www.weather.gov/jetstream/climate_max.
28. "IPCC, 2018 Summary for Policymakers" in *Global Warming of 1.5°C. An IPCC Special Report on the Impacts of Global Warming of 1.5°C above Pre-Industrial Levels and Related Global Greenhouse Gas Emission Pathways, in the Context of Strengthening the Global Response to the Threat of Climate Change, Sustainable Development, and Efforts to Eradicate Poverty*, eds. V. Masson-Delmotte, et al.(Geneva: World Meteorological Organization, 2018).
29. "Climate Change: How Do We Know?" *National Aeronautics and Space Administration*, https://climate.nasa.gov/evidence/.
30. "JetStream Max."
31. "Delaware Climate Information," Office of the Delaware State Climatologist, http://climate.udel.edu/delawares-climate.html.
32. Ibid.
33. "State Climate Extremes Committee," *NOAA National Centers for Environmental Information*, https://www.ncdc.noaa.gov/extremes/scec/records/de.
34. Ibid.
35. DNREC, "Delaware Climate Change Impact Assessment" (Dover: Delaware Department of Natural Resources and Environmental Control Division of Energy and Climate, 2014).
36. Ibid.
37. DNREC, "Preparing for Tomorrow's High Tide, Sea Level Rise Vulnerability Assessment for the State of Delaware," Prepared for the Delaware Sea Level Rise Advisory Committee by the Delaware Coastal Programs of the Department of Natural Resources and Environmental Control (Dover: DNREC Division of Energy and Climate, 2012).
38. John A. Callahan, Benjamin P. Horton, Daria L. Nikitina, Christopher K. Sommerfield, Thomas E. McKenna, and Danielle Swallow, "Recommendation of Sea-Level Rise Planning Scenarios for Delaware: Technical Report," Prepared for Delaware Department of Natural Resources and Environmental Control Delaware Coastal Programs (Newark: DNREC Division of Energy and Climate, 2017).
39. Jerry M. Melillo, Terese (T. C.) Richmond, and Gary W. Yohe, eds., *2014: Climate Change Impacts in the United States: The Third National Climate Assessment. U.S. Global Change Research Program*, doi:10.7930/J0Z31WJ2.
40. L. H. Ziska, R. C. Sicher, K. George, and J. E. Mohan, "Rising Atmospheric Carbon Dioxide and Potential Impacts on the Growth and Toxicity of Poison Ivy (Toxicodendron Radicans)," *Weed Science* 55(4) (July 1, 2007): 288–92.
41. Peter Stiling and Tatiana Cornelissen, "How Does Elevated Carbon Dioxide (CO_2) Affect Plant-Herbivore Interactions? A Field Experiment and Meta-Analysis of CO_2-Mediated Changes on Plant Chemistry and Herbivore Performance," *Global Change Biology* 13 (2007): 1823–42, doi: 10.1111/j.1365–2486.2007.01392.x.

42. Jerry L. Hatfield and John H. Prueger, "Temperature Extremes: Effect on Plant Growth and Development," *Weather and Climate Extremes* 10 (2015): 4–10.
43. "IPCC, 2018 Summary for Policymakers."
44. E. M. Morton and N. E. Rafferty, "Plant-Pollinator Interactions under Climate Change: The Use of Spatial and Temporal Transplants," *Applications in Plant Science* (2017), 5(6):apps.1600133 doi:10.3732/apps.1600133.
45. "Climate Change Increases the Number and Geographic Range of Disease-Carrying Insects and Ticks," *Centers for Disease Control*, https://www.cdc.gov/climateandhealth/pubs/VECTOR-BORNE-DISEASEFinal_508.pdf.
46. DNREC, "Draft State of Delaware 2018 Combined Watershed Assessment Report (305(b)) and Determination for the Clean Water Act Section 303(d) List of Waters Needing TMDLs (The Integrated Report)" (Dover: Delaware Department of Natural Resources and Environmental Control, 2018).
47. DWSCC, "Twelfth Report to the Governor and General Assembly Regarding the Progress of the Delaware Water Supply Coordinating Council—Estimate of Water Supply & Demand for Kent County and Sussex County through 2030," Prepared by the Delaware Department of Natural Resources and Environmental Control Division of Water, Delaware Geological Survey, and University of Delaware Institute for Public Administration—Water Resources Agency (Dover: DNREC Division of Water, 2014).
48. Matthew L. Kirwan and Keryn B. Gedan, "Sea-Level Driven Land Conversion and the Formation of Ghost Forests," *Nature Climate Change* (2019): 450–57
49. "Natural and Working Lands—An Initiative of the US Climate Alliance," *US Climate Alliance*, https://www.usclimatealliance.org/nwlands. The Paris Agreement is a 2015 agreement by the parties to the United Nations Framework Convention on Climate Change to strengthen the global response to climate change.
50. DFS, "Delaware Forest Resource Assessment" (Dover: Delaware Forest Service, Delaware Department of Agriculture, 2017).
51. Susan Love, Tricia Arndt, and Molly Ellwood, "Preparing for Tomorrow's High Tide: Recommendations for Adapting to Sea Level Rise in Delaware, Final Report of the Delaware Sea Level Rise Advisory Committee" (Dover: Delaware Department of Natural Resources and Environmental Control, Delaware Coastal Programs, 2013).
52. A. Gustafson, P. Bergquist, A. Leiserowitz, and E. Maibach, "A Growing Majority of Americans Think Global Warming Is Happening and Are Worried" (New Haven, CT: Yale Program on Climate Change Communication, 2019).
53. A. Leiserowitz, E. Maibach, C. Roser-Renouf, S. Rosenthal, M. Cutler, and J. Kotcher, "Climate Change in the American Mind" (New Haven, CT: Yale Program on Climate Change Communication, 2017).

CHAPTER TWELVE

Invasive Species and Habitat Management

DOUG TALLAMY

THE TERM *invasive species* has only recently entered our environmental lexicon and, like so many aspects of today's culture, has become curiously controversial. There is no public outcry demanding the introduction of new diseases or pest insects to kill our plants, new mussels to fowl our freshwater ecosystems, or new species of Asian fish to wreak havoc on North American fisheries. Yet, for more than a century, there has been a powerful public demand for ornamental plants that evolved on other continents, and this demand has brought us 86 percent of our worst invasive plants. And because invasive plants are what the public argues most about regarding the fulfillment of this demand, I will focus on them in this chapter.

NOVEL ECOSYSTEMS

For as long as we *Homo sapiens* have been on the move around the globe, we have carried plants, and—to a lesser extent—animals, with us. Modern modes of transportation, international trade, and a keen desire to display unusual plants in our yards have turned what was once a trickle of introductions of such plants in past centuries into a torrent of new species entering North America from foreign lands. So great, so sudden, and so disruptive has the influx of new species been that ecologists now call the majority of today's ecosystems "novel"—literally, "new things" under the sun.[1] They are considered "novel" because many of the species within these ecosystems are just meeting each other for the first time in evolutionary history. That is, their interactions with each other are occurring without the tempering effects of long periods of coevolution.

Although some scientists are excited about the evolutionary potential of novel ecosystems, such potential will not be realized for eons and, in the meantime, many of these introductions have been devastating to native populations, and thus to ecological productivity, within existing ecosystems. When a novel predator is introduced, for example, prey rarely have appropriate adaptations to defend themselves, and quickly fall victim.[2] Our cute and cuddly house cats kill between two and three *billion* birds in the United States alone each year.[3] The largest extinction event in the Holocene Epoch occurred when humans colonized the Polynesian Islands some four thousand years ago.[4] An astounding eight hundred to two thousand species of birds were lost to human hunting, but were also devastated by the introduction of rats to the islands. The same type of ecological ruin has occurred time and again when diseases such as chestnut blight and white pine blister rust, or non-native insects like the hemlock wooly adelgid, emerald ash borer, and gypsy moth, have been introduced to North America from other lands.

But the most widespread and underappreciated consequences of creating novel ecosystems are those that occur when introduced plants replace native plant communities.

NATIVE VS. INTRODUCED

Nearly all of us get our plants from nurseries, but the plants in most nurseries fall into two very distinct categories: they are either *native* to your area (that is, they share an evolutionary history with the plant and animal communities in your ecoregion or biome), or they developed the traits that make them unique species somewhere else. In Delaware, that "somewhere else" is typically from East Asia, although our nurseries carry many plants from Europe as well.

In the past, we didn't care much about where a plant came from; we chose our ornamentals for the sole purpose of meeting a specific aesthetic taste. Maybe we desired a particular color or habit to complete a landscape design, or perhaps we sought a plant with a striking bloom to serve as an accent or focal point in our yard, or maybe we wanted a dense evergreen to serve as a screen, or—more often than not—we chose our plants so that our yard would look just like our neighbor's, and their neighbor's, and so on. When these were our only goals, the geographic origins of the plant mattered little.

However, if your goal is to create a landscape that enhances your local ecosystem rather than degrades it, then geographic origin is the very first attribute you must consider. Plants native to your region are almost always far better at performing local ecological roles than plants introduced from somewhere else are. It is important to recognize that all invasive plants are introduced from somewhere else, but all introduced plants are not invasive. That is, they do not spread into natural areas and displace native plant communities. This is an important ecological distinction, unless, as with noninvasive crepe myrtle, we plant so many of them in so many places that they might as well be invasive in terms of their impact on local biota.

INVASIVE SPECIES

Invasive species create novel ecosystems, and there are more species of invasive plants than of all other invasive organisms combined.[5] *Invasive* plants, defined as non-native species that displace native plant communities, should not be confused with fast growing, "aggressive" native plants for one simple reason: native plants, aggressive or otherwise, have been duking it out with each other, competing for space, light, water, and nutrients, for millions of years. Over the eons, native species have evolved ways to cope with each other, and the results of their interactions define the highly diverse species composition of most native plant communities all over the world. Invasive plants, in contrast, have just arrived within a community (and "just" can be defined as within the last several hundred years, a blink of an evolutionary eye). They also have arrived without their suite of natural enemies: the insects, mammals, and diseases that keep them in check in their homeland. Invasive plants thus have an enormous competitive advantage over most native plant species, which enables many of them to run amuck across the landscape. And so interactions between invasive plants and our native species are anything but tried and true. Invasives and natives are just starting to negotiate what their future coexistence will look like, and it will take

FIGURE 12.1. The invasive Japanese stiltgrass all but eliminates plant diversity near the ground in Delaware. (Photo courtesy of Doug Tallamy.)

hundreds or thousands of generations for these negotiations to reach a compromise.

You might think monocultures of Japanese stiltgrass, autumn olive, privet, bush honeysuckle, or Phragmites would easily convince anyone that fighting such invasions is a good idea. Yet questions about the wisdom of attempts to curb vegetative incursions have been raised ever since invasive ornamentals started moving from our gardens into natural areas. People concerned about the impacts of invasive plants on ecosystems have been accused of being too emotional about plant invasions, of not letting nature take its course, of ignoring the beneficial side of introduced plants, and of trying to return our ecosystems to some pristine state that has not existed for at least fourteen thousand years. Some criticism of efforts to control invasive plants is understandable: we spend billions each year trying to manage invaded ecosystems, and though local success is common, eradication of invasives is nigh impossible. Wide-scale control of these plants is extraordinarily difficult, particularly when we continue to sell them in our nurseries. Fighting what many see as a losing battle might seem a fool's errand if there were not compelling reasons to do so. But there are good reasons to keep introduced plants off our properties.

Invasive plants can disrupt long-standing ecosystem dynamics in several ways. They can change soil hydrology, increase fire frequency, competitively exclude and hybridize with native plants, and degrade aquatic habitats. Critics of invasive control efforts argue that all ecosystems are in a constant state of flux; if an introduced plant enters an ecosystem and changes the diversity and abundance of native species, supporters of this argument say, then that change is a "natural process" that should not be challenged by humans (even though humans were responsible for it). They further argue that plant invasions are not "bad" for ecosystems because there are no records of a plant invasion causing a continent-wide extinction of a native species, and that there are more plant species in North America after an invasion than before.

Are these claims valid? Let's take a close look at the evidence. First, are there really more species present after an introduced plant displaces native plant communities? The answer depends on the geographic scale being considered.[6] If we have introduced thirty-three hundred new plant species to North America, then species diversity should now be higher. And on a continental scale, it is. But ecosystems don't function on a continental scale; they function locally, and there are oodles of studies documenting the reduction (or complete elimination) of one or more plant species on a local scale after the arrival of an invasive plant.[7]

Moreover, when counting species impacted by introduced plants, we should not just count plant species. We need also to consider the animals that eat or pollinate plants, and, in turn, the predators of those animals. Every time a native plant is removed from an ecosystem, or even diminishes in abundance, populations of all of the animals that depend exclusively on that plant are also removed or diminished, as are the natural enemies of those species. In sum, then, at the local scale—the scale that counts ecologically—invasive plants typically shatter local species diversity, and claims to the contrary have not been supported by rigorous field studies.

What about the assertion that invasive plants have not caused any native plant extinctions on continents? The qualifier "on continents" has to be added to this statement because the threat of extinctions caused by invasive species on islands is quite high.[8] In contrast, on continents, native plant populations are larger and more dispersed, and there are more places for natives to escape invasive species; even if a native plant population is clobbered in one place, other populations of it may persist in another. However, there is one biological phenomenon associated with some plant invasions that is so pernicious, even continental scales are not protecting natives from it. I speak of *introgression* or *introgressive hybridization*, where the invasive species hybridizes with a closely related native in repeated back crosses. Through that hybridization and directional gene flow, the gene pool moves closer and closer to that of the invader. This is the process by which Africanized bees have replaced the European honey bee genotypes wherever the two have come into contact, and all in just a few generations. American bittersweet and red mulberry are two examples of native plants that are rapidly disappearing from their native range in a similar manner, through directional introgression with oriental bittersweet and white mulberry, respectively.[9] Both species of native plants have been entirely replaced by their non-native relatives in Delaware. Unfortunately, the introgression is proceeding at such a rapid pace that their extinction seems imminent.

Even though the claim that there are no records of extinctions caused by invasive plants may not remain true for long, it is certainly true that the introduction of non-native plants has not (yet) created an extinction threat for most native species. That, however, does not mean that ecosystem function has not been compromised at the site of each invasion. Besides, using global extinction as the only indication of harm is like saying the only symptom that warrants a visit to the doctor is death. When invasive plants like autumn olive, barberry, bush honeysuckle, and Phragmites invade Delaware plant communities, they completely replace the local native species at that site, causing a decline in the plants upon which that ecosystem has depended for eons. If introduced plants were the ecological equivalents of the native species they replaced, then ecosystems would look different after an invasion, but they would be just as productive (though less stable). Introduced plants may, in fact, be equal to natives in their ability to produce ecosystem services like carbon sequestration, but they pale in comparison to natives in perhaps the most critical role plants play in nature. That is, introduced plants are poor at providing food for the animal life that runs our ecosystems.

Plants, in essence, enable animals to eat sunlight; by capturing energy from the sun and storing it through photosynthesis in the carbon bonds of simple sugars and carbohydrates, plants are the basis of every terrestrial and most aquatic food webs on the planet. But animals benefit from the energy captured by photosynthesis only if they can eat plants, or eat something that ate plants previously. And there's the rub: the group of animals that is best at transferring energy from plants to other animals is insects. Unfortunately, most insects are very fussy about which plants they eat.

THE CURSE OF SPECIALIZATION

How easy conservation would be if all plants delivered the same ecological benefits, especially to all insects. We could plant eucalyptus around the world and plant-eaters everywhere would be as happy as koalas. The nectar-filled butterfly bush so many people plant "to help the butterflies" would actually serve as a larval host for all butterflies (instead of only for one species in the Southwest) and deliver pollen and nectar to all four thousand species of native bees rather than to just a few generalist bees. We could rename the evening primrose moth the "every plant" moth, and the ornamental bamboos that are consuming yards and road shoulders throughout Delaware would feed monarch butterflies as well as milkweeds do.

But, alas, specialized relationships between plants and animals are the rule rather than the exception in nature, and they are far more common than generalized ones. This is particularly true for specialized relationships involving food webs, those interconnected relationships that facilitate the transfer of energy harnessed from the sun by plants to animals that eat plants, and then to animals that eat animals. Many people refer to this transfer of energy as a food "chain," but if you were to make a diagram of a plant and then all of the species that eat that plant, as well as all of the species that eat each of those plant eaters, the result would look far more like a very complex food "web" than a linear "chain."

By far, the most important and abundant specialized relationships on the planet are the relationships between the insects that eat plants and the plants they eat. Most insect herbivores, some 90 percent, in fact, are diet specialists, or what we call "host plant specialists," restricted to eating one or just a few plant lineages.[10] Host plant specialization has been well-documented by entomologists since the early 1960s, but scientists have never been very good at talking to each other, so the importance of host plant specialization in gluing ecosystems together is still underappreciated by many ecologists, restoration biologists, and particularly by conservation biologists. This is why proposals to reforest tropical areas of the world with eucalyptus have not been met with what should be jaw-dropping outrage. Blue gum eucalyptus from Australia and the African rubber tree are now the most abundant trees in Portugal and Puerto Rico respectively.[11] More and more frequently, shade coffee marketed as "good for the birds" is being grown under the shade of eucalyptus, citrus, and mango, even though these plants make little to no food for birds.

The specialized relationships between insects and plants are so important in determining ecosystem function and local carrying capacity that it is worth spending a little time to explain why this is so, and how these relationships have come about. Plants, of course, don't want to be eaten; they want to capture the energy from the sun and use it for their own growth and reproduction. So, in an attempt to deter plant eaters, they manufacture nasty-tasting chemicals and store them in vulnerable tissues like leaves. These chemicals are secondary metabolic compounds that do not contribute to the primary metabolism of the plant. That is, they are not a necessary part of the everyday jobs of living and growing. Instead, their job is to make various plant parts distasteful or downright toxic to insect herbivores. Some well-known plant defenses include toxic compounds like cyanide, nicotine, cucurbitacins, and pyrethrins, heart-stoppers like cardiac glycosides, and digestibility inhibitors like tannins.[12]

But if plants are so well defended, then how can insects eat them without dying? This question dominated studies of plant-insect interactions for three decades, but at this point, the answer has been thoroughly delineated. Caterpillars and other immature insects are eating machines. Some species increase their mass seventy-two thousand-fold by the time they reach their full size.[13] Because caterpillars necessarily ingest chemical deterrents with every bite, there is enormous selection pressure to restrict feeding to plant species they can eat without serious ill effects. Thus, a *gravid* (pregnant) female moth attempts to lay eggs only on plants with chemical defenses their hatchling caterpillars are able to disarm.

There are many physiological mechanisms by which caterpillars can neutralize plant defenses, but they all involve some combination of sequestering, excreting, and/or detoxifying defensive phytochemicals before they interfere with a caterpillar's health. Caterpillars have typically come by these adaptations through thousands of generations of exposure to the plant lineage in question. In short, by becoming host plant specialists, insect herbivores have developed the capacity to circumvent the defenses of a few plant species well enough to make a living, while ignoring the rest of the plants in their ecosystem. For our purposes, the key point regarding host plant specialization is that it does not happen overnight. Although every once in a while an insect species coincidentally possesses enzymes that are able to disarm the defenses of a plant species that it has never before encountered in its evolutionary history, it usually takes many eons for an insect to adapt to a new host plant, if it can adapt to it at all.

Does this mean insect specialists have won the evolutionary arms race with plants? Somewhat, but only in relation to the plant lineage on which they have specialized. When viewed across all lineages, plant defenses are very effective at deterring most insects. The monarch butterfly provides a great example of this. This species is a specialist on milkweeds that use various forms of toxic cardiac glycosides to protect their tissues. Very few insects can eat plants containing cardiac glycosides, but over the ages, monarchs have developed enzymes that can make cardiac glycosides less toxic. They also have found a physiological mechanism for storing these distasteful compounds in their wings and blood, rendering their own bodies unpalatable to predators. And they have gone one step further. Tasting bad after eating milkweed plants does not help a monarch if a bird has to eat it in order to discover its unpalatability. So monarchs, like many other distasteful insects, advertise their bad taste with an orange and black pattern that serves as a universal warning signal to would-be predators: "Don't eat me. I taste bad."

Milkweeds are so named because, in addition to using cardiac glycosides, they defend their tissues with a milky latex sap that jells on exposure to air. Insects that attempt to eat milkweed leaves soon find their mouthparts glued permanently shut by the sticky sap. Yet monarchs have found a simple but amazing way to defeat this defense: they block the flow of sap to milkweed leaves.[14]

FIGURE 12.2. Monarch butterflies have adapted to survive and reproduce on milkweed. But such specialization has come with a cost: even if adult monarchs can nectar on multiple plants (like this one on a viburnum), their caterpillars can only eat the leaves of milkweeds. (Photo courtesy of Doug Tallamy.)

This is an example of a behavioral adaptation (as opposed to a physiological adaptation), and you can easily watch it in action right in your yard. When a monarch caterpillar first walks onto a milkweed leaf, it usually moves to the tip of the leaf and starts to eat. If any latex sap starts to ooze from the wound, the caterpillar immediately stops eating, turns around, and crawls two-thirds of the way back up the leaf. There it chews entirely through the large midrib of leaf. That simple act severs the main latex canals that move the sap throughout the leaf. With the canals blocked, all of the leaf tissues below the midrib wound become latex-free and the monarch can eat them without gumming up its mouthparts. If the monarch decides to chew through most of the leaf petiole (leafstalk) instead of the leaf midrib, latex is blocked from the entire leaf. Incidentally, this behavior provides monarch hunters with a convenient tool for finding monarch caterpillars, for the leaf flags at the point where the monarch weakened the midrib. Any milkweed plant with a flagged leaf is or has been the home of a monarch.

The advantage of these adaptations is obvious for the monarch, but there are also disadvantages to such specialization, especially in today's world. Unfortunately for the monarch, the ability to detoxify cardiac glycosides and block latex sap in milkweeds does not confer the ability to disarm the chemical defenses found in other plant lineages. This means that of the 2,137 native plant genera in the U.S., the monarch can develop (with very minor exceptions) on only one, the milkweed genus *Asclepias*. The evolutionary history of this butterfly has locked it into a dependent relationship with milkweeds, and if milkweeds should disappear from a landscape, so would the monarch.

And this is exactly what has happened across the U.S. in recent years. A growing culture that favors neat, turf-lined (and chemically sanitized) agricultural fields and suburban lawns, combined with a broad unwillingness to share designed domestic landscapes with milkweeds, helped reduced monarch populations 96 percent in just forty years, between the 1970s and 2013.[15] Can monarchs adapt to other plant species? In theory, yes, but in reality, no. The monarch lineage has been genetically locked into a relationship with milkweeds for millions of years. Adaptation could conceivably modify this relationship—very slowly and over enormously long periods—but asking monarchs to suddenly (within a few decades!) switch their dependence on milkweeds to an entirely different plant lineage is like asking humans to learn to eat grass or oak leaves. The number of genetic changes required to make such a switch reduces the probability of it happening before monarchs disappear to near zero.

Please note that monarchs are not exceptions, either in their specialized relationship with milkweeds, or in their current plight. They are typical of 90 percent of the insects that eat plants, whose evolutionary history has restricted their development and reproduction to occur on plants of the lineage on which they have specialized. And as we have homogenized plant life in Delaware by replacing diverse native plant communities with a small

FIGURE 12.3. 90 percent of the insects that eat plants, like this exquisite curve-lined owlet caterpillar, can only develop on one or two plant lineages. (Photo courtesy of Doug Tallamy.)

palate of ornamental favorites from other lands, the insects that depend on local native species have declined. We have caused these declines by the way we have designed landscapes in the past. But we can and must reverse the declines by the way we design landscapes in the future, for such decisions will determine how well our ecosystems function.

ECOSYSTEM FUNCTION

I talk a lot about *ecosystem function*, and usually draw blank stares when I do. I use the term "function" because, in some ways, ecosystems are like well-oiled machines. They are built from many interacting parts that combine to perform different functions. Creating the life support systems that keep us fed, healthy, buffered from severe weather, and supplied with plenty of clean air and water—that is, the ecosystem services on which we all depend—is some of what functioning ecosystems do every day.

When we look more closely at how ecosystems function, though, my analogy breaks down. Machines have a specified number of parts, and taking some parts away almost always impairs or destroys the ability of a machine to run, while adding more parts than the design calls for does not make the machine function better. In contrast, research over the past sixty years has shown that ecosystems are far more flexible than machines in the number of parts that run them. To me, the most fascinating result of this research has been to discover the relationship between the number of species (parts) in an ecosystem and how well that ecosystem performs its various functions.

In 1955, at the very dawn of critical ecological thinking, Robert MacArthur published a paper in which he suggested that *ecosystem productivity*, the stability of ecosystems and their ability to function, was inseparably related to the number of species residing within that ecosystem. As the number of species increased, so did both ecosystem stability and productivity. MacArthur was perhaps the most brilliant theoretical ecologist of the twentieth century, so for him this was an unusually simple hypothesis. Testing it, however, would prove to be extraordinarily difficult, so he didn't. He just called this relationship the "law of nature" and left it to future generations of ecologists to test.[16]

Other hypotheses followed, including the famous *rivet* hypothesis proposed by Ann and Paul Ehrlich,[17] in which each species in an ecosystem is compared to a rivet holding a plane together; and the *redundancy* hypothesis, in which Brian Walker[18] suggested that species doing the same job in an ecosystem were functionally "redundant" and thus could be lost without harm to the ecosystem. These ideas too were largely theoretical exercises that went untested in the field. Slowly, however, as the resources required for long-term ecosystem studies became available and ecologists learned how to manipulate simplified ecosystems, direct measures of the relationship between species richness and ecosystem function began to appear in the literature.[19] Surprisingly, all of the studies were largely in agreement. MacArthur had been right after all.

Like machines, ecosystems run more smoothly, longer, and more productively when they contain all of their parts. Also, like with machines, some parts (species) are more vital to ecosystem function than others, and losing the most vital parts can shut down the entire system. But unlike machines, there may be no upper limit to the number of parts that help run ecosystems; every time a new species joins the ecosystem, it runs better than it did before. Unfortunately, the opposite is also true: every time we remove a species, or diminish its numbers to the point where it can no longer perform its role effectively, its ecosystem becomes less functional and less stable. And that in a nutshell explains why replacing native plant communities with introduced plants compromises ecosystem function. Not only do such plants often reduce the number of species in the ecosystem, but they always reduce the number of interacting species. Establishing a presence but not interacting with other local species is akin to throwing a monkey wrench in a machine. The monkey wrench is a new part added to the machine, but not only does it not interact in a positive way with the other parts of the machine, it actually prevents the other parts from interacting effectively with each other.

Introduced species occupy space in an ecosystem, space once occupied by contributing native species, but they have not been present for the thousands of generations required to form the specialized relationships that run that ecosystem.

INTERACTION DIVERSITY

Friedrich Wilhelm Heinrich Alexander von Humboldt, an eighteenth-century philosopher, linguist, botanist, biogeographer, and gifted naturalist, accumulated a list of scientific contributions during his ninety years that was even longer than his name. Von Humboldt spent much of the time between 1799 and 1804 exploring the tropical regions of the Americas. Among other things, he was keenly interested in the way the natural world worked. Not only was he the first to propose that Africa and South America had once been attached to each other, and that human activities would change the Earth's climate if they continued unabated, but—more to our point here—he was also the first to formally recognize that it is the way species *interact* with each other (and not the species themselves) that provides the glue that holds nature together. Humboldt reasoned that nature was not an abstract idea, but a living, interconnected entity.[20] Yes, the species that comprise nature are important, but it is how these species interact and the diversity of these interactions that make nature a living force.

Sadly, like his ideas on continental drift and climate change, the importance of the idea what we now call *interaction diversity* was forgotten soon after von Humboldt's death in 1859. What replaced it was a species-centric view. Recently, this view of the world has intensified as more and more species have become threatened with local or global extinction in the twentieth century. And despite growing calls from ecologists to preserve entire ecosystems, what still dominates the public's perception of conservation is the plight of individual (typically "charismatic") species; our mailboxes are full of pleas to save the whales, elephants, rhinos, tigers, and jaguars. The importance of interacting suites of species has been too abstract for the public, and even for many scientists, to appreciate.

Fortunately, this is changing, and the change is being led by tropical ecologists. While lamenting the local extinction of countless species throughout Central America, Dan Janzen, perhaps the most perceptive ecologist of the last fifty years, noted that the most insidious form of extinction was not the loss of individual species, but the extinction of ecological interactions.[21] This thinking has spawned a new way of measuring nature called *network analysis*, and Lee Dyer, an ecologist at the University of Nevada, thinks such analysis will soon show that interaction diversity is an even better predictor of ecosystem function than MacArthur's species diversity is.[22] After all, Dyer argues, it is interactions among species that affect all aspects of ecosystems, from primary productivity, to the way populations fluctuate, to the survival and reproductive success of individual species. In the few cases in which it has been studied, interaction diversity has shown to have been devastated by the introduction of non-native plants.

I will close this section with some numbers, not just because they demonstrate that introduced plants reduce both species and interaction diversity, but because they hammer home how large these reductions are. A few years ago, my students Melissa Richard and Adam Mitchell set out to measure what happens to caterpillars when invasive plants create a novel ecosystem. Finding habitats that were thoroughly invaded by introduced plants such as autumn olive, multiflora rose, Callery pear, porcelain berry, burning bush, and bush honeysuckle was easy. These plants now typify the "natural" areas near the University of Delaware where we did our study. The trick was finding places that were still relatively free of invasive plants. Using a combination of restored sites and areas not easily accessed by deer (which exacerbate the spread of invasive plants), we finally found what we were looking for: four invaded sites and four primarily native sites of similar size.

Using standard methods, we counted and weighed caterpillars at each site, once in June and again in late July. By every measure, the caterpillar community and, by extension, the community of insectivores that relied on the caterpillars for food, were seriously diminished when introduced plants replaced native plants. Even though there was more plant biomass along the invaded transects, there were 68 percent fewer caterpillar species, 91 percent fewer caterpillars, and 96 percent less caterpillar biomass than what we recorded in native hedgerows.[23]

To summarize these numbers in terms of the everyday needs of the animals that eat caterpillars, we found 96 percent less food available in the invaded habitats. Interactions between caterpillars and hedgerow plants were also significantly impacted by introduced plants, with invaded hedgerows supporting 84 percent fewer interactions and 2.3 times less interaction diversity than coevolved hedgerows did. Had we compared the dozens of species that depend on caterpillars or their adult moths in invaded and native transects—the parasitoids, as well as the assassin bugs, damsel bugs, predatory stink bugs, spiders, toads, and birds—the impact of plant invasions on interaction richness and diversity would have been shown to be many times larger.

CONSEQUENCES

My students and I did our study in unmanaged hedgerows, what passes for "natural" areas where I live. But would we have seen the same impact on insect populations if we had conducted the study in a typical suburban neighborhood? The answer would depend entirely on the percentage of the plant life in the study landscapes that was introduced. Unfortunately, in most urban/suburban, and even exurban landscapes, the majority of plants are from somewhere else.[24] My students and I have measured this in twenty-five-year-old suburban developments in Delaware, Northeast Maryland, and Southeast Pennsylvania. We didn't have to work too hard, because these landscapes contained very few plants at all. They were 92 percent lawn! But of the plants that were there, 79 percent (on average) were introduced not just from other parts of the North America but from other continents. Moreover, they were largely the same species we had studied in the invaded hedgerows: Callery pear, bush honeysuckle, privet, burning bush, Oriental bittersweet, barberry, and Norway maple. Unfortunately, homeowners are still legally allowed to landscape with invasive species throughout the country.

This should change.

For years I have speculated about the consequences of such landscaping choices for the birds with whom many of us would like to share our yards. I had to speculate because no one had directly measured what happens to bird populations in landscapes that favor introduced plants. I was pretty safe in my speculations because logic dictates that if you take away the food birds need, they won't do well. This, as the saying goes, is not rocket science. Nevertheless, I need speculate no longer. In the first study of its kind, my student Desiree Narango has measured what happens to Carolina chickadee populations and the caterpillars that support them when native plants are replaced by introduced ornamentals in suburban settings.[25]

For three years, Narango and a team of field assistants followed breeding chickadees in the suburbs of Washington, D.C., during the nesting season. Using video cameras at the nests, radio isotope analyses, territory mapping, vegetation analyses, and foraging observations, and employing citizen scientists,

FIGURE 12.4. As the percentage of introduced plants increases in suburban yards, the number of chicks fledged from chickadee nests decreases. (Graph courtesy of Doug Tallamy.)

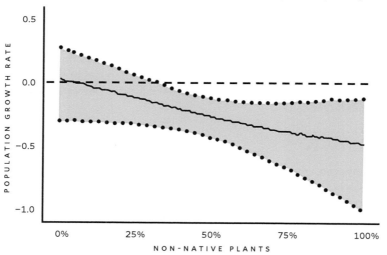

FIGURE 12.5. Although they depend on seeds much of the year, Carolina chickadees must find thousands of caterpillars to rear one clutch of young. (Photo courtesy of Doug Tallamy.)

Narango was able to quantify all of the variables required to model the population growth of chickadees as a function of the percentage of introduced plants within the chickadee's breeding territories.

Narango found out far more than I have space to describe, but here are some of the highlights of her research. Throughout her study, parent birds foraged for food on native plants 86 percent of the time. Compared to primarily native landscapes in her suburban study sites, yards dominated by introduced plants produced 75 percent less caterpillar biomass and were 60 percent less likely to have breeding chickadees at all. Apparently, chickadees were able to assess the quality of the landscape before they decided whether or not to set up house in one of Narango's chickadee boxes. If a chickadee did build a nest in a yard with many introduced plants, it contained 1.5 fewer eggs than nests in yards dominated by natives and those nests were 29 percent less likely to survive. Chickadees that nested where there were not enough caterpillars fed their young spiders and aphids but these food items did not compensate nutritionally for the lack of caterpillars in their chicks' diets. Nests in yards dominated by introduced plants produced 1.2 fewer chicks, and delayed chick maturation by 1.5 days compared to nests located in yards with lots of native plants.

Some of these differences may not sound very big to you, but cumulatively they are making a huge, negative impact on suburban chickadee populations. Chickadee populations achieved replacement rate (that is, produced enough chicks each year to replace the adults lost to old age and predation) only in yards with less than 30 percent introduced plants. Unfortunately for the chickadees in Washington, D.C., suburbs, Narango found that, on average, 56 percent of the plants in yards are introduced.[26]

Although these results are not a surprise, they remove the guesswork from understanding how much our plant choices impact the life around us. Here is solid evidence that, at least for Carolina chickadees, introduced plants are not the ecological equivalents of the native plants they replace. It is hard to imagine that other insectivorous birds would not be similarly affected by introduced plants. Narango's research helps us understand that it is the plants we have in our yards that make or break bird reproduction, and not the seeds and suet we so dutifully buy for our feathered friends, although those supplements certainly help our birds after they have successfully reproduced.[27] Her results also give us insight into what is happening beyond our yards in the natural areas that have been invaded by the ornamental plants we have brought from Asia and Europe. We can now better understand one of the factors that has caused 432 species of birds to decline at such a perilous rate.[28]

DO BIRDS CARE IF A BERRY IS NATIVE OR NOT?

These studies are not the only ones that show how seriously insect populations, and thus birds, are affected by introduced plants. Dozens of other labs have found the same effect over and over again. But what about berries? Do introduced plants make up for the loss of insects by producing lots of berries that sustain birds? Mark Davis, a botanist at Macalaster College, asked this interesting question in 2011 when the controversy over fighting invasive plants was at its peak. He concluded, quite logically I might add, that if birds readily eat berries from introduced plants, they must not care about the evolutionary origin of those berries.[29] As anyone who has witnessed a flock of cedar waxwings strip a European crabapple of its fruit in minutes can attest, many birds do eat berries produced by introduced plants. In fact, it is birds that make so many introduced plants highly invasive; they eat seed-laden fruit, fly some distance away, and defecate the seeds that will start another generation of those plants. So, if introduced plants are providing lots of good food for birds, maybe they aren't so bad after all?

The heart of this question focuses on the nutritional value of berries produced by different species of shrubs. We can assume that all berries are equal in what they deliver to birds, but assuming things is not science. Science, as Tara Trammell writes in her chapter in this volume, is all about hypothesis testing, and that is exactly what Susan Smith Pagano and her collaborators at the Rochester Institute of Technology have been up to lately. They tested the hypothesis that berries are nutritionally equivalent regardless of whether they are produced by native shrubs or introduced shrubs.[30]

Pagano focused on berries produced in the fall for two reasons: nearly all of our invasive shrubs produce their berries in the fall (I can't think of any that don't), and both migrating and overwintering birds depend on fall berries

FIGURE 12.6. Birds like this red-eyed vireo are the primary dispersers of seeds from within berries. (Photo courtesy of Doug Tallamy.)

FIGURE 12.7. In the fall, birds like downy woodpeckers need berries high in fat to help them make it through the winter or to give them the energy required for migration. Poison ivy berries are among the best in this regard. (Photo courtesy of Doug Tallamy.)

for the fats they need to either fuel their migration or build fat reserves for the long winter months if they don't migrate.

Pagano's work has revealed a surprising and distressing pattern. Berries from introduced Eurasian plants like autumn olive, glossy buckthorn, bush and Japanese honeysuckle, and multiflora rose contain very little fat, typically less than 1 percent, while berries from natives like Virginia creeper, wax myrtle, arrowwood viburnum, spicebush, poison ivy, and gray dogwood are loaded with valuable fat, often nearly 50 percent by weight. There is variation among species, but the pattern is clear: introduced berries are high in sugar at the time of year our birds need berries high in fats.

Why would a bird eat a berry that does not meet its nutritional needs? I can think of two reasons. First, when an invasive shrub moves into a habitat, it typically eliminates the native shrubs that used to grow at that site. Thus, the berries produced by the invasive shrub are now the only ones available for birds. With no other option, hungry birds eat them. I can suggest a second reason birds might eat nutritionally bereft berries by asking why you would eat a sugar-coated donut when you ought to be eating a calorie-dense piece of cheese. If we humans, with the most developed reasoning capacities in the animal kingdom, are unable to distinguish what is good for us and what is not, then why would we expect birds to be able to bypass a sugary treat? Despite this logic, new research suggests that birds *do*, in fact, "care" whether a berry is native or not and that they discriminate against introduced berries whenever they have the option. When fall migrants stop to rest and eat in a habitat loaded with invasive shrubs, they do not stay long. Instead they seek a habitat with plenty of the spicebush and arrowwood viburnum berries they need to fuel their migration.[31]

COSTS VS. BENEFITS

Every now and then a paper appears in the scientific literature that points out the ecological benefits delivered by introduced plants. The conclusion drawn by the authors is always the same: if introduced plants are doing good things in local ecosystems, perhaps we should tolerate, or maybe even encourage, their presence. This logic has been used to justify planting more eucalyptus in California (people love trees more than dry grasslands) and to discourage municipalities from spending money to fight invasive species almost everywhere. I would agree with this line of thinking on one condition: the net effect of the plant in question must be positive. Benefits cannot be viewed in ecological isolation; demonstrating the benefits that a plant delivers is meaningless unless we also measure the ecological costs that plant brings to the system. Only in this way can we estimate whether the benefits a plant provides outweigh the costs, or vice versa. If the net effect of an introduced plant improves ecosystem function, then yes indeed, let's rethink our bias against it.

Here is a typical scenario: kudzu, the Asian plant that now exclusively occupies over seven million acres in the Southeast U.S.,[39] serves as a host plant to our native silver spotted skipper. The skipper, a legume specialist, has found the defensive chemicals of kudzu to be within the range of its detoxification abilities, and so it can reach maturity on the nutritious leaves of this introduced species. We can put the new host association between the sliver spotted skipper and kudzu in the benefits column, as kudzu is creating a food web option that did not exist before its introduction. Sticking to the food web theme, though, we now have to consider whether kudzu is having any negative impacts. The answer, of course, is yes. When kudzu smothers young oak trees in an area of Camden County, Georgia, for example, the oaks—host options for 454 species of caterpillars—disappear. Similarly, if black cherry is eliminated from this kudzu patch, 324 caterpillar species will be lost. If willows, hickories, and maples are covered by kudzu, 247, 229, and 223 species of caterpillars will be lost, respectively.[33] Such losses will hold for all of the woody plant genera lost to kudzu at this single site in Georgia.

What about herbaceous plants? If kudzu smothers goldenrod, as it surely does, ninety-four species of caterpillars are then lost. If it covers native asters, another eighty species are lost. Sunflowers lost to kudzu host sixty-seven species of caterpillars, horsenettle sixty-seven more species, and so on. By allowing kudzu to invade an area in Camden County, Georgia, we have gained a host plant for silver spotted skipper, but lost host opportunities for the caterpillars of over a thousand moths and butterflies. And, as with all food webs, removing food from the base of the web reverberates throughout the entire web, impacting all of the species, especially birds, that eat the caterpillars lost to kudzu. No doubt, the net effect of this kudzu invasion is not just slightly negative, it is hugely negative in terms of supporting local biodiversity.

WHY NOT LET NATURE TAKE ITS COURSE?

Some might actually use our kudzu example as evidence that nature is, in fact, taking its course to bring introduced plants into functional relationships with their new ecosystems. After all, the silver spotted skipper has already adapted to kudzu (or, more likely, already possessed the necessary physiological adaptations required to eat kudzu when kudzu first arrived in Georgia). Isn't this evidence that nature is repairing itself without our help? In the long view, I suppose it may be, but unfortunately, the rate at which nature is repairing the damage we have inflicted is so incrementally slow compared to the rate at which we keep inflicting damage that true repair will not occur fast enough to prevent the loss of what we now know as nature.

Journalist Michael Pollan once asked whether there should be a statute of limitations on being "native." In other words, if a plant or animal has been in North America long enough, shouldn't we consider it ecologically equal to the organisms that evolved here? It's a good question, but it reflects how difficult it is for most people to comprehend the immense periods of time required to build evolutionary relationships. In my view, *native* is not a label a species earns after a given period of time. It is a term that describes function. For example, a plant should be considered a native when it acts like a native: that is, when it has achieved the same ecological productivity that it had in its evolutionary homeland; when it has accumulated the same number of specialized relationships that had been nurtured by the native plants it displaced; and when it has accumulated the same number of diseases, predators, and parasites that species which evolved in North America must endure. Time in residence is not the variable to be measured here, but the rate at which local organisms adapt to the plant's presence is.

The common reed, *Phragmites australis,* provides a great example of this. The European genotype that has displaced wetland vegetation from the Atlantic Coast to the shores of Lake Michigan has been in North America for hundreds of years. There is good evidence that it was used as packing material in the holds of the earliest sailing ships some five hundred years ago. In Europe, Phragmites supports 170 species of insects. After hundreds of years of residence in North America, only five insect species have started using Phragmites as a nutritional resource.[34] Adaptation is happening, but at a glacial rate typical of evolutionary change.

And Phragmites is not an exception. Very slow rates of adaptation have been recorded for a number of introduced plants. *Melaleuca quinqueneryia* has been in the Florida Everglades for over 130 years. In Australia, where it evolved, *Melaleuca* supports 409 species of insects, but in Florida, it only supports eight North American insects.[35] Only one species has adapted to *Eucalyptus* in California after 110 years,[36] and no species have adapted to the cactus *Opuntia ficus-indica* after 260 years.[37] In short, it takes enormous periods of time before introduced plants act like the natives they replace.

There are other reasons "letting nature take its course" after the introduction of hundreds of non-native plants is not a good idea. First, these types of introductions are anything but natural. We have perpetrated a biological exchange of species across the globe so rapidly—instantaneously on an evolutionary time scale—that it has created a phenomenon that ecosystems have never before encountered in the history of life on earth. There is no "natural" response to counter our meddling. Moreover, by moving introduced plants beyond the suite of natural enemies—the insects, mammals, and diseases—that keep them in check in their homelands, we have stacked the competitive deck against native plants that do have to contend with hundreds of herbivores and diseases. Expecting native species to successfully duke it out with introduced plants is ecologically unrealistic. And that is why natural succession from one type of plant community to another is essentially dead when invasive plants enter the scene. Most disturbed communities these days no longer progress from grassland to meadow to scrub to forest. Instead, they

FIGURE 12.8. White Clay Creek State Park in Newark has been so invaded by introduced plants that now in many places within the park more than one-third of vegetation is from Asia. (Photo courtesy of Doug Tallamy.)

become frozen in a perpetual tangle of invasive vines and shrubs. At least, that has been the case during the last thirty years. Will our native species prevail in the end? Some think they will, and I hope those optimists are right. But I wonder.

To those who claim it is a fool's errand to try to restore nature to some mythical pristine state, I say, "That's not the goal!" The natural world that existed before humans entered North America at least fourteen thousand years ago is gone, along with the Pleistocene-Epoch megafauna that shaped it. The modified natural world that Europeans encountered in the Americas in the fifteenth century is gone. Since then we have dramatically altered our watersheds, soils, forests, wetlands, and grasslands to meet first our agricultural needs and then our industrial needs. And we continue to change the land today. But none of these changes means we have to destroy (or can afford to destroy) ecosystem function. Wherever and whenever we can, we must repair the coevolved relations between plants and animals and among animals themselves that enable ecosystems to produce the life support systems we all need. Introduced plants only hinder our efforts to do so.

NOTES

1. Richard Hobbs, Eric Higgs, and Carol Hall, *Novel Ecosystems* (West Sussex, UK: Wiley-Blackwell, 2013).
2. William Stolzenburg, *Rat Island* (New York: Bloomsbury Publishing, 2011).
3. Peter Marra and Chris Santella, *Cat Wars* (Princeton, NJ: Princeton University Press, 2017).
4. Richard Duncan, Alison Boyer, and Tim Blackburn, "Magnitude and Variation of Prehistoric Bird Extinctions in the Pacific," *Proceedings of the National Academy of Sciences* 110, no. 16 (April 2013): 6436–41.
5. Hong Qian and Robert Ricklefs, "The Role of Exotic Species in Homogenizing the North American Flora," *Ecology Letters* 9, no. 12 (October 2006): 1293–98.
6. Kristin Powell, Jonathan Chase, and Tiffany Knight, "Invasive Plants Have Scale-Dependent Effects on Diversity by Altering Species-Area Relationships," *Science* 339, no. 6117 (January 2013): 316–18.
7. Matthew Collier, John Vankat, and Michael Hughes, "Diminished Plant Richness and Abundance below *Lonicera maackii*, an Invasive Shrub," *The American Midland Naturalist* 147, no. 1 (January 2002): 60–72; Kathleen Knight, et al., "Ecology and Ecosystem Impacts of Common Buckthorn (*Rhamnus cathartica*): A Review," *Biological Invasions* 9, no. 8 (December 2007): 925–37; Kristina Stinson, et al., "Impacts of Garlic Mustard Invasion on a Forest Understory Community," *Northeastern Naturalist* 14, no. 1 (March 2007): 73–89.
8. Paul Downey and David M. Richardson, "Alien Plant Invasions and Native Plant Extinctions: A Six-Threshold Framework," *AoB Plants* 8 (August 2016): plw047, https://doi.org/10.1093/aobpla/plw047.
9. David Zaya, et al., "Genetic Characterization of Hybridization between Native and Invasive Bittersweet Vines (*Celastrus* spp.)," *Biological Invasions* 17, no. 10 (October 2015): 2975–88; Kevin Burgess and Brian Husband, "Habitat Differentiation and the Ecological Costs of Hybridization: The Effects of Introduced Mulberry (*Morus alba*) on a Native Congener (*M. rubra*)," *Journal of Ecology* 94, no. 6 (July 2006): 1061–69.

10. Matthew Forister, et al., "The Global Distribution of Diet Breadth in Insect Herbivores," *Proceedings of the National Academy of Sciences* 112, no. 2 (December 2015): 442–47.
11. Michaela McGuire, "The Eucalypt Invasion of Portugal," *The Monthly* (June 2013), https://www.themonthly.com.au/issue/2013/june/1370181600/michaela-mcguire/eucalypt-invasion-portugal.
12. Douglas Tallamy, "Do Alien Plants Reduce Insect Biomass?" *Conservation Biology* 18, no. 6 (December 2004): 1689–92.
13. O. W. Richards and R. G. Davies, (eds.), *Imms' general textbook of entomology* (Dordrecht, Netherlands: Springer, 1977).
14. David Dussourd and Thomas Eisner, "Vein-Cutting Behavior: Insect Counterploy to the Latex Defense of Plants," *Science* 237, no. 4817 (August 1987): 898–901.
15. Lincoln Brower, et al., "Decline of Monarch Butterflies Overwintering in Mexico: Is the Migratory Phenomenon at Risk?" *Insect Conservation and Divers* 5, no. 2 (March 2012): 95–100.
16. Robert MacArthur, "Fluctuations of Animal Populations and a Measure of Community Stability," *Ecology* 36, no. 3 (July 1955): 533–36.
17. Paul Ehrlich and Anne Ehrlich, *Extinction* (New York: Random House, 1981).
18. Brian Walker, "Biodiversity and Ecological Redundancy," *Conservation Biology* 6, no. 1 (March 1992): 18–23.
19. Oswald Schmitz, Peter Hambäck, and Andrew Beckerman, "Trophic Cascades in Terrestrial Systems: A Review of the Effects of Carnivore Removals on Plants," *The American Naturalist* 155, no. 2 (February 2000): 141–53; José Rey Benayas, Adrian Newton, Anita Diaz, and James Bullock, "Enhancement of Biodiversity and Ecosystem Services by Ecological Restoration: A Meta-Analysis," *Science* 325, no. 5944 (August 2009): 1121–24; Peter Reich, et al., "Impacts of Biodiversity Loss Escalate through Time as Redundancy Fades," *Science* 336, no. 6081 (May 2012): 589–92.
20. Andrea Wulf, *The Invention of Nature* (New York: Knopf, 2015).
21. D. H. Janzen, "The Deflowering of Central America," *Natural History* 83, no. 4 (1974): 49–53.
22. Lee Dyer, et al., "Diversity of Interactions: A Metric for Studies of Biodiversity," *Biotropica* 42, no. 3 (May 2010): 281–89.
23. Melissa Richard, Douglas Tallamy, and Adam Mitchell, "Introduced Plants Reduce Species Interactions," *Biological Invasions* 21, no. 3 (March 2019): 983–92.
24. Michael McKinney, "Urbanization, Biodiversity, and Conservation: The Impacts of Urbanization on Native Species Are Poorly Studied, but Educating a Highly Urbanized Human Population about These Impacts Can Greatly Improve Species Conservation in All Ecosystems," *Bioscience* 52, no. 10 (October 2002): 883–90.
25. Desiree Narango, Douglas Tallamy, and Peter Marra, "Native Plants Improve Breeding and Foraging Habitat for an Insectivorous Bird," *Biological Conservation* 213 (September 2017): 42–50; Desiree Narango, Douglas Tallamy, and Peter Marra, "Nonnative Plants Reduce Population Growth of an Insectivorous Bird," *Proceedings of the National Academy of Sciences* 115, no. 45 (October 2018): 11549–54.
26. Ibid.
27. John Marzluff, *Welcome to Subirdia* (New Haven, CT: Yale University Press, 2014).
28. NABCI, "State of the Birds Report 2016" (Washington, DC: U.S. North American Bird Conservation Initiative, 2017), http://www.stateofthebirds.org/2016.
29. Mark Davis, et al., "Don't Judge Species on Their Origins," *Nature* 47 (June 2011): 153–54.
30. Susan Smith, Samantha DeSando, and Todd Pagano, "The Value of Native and Invasive Fruit-Bearing Shrubs for Migrating Songbirds," *Northeastern Naturalist* 20, no. 1 (March 2013): 171–85.

31. Susan Smith, et al., "Local Site Variation in Stopover Physiology of Migrating Songbirds near the South Shore of Lake Ontario is Linked to Fruit Availability and Quality," *Conservation Physiology* 3, no. 1 (August 2015), https://doi.org/10.1093/conphys/cov036; Yushi Oguchi, Robert Smith, and Jennifer Owen, "Fruits and Migrant Health: Consequences of Stopping over in Exotic- vs. Native-Dominated Shrublands on Immune and Antioxidant Status of Swainson's Thrushes and Gray Catbirds," *The Condor: Ornithological Applications* 119, no. 4 (October 2017): 800–16.

32. Irwin Forseth and Anne Innis, "Kudzu (*Pueraria montana*): History, Physiology, and Ecology Combine to Make a Major Ecosystem Threat," *Critical Reviews in Plant Sciences* 23, no. 5 (August 2004): 401–13.

33. "Native Plant Finder," National Wildlife Federation, 2015–2019, https://www.nwf.org/NativePlantFinder/Plants.

34. Lisa Tewksbury, et al., "Potential for Biological Control of *Phragmites australis* in North America," *Biological Control* 23, no. 2 (February 2002): 191–212.

35. Sheryl Costello, et al., "Arthropods Associated with Above-Ground Portions of the Invasive Tree, *Melaleuca quinquenervia*, in South Florida, USA," *Florida Entomologist* 86, no. 3 (September 2003): 300–23.

36. Donald Strong, John Lawton, and Sir R. Southwood, *Insects on Plants* (Cambridge, MA: Harvard University Press, 1984).

37. D. P. Annecke and V. C. Moran, "Critical Reviews of Biological Pest Control in South Africa 2. The Prickly Pear, *Opuntia ficus-indica* (L.) Miller," *Journal of the Entomological Society of Southern Africa* 41, no. 2 (1978): 161–88.

CHAPTER THIRTEEN

Environmental Justice

VICTOR W. PEREZ

INTRODUCTION

One of the most important (and long overdue) developments in recent environmental restoration work has been the addition of social, economic, and racial factors to the traditional ecological indicators of a region's environmental health. As a professor of sociology, I consider my most challenging and important work to be conducting local research and teaching about *environmental justice*. Many students at the University of Delaware have never heard the term before, let alone related terms like *environmental racism* or *environmental inequity*. Examining these concepts offers an excellent analytical framework to investigate communities that continue to bear a disproportionate level of exposure to environmental hazards.[1]

It is through this lens that we can begin to understand just how personal pollution can be for our own neighbors, especially for people with the least economic or political power. For example, in the classroom I use Geographic Information Systems (GIS)-based maps to show where *brownfields*, defined as "abandoned, idled, or underutilized industrial and commercial facilities where expansion or redevelopment is complicated by real or perceived environmental contamination,"[2] are in relation to the racial composition of a community. Often, brownfields are situated nearby and in areas of high concentrations of minoritized groups. This exercise very quickly and vividly shows students that this is rarely a coincidence. I further present factors that typically drive (or delay) the cleanup of some of these contaminated sites, noting the complexities of urban revitalization and redevelopment through such environmental cleanups and the market-driven economics that can often spur (or delay) those efforts. We ask questions like, "Which brownfields are being cleaned up first and why?"

My students and I also examine the frequent contrast between the self-reported health experiences of residents in environmentally degraded communities and official public health statistics, pondering how we explain this difference. I also bring undergraduates and graduate students into local communities to directly engage these issues and work with community members who volunteer their stories and their perspectives. Whenever community members share their lives with us, it is a profound act of vulnerability and empowerment on their part. Just as thrilling is when community members come into the classroom and open up a new world to students that only residents' voices can illuminate.

So, let us begin by defining three important terms: environmental justice, *environmental injustice*, and environmental racism. From there, we will examine a brief history of the environmental justice movement and then segue into several related features in the study of environmental justice, including: zoning; racial residential segregation; proximity to environmental hazards; the impact of environmental hazards on human health (and what I call "grey areas" of environmental justice); the role of academic-citizen/community partnerships in addressing community-level environmental justice issues and the complexities of community dynamics; and community efforts called citizen science and "popular epidemiology." We will then take a look at how climate change impacts promise to have a disparate impact on communities with already heightened social vulnerabilities and existing environmental hazards.

We will also look at some initiatives to address environmental injustice at the community level and their subsequent potential for enhancing human health and quality of life. As a rule, none of these initiatives do well by communities unless residents themselves are intimately involved from the very beginning in getting others to recognize their own environmental trauma *and* create equitable solutions. Well-intentioned policies to help clean up environmental hazards in environmentally degraded communities can also have the long-term (and in many cases unintended) effect of increasing property values and displacing current residents if things like affordable housing and rent control are not in place. Needless to say, the complexities and of the politics of environmental justice abound.

Throughout the chapter, I will reference environmental justice issues in Northern Delaware as examples. In recent years, Delaware communities impacted by environmental hazards—like air pollution, toxic dust, and brownfields—have gotten increased attention in the news media. Local and regional activists have coalesced to form groups dedicated to community organizing and the empowerment of residents impacted by these issues. This collective mobilization is being shaped by the myriad social, political, and economic dynamics within and across communities that share environmental hazards. The mobilization is also shaped by research and the multiple community narratives increasingly available to the public, as well as by the relationships between communities and local environmental and health regulatory agencies.

THE CONCEPTUAL TRIAD

Robert Bullard, arguably the most influential academic sociologist, notes that environmental justice is the "right to live and work and play in a clean environment."[3] The Environmental Protection Agency (EPA) has defined environmental justice as the "fair treatment and meaningful involvement of all people regardless of race, color, national origin, or income with respect to the development, implementation, and enforcement of environmental laws, regulations, and policies."[4] In essence, this "fair treatment" is meant to ensure that no social group bears a disproportionate health or environmental burden as a result of industry, or of municipal or commercial activity, either directly or indirectly.[5] This is important because this focuses on the social structural conditions that, historically, have positioned some communities to be at a greater risk for negative health impacts and exposure to environmental hazards from active industrial activity (like waste incineration), as well as the historical remnants of industry (like soil contamination and contaminated brownfields). These communities generally lack social, political, and economic power, and are often communities of minoritized groups. At its root, the emphasis on acknowledging social inequalities has taken a page from ongoing struggles over civil rights to prohibit discrimination and attempt to reassign responsibility for communities' toxic exposures to the federal government.[6]

Environmental injustice is a reality that many communities face on a day-to-day basis. It may stem from exposure to particulate matter in the air that exacerbates asthma, or to soil and water pollution from runoff of a toxic waste site, or from a landfill that might impact locally grown food or pollute groundwater. It may result in cumulative *body burdens* (the accumulation of toxic chemicals in one's body) that negatively impact the health of residents in these communities. It may be related to ill health and generational trauma, whereby families live decade after decade in the presence of smokestacks, or with the uncertainty of not knowing what may lie in the soil or air. In many cities around the world—including cities like Wilmington—this is the reality for many "fenceline" communities living near industrial sites.[7] A substantial body of research literature points to the disproportionate burden created by environmental hazards that poorer and racially minoritized communities contend with and their frequently negative cumulative health consequences.[8] For example, a recent report shows that "of the 73 [waste] incinerators remaining in the U.S., 79% are located in low-income communities and/or communities of color."[9]

This is not without scientific uncertainty or controversy. The responsibility of proving negative environmental health consequences is often laid not upon polluting industries but upon the residents of communities themselves. Though there are some historical cases where exposure to local chemicals has been clearly illustrated in body burden tests among residents,[10] usually concerns over environmental exposures involve aggregate health measures at the community level, coupled with data on such issues as air pollution or frequent flooding. It is difficult to quantify the direct, individual-level impacts of environmental hazards with these aggregate metrics, and this uncertainty can lead to two very different things: it can allow for inaction, or it can catalyze preventative action.[11] Emissions from waste incinerators, for example, may have an impact on local residents' health, but the uncertainty lurks in the difficult task of proving a direct link between exposure and negative human health outcomes. Most often, the focus of environmental injustice is on examining the health toll on communities disproportionately exposed to these hazards, but the state of epidemiology and the inherent uncertainties can play a powerful role in the ongoing debate around exposure and human health. Often, this scenario plays out in long, incremental stages of activism, corporate and political resistance, and ongoing campaigns for change.

Environmental racism is arguably the most powerful conceptual lens for examining which communities tend to face the disproportionate impact of environmental hazards. Though both social class and race play important, interacting roles in determining the hazard exposure of a community, many studies show that a concentration of racially minoritized people in a community is the best predictor of that community's proximity to and experience with environmental hazards. For example, hazardous waste facilities are significantly more likely to be located in Black, Latinx, and poor communities than in white communities.[12] For my students, this brings to mind many questions, including: "Why don't people just leave?" or "Were they targeted because of their race, or because the property values were generally cheaper?" Both of these questions allude to the historical importance of racial segregation, isolation, and lack of resources, and to the legacy of racially segregated communities in the United States in creating these outcomes. Recent research shows that *redlining*, the longtime practice of imposing racially segregated restrictions on access to housing and homeownership in American communities, still has a lingering impact today on economic disempowerment and isolation. Redlining involved coding certain areas in cities on a map with the color red to indicate that it was "hazardous" (i.e., financially risky) for banks to give mortgages to people living there. These areas were also typically communities with heavy concentrations of minorities.

Put simply, bank lending practices were often designed to prevent African Americans from purchasing homes in many city areas. This not only included traditionally affluent or white neighborhoods, but even those designated as "hazardous"—communities where people of color often already lived. This prevented residents from investing in these neighborhoods, created a gap in homeownership and long-term wealth accumulation, and perpetuated structures of economic distress and racial segregation.[13] These practices were in place for decades in cities all over the United States, and their legacy still dramatically compromises the upward economic mobility—to say nothing of the health—of many of our citizens. A significant number of the communities redlined in the 1930s are still economically struggling today.[14]

A BRIEF HISTORY OF ENVIRONMENTAL JUSTICE IN THE UNITED STATES

So, how did we get here? I provide here some history of the modern environmental justice movement and background on the framing of these issues.[15]

Scholars tend to point to a handful of events and research reports, dating back to about 1980, that set the tone and character for the contemporary environmental justice movement, particularly as it has come to emphasize racial and economic inequities. But it is important to note that other forms of environmental awareness, environmental social movements, and environmental justice initiatives have also emerged over a longer period of time. Community organizing to protest environmentally hazardous threats in the United States has included the efforts of Cesar Chavez to protect Latinx farmers from harmful agricultural chemicals and abusive labor practices in the 1960s; of African American students protesting a local garbage dump in Houston in 1967; and of West Harlem residents' opposition to a sewage treatment plant in 1968, among others.[16]

In the late 1970s and early 1980s, Warren County, North Carolina, was the site of a proposed landfill that locals argued would be situated dangerously close to residences and potentially release polychlorinated biphenyl (PCB), a known human carcinogen. This development caught the eye of national media when locals, activists, and a U.S. congressman protested the movement of PCB-contaminated soil into the landfill, while some literally tried to block the trucks of soil from coming into the area.[17] Because Warren County was 60 percent African American in a state where African Americans made up only 22 percent of the population, and because there was sufficient media coverage of the controversy, it helped to fashion the idea of environmental racism in the minds of the public.[18] The Warren County site was the subject of numerous lawsuits, including one by the National Association for the Advancement of Colored People (NAACP) that further made the case for environmental racism when it argued that "the driving force behind the decision to construct the landfill near Afton [community in Warren County] was the fact that the community was predominantly Black, rural, and poor."[19]

The Warren County episode helped to usher in a handful of studies that set the landscape for illustrating the issues of environmental racism and environmental inequity for decades to come.[20] Studies by the General Accounting Office in 1983, by Robert Bullard in 1983, and by the United Church of Christ in 1987 all pointed to substantial evidence that communities of color disproportionately bore the burden of exposure to environmental hazards. Much of this research was predicated on the way that environmental injustices were being examined, such as the process of siting environmentally hazardous waste in majority Black communities.[21] For example, Bullard's 1983 study showed that "the siting pattern of waste dumps in Houston ... 'were not randomly scattered over the Houston Landscape.'"[22] The study *Toxic Waste and Race in the United States*, published by the United Church of Christ in 1987, "documented that three of every five blacks live in communities with abandoned toxic waste sites."[23] This was soon followed by the seminal publication by Robert Bullard in 1990, *Dumping in Dixie: Race, Social Class, and Environmental Quality*, showcasing a pattern of environmental racism throughout the South.[24]

More recent research efforts not only focus on community exposure to hazards through active and former industry, but also on the toxicity of consumer products in convenience stores that are disproportionately located in poor and minoritized communities. Activists continue to illuminate the many forms of injustice, framing this toxic exposure as a justice issue because these stores—often the only ones available for residents—create a cycle of "subprime groceries"[25] and sell products containing toxics harmful to human health.[26]

Ultimately, the environmental justice movement has articulated five basic principles for addressing social injustices and pursuing social change, including:[27]

1. The right of all to be protected from environmental harms and degradation
2. A precautionary public health approach that focuses on prevention
3. Shifting the responsibility for proving harm away from affected communities and onto polluters or those who fail to protect citizens
4. Using disparate impacts of environmental hazards, as opposed to the intent of polluters, as evidence of discrimination
5. Taking direct action to remedy disproportionate environmental risks and burdens

The term *environmental justice community* is used fairly often to identify specific communities that have some form of disproportionate exposure to environmental hazards. That said, the social dimensions of interest, which generally frame the analytical lens on a community's vulnerability to environmental hazards, are many. For example, the social dimensions may include race, ethnicity, class, income, deprivation, gender, age, indigenous status, disability, as well as considerations involving future generations, such as single-parent households.[28] Further, the types of environmental hazards that a community may encounter include air pollution, accidental hazardous releases, waste landfills, waste incinerators, contaminated land, brownfield land, lead in paint and pipes, flooding, noise, drinking water quality, the hog industry, oil drilling and extraction, access to healthy food, climate change, emissions trading, and land reform, to name just several.[29]

When it comes to siting hazardous facilities and zoning, industry may target communities that cannot effectively mount a resistance to their local development due to a lack of social, political, and economic power.[30] Such a dynamic has been happening recently in South Wilmington, where there has been local debate about the siting of a slag grinding facility in the community of Southbridge. Though slated for an area adjacent to the community already zoned for manufacturing, some local residents saw that, by building and operating next to their already socially, economically, and environmentally vulnerable community, this facility would add to the cumulative health burdens of residents.[31]

So, how do we define an "environmental justice community"? If we base the definition on the social indicators presented above, it would typically include communities of persons of color and the poor that are exposed to a higher level of environmental hazards than those in neighboring communities. This is particularly true when you consider that some forms of pollution and environmental hazards are more toxic than others and carry varying degrees of exposure potential.[32]

RACIAL RESIDENTIAL SEGREGATION AND ZONING

It is worth introducing the role of racial residential segregation and zoning practices that have contributed to some of the disparate exposures to environmental hazards that minoritized communities have endured. Studies find that residential segregation by race and environmental hazard exposure are related, and that increased incomes do not necessarily protect African Americans and Latinx persons from segregation, nor do they protect African Americans from pollution.[33]

Zoning policies (such as redlining) also have a historical influence on why minority communities face disproportionate environmental hazard exposure,

FIGURE 13.1. An example of Route 9 zoning, New Castle County. (Google Earth.)

because zoning is what dictates how land can be used in an area. If zoning permits residential and industrial land use in close proximity to each other, then communities adjacent to industry may be impacted by whatever environmental hazards and pollution that industry (or industries) may produce. This type of zoning has been referred to as "intensive," and some scholars point to how poor and racially minoritized communities have less power to forestall this type of land use planning. In communities where zoning permits residential, commercial, and industrial land use, community members who can afford to leave might go off in search of other areas.[34]

A local example of intensive zoning can be seen in Figure 13.1. This is the Route 9 corridor in New Castle County (i.e., New Castle Avenue), toward the northern tip of the corridor, before entering the city limits of Wilmington. Here, you can see the close proximity between residential and industrial zoning, positioning some communities directly next to different types of industrial activity and land use. This type of zoning can impact the quality of life of residents, and may also be a contributor to some health issues that communities face. Recently, there have been several news stories about the history of mobilization by two communities on the northernmost point of Route 9, Eden Park and Hamilton Park, to address these issues, including complaints about fugitive dust and other air quality problems.[35]

HEALTH IMPACTS OF RESIDENTIAL SEGREGATION AND CHRONIC, CUMULATIVE EXPOSURE

Arguably, the driving focus of the modern environmental justice movement has been human health impacts on local communities. This is also, in my view, the most contentious arena for debate because of the complexities of illustrating how different types of pollution directly impact human health. This contentiousness builds, in part, from the stark contrasts between health experiences reported by many residents in these communities and official public health statistics. In other words, public health statistics often do not support the negative health-related claims that residents make, nor do they support a direct connection to local pollution and/or contaminants. This is often related to the methodological complexities for public health professionals in measuring toxics and pollution, as well as health outcomes. Further, many environmental regulations and thresholds are based on tests of a single pollutant at a time, and not cumulative tests of multiple pollutants that might be present in an area. Thus, the potential cumulative impact is lost when tests are done in isolation to determine a pollutant's toll on the health of residents in adjacent communities. Some scholars have noted this as a limitation in connecting environmental hazard exposure and poor health outcomes, and state that: "Activists have pushed environmental health researchers and regulatory authorities to move beyond assessments focused on single chemicals or facilities and toward a cumulative-exposure approach that takes into account where affected populations live, work, and play in order to elucidate how race and class discrimination increase community susceptibility to environmental pollutants."[36]

Thus, when discussing disparities in health that may be attributed to local environmental hazards, it is important to start with differences in air, soil, and water quality. These differences are partly created by a community's proximity to land use and industry that can produce toxic emissions, such as an oil refinery, high volumes of truck and automobile traffic, as well as waste incinerators or landfills. These types of land uses, research shows, are disproportionately located in communities of racially minoritized groups and poorer communities.[37] Consider the additional factors that may contribute to chronic exposure and poor health, such as the role of the larger social space. The broader social context may also include poor-quality housing, lack of access to green space, lack of access to nutritious food, and other social challenges. It is important to note, too, that children are highly susceptible to environmental conditions that may contribute to poor air quality. These conditions, taken together, can cause things like childhood asthma, one of the more robust measures of a cumulative impact of local environmental hazards, lower-quality housing, and mold. Overall, research suggests that residents of communities that are disproportionately located near environmental toxics experience an increased risk of adverse perinatal outcomes, respiratory and heart diseases, psychosocial stress, and mental health impacts. Members of racial or ethnic minority groups and people of low socioeconomic status are also more likely than others to live near busy roads, where traffic-related air pollutants concentrate. Research has linked a wide array of adverse health outcomes to residential proximity to traffic, including asthma, low birth weight, cardiovascular disease, and premature mortality.[38]

People living in communities zoned with industry and residents situated side-by-side unequivocally face daily quality of life issues and very likely experience cumulative health problems as a result of their proximity. The trouble is, traditional epidemiological assessments that rely on population size, differential types of measurements of toxics, and documented health outcomes to produce risk estimations may not be able to represent a community's experiences. I call these places "grey areas" of environmental justice, or communities where "isolating the effects of environmental factors is very difficult as the populations that are exposed are also affected by a myriad of other suboptimal conditions, e.g., poor housing, poor schools, lack of access to health care, insufficient nutritious food, lack of outdoor recreational opportunities, neighborhood crime, psychological stressors, and others."[39]

There are numerous indigenous, civic, and environmental activists in New Castle County, representing various communities and corridors. Recently, in attempting to raise awareness of environmental conditions for residents of some of the communities along the Route 9 corridor, the news media has showcased some of the activism in response to local industry over potentially hazardous toxic releases. Some environmental justice activists in New Castle County have become more organized and more prominently featured in local news media.[40] A recent report[41] co-sponsored by some of these activists suggests that local environmental hazards increase the risk of cancer and respiratory illness for several socially disadvantaged communities along the Route 9 corridor, relative to the community of Greenville. Though the report's methods and the conclusions that the average reader could extrapolate from them are debated by local health officials,[42] the report nonetheless can serve a rhetorical role for promoting a narrative of environmental inequity in the area. Reports like these, though sometimes controversial, illustrate one of the most challenging impasses in the environmental justice arena: debate over the causality between environmental hazard exposure and negative health impacts.

CITIZEN SCIENCE ALLIANCES AND POPULAR EPIDEMIOLOGY

Gathering and articulating the health experiences of residents in environmental justice communities are the first steps in mobilizing people to address their concerns. It may be that numerous community members have similar experiences with cancer, or that a seemingly inordinate number of children in the area have asthma. It is when residents begin to share these collective experiences that their interpretations of the relationship between local environmental hazards and health can crystalize into something more substantial, catalyzing a movement to address their concerns. This is what sociologist Phil Brown and colleagues call an "embodied health movement" (EHM), a social movement that emphasizes "the embodied experience of illness—that is, people's everyday lived experience ... [embodied health movements] identify the collective experience of health and illness as their core—a source of solidarity, motivation, and urgency. At the same time, EHMs link the personal experience of illness with the collective experience and with the institutional and political-economy structures that can cause disease as well as treat it."[43] The experiences of some residents with the environmental conditions in Eden Park, New Castle County, can be understood using this idea, as some residents have health concerns as a result of living near the industry in the area.[44] Awareness of such experiences has led the state Department of Natural Resources and Environmental Control (DNREC) to conduct further air quality monitoring and testing.[45]

The issue of *fugitive dust* is not new in the area—locals have long discussed how dust covers cars and homes—and it has recently has been measured via a Moveable Monitoring Platform by DNREC to assess when the dust exceeds legal thresholds for quality of life and health protection. But what is particularly interesting is DNREC's recent efforts to see where the dust is coming from and what it might be made of. If a clear source for contaminants could be found, then cleanup and regulation, presumably, would be far easier to pursue. This example shows how gathering stories about the compromised quality of life and the lived experience of community members can attract greater involvement of regulatory agencies and deeper scientific testing, prompting the generation of more scientific questions that might not have been asked without the voices of the residents. Indeed, residents telling their stories help show not only the differences between their lived experiences and the data that scientific studies of environmental conditions capture, but also how those stories and data can complement one another and be stronger in combination with attention to the limitations and value in both.

The next steps that community members may take in addressing their health concerns over local environmental conditions is part of a process called "popular epidemiology." In short, *popular epidemiology* is a form of citizen-science alliance, where local citizens in an environmentally hazardous area use their own expert knowledge of the area, combined with scientific tools, to measure things like air, soil, and water quality, performing their own environmental testing and health assessments in consultation with scholars and scientists. It is "the process by which laypersons gather scientific data and other information, and also direct and marshal the knowledge and resources of experts in order to understand the epidemiology of disease."[46]

Let's take a look at a prominent example that may be understood more clearly through the lens of popular epidemiology: air quality testing done by a group called the Delaware City Environmental Coalition in Delaware City, a community in New Castle County located next to the Delaware City Refinery. Those involved included at least one local resident at the time and a local chapter of the Sierra Club. They collected samples of air quality to compare them before and after the reopening of the refinery and had them reviewed by an outside scientific expert, ultimately creating a report on their findings and incorporating it into their efforts to address air quality concerns. Their report suggested the presence of some high levels of various toxics in the air that were capable of compromising human health, and showed along Clinton Street, a major corridor in town downwind of the oil refinery, the highest levels of air quality issues.[47] Though some other possible sources of air pollution were noted, such as nearby ships at the Delaware City Port, this research nonetheless serves as an example of the way that local Delaware citizens have used scientific assessments in their efforts to document and address some shared environmental and health concerns.

Popular epidemiology is one way that concerned local residents can coalesce into an organized social movement and push for change, such as stronger environmental regulations and validation of their concerns. This form of resident-scientist activism pushes back against the dominant, traditional way that science can often overlook (or refuse to validate) the health concerns of people in potentially environmentally hazardous communities.

As Tara Trammell points out elsewhere in this volume, citizen science can take many forms, including local citizens simply gathering data on rainfall or high tides, but I argue that it has its most profound implications when residents of communities facing environmental threats take up scientific methods to measure and document their shared experiences. It is within this context that citizen science serves to realign the agency of residents to dictate the future of their own health and the health of their community and of future generations.

The final sections of this chapter will begin with a few observations about larger environmental issues, such as climate change, and how they may interact with local environmental justice issues in Northern Delaware. Additionally, the role of academic-community partnerships and community-based participatory research will be mentioned. Further, the role of citizens who are not residents of environmental justice communities, but who want to assist these communities, will be briefly explored. Lastly, I will illustrate a handful of initiatives that can benefit communities dealing with local industrial pollutants, like community benefits agreements.

CLIMATE CHANGE AND ENVIRONMENTAL JUSTICE

Climate change impacts will exacerbate current environmental and health vulnerabilities.[48] Communities will face a dual impact, with increased storm surges, increased heat, and sea level rise, which will almost certainly exacerbate existing environmental justice challenges. This is especially true of coastal areas, where sea level rise may contribute to floods during high tides and storm surges.[49] Take a look at Figure 13.2.

This image shows the area of part of the East Side of Wilmington and South Wilmington, with current mean high tide illustrated. Green and blue indicators show existing water entities, like the wetland in South Wilmington (green) and the Christina River (blue). This is what water in the area looks like with the average high tide. Now take a look at Figure 13.3. This shows the mean high tide with a sea level rise scenario of 1.5 meters by the year 2100. As you can see, large parts of communities, all things left the same and using this "bathtub model" of sea level rise predictions, would be inundated with water.[50]

Southbridge is a socially vulnerable community that deals with flooding and has some contaminated properties and brownfields, which may be impacted by repeated water inundation.[51] The community has experienced flooding for years, and residents have been vocal about its impact on quality of life, housing, and community vitality overall. It is also bracing for sea level rise impacts and working with the local government to try and address the threat.[52] One promising initiative to help control flooding is the revitalization of an existing wetland in the community to hold excess stormwater, and to create separate systems for stormwater and sewage to prevent the two from mixing during flooding events.[53] This wetland will become a city park that allows residents to walk and bike across it, helping to connect Southbridge to the western part of South Wilmington.

FIGURE 13.2. Christina River current mean high tide.
(Courtesy of the National Oceanic and Atmospheric Administration.)

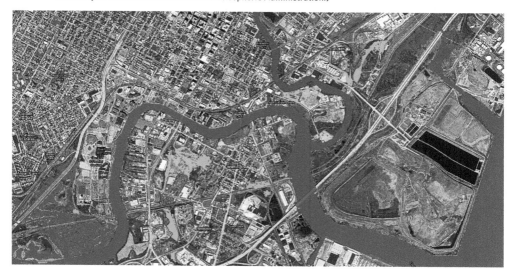

FIGURE 13.3. Christina River mean high tide with a 1.5 meter increase in sea level rise by 2100. (Courtesy of the National Oceanic and Atmospheric Administration.)

The City of Wilmington is planning to help fortify the area with resilience measures through building dikes, creating a floodable levee or waterfront park, and elevating or floodproofing structures in the area, and these measures will likely have a tremendous impact on the quality of life and the vitality of the community and of the nearby Riverfront area.[54] However, it remains unclear how these efforts will enhance resiliency over the long term.[55] Short-term and long-term adaptation to flooding and sea level rise predictions should be in alignment, but if adaption in the short term does not help to mitigate these very real long-term problems, at some point in the future the area will likely be dealing with significant sea level rise complications.

ACADEMIC-COMMUNITY PARTNERSHIPS

If you're reading this chapter and you live in an environmental justice community, you may very well know the complexities we've examined. However, for those of you who might want to get involved and engage residents, advocates, government, and academia in these issues, I hope the framework provided here is a useful guide to understanding the complexities. Also, I think it is important to take note of a few pointers I have picked up along the way.

First, communities are nuanced, intricate places, with a variety of stakeholders, voices, and perspectives on environmental justice issues. I have never worked with a community where everyone shared the exact same view on their local environmental conditions and what the consequences of those conditions were (e.g., health, housing value, quality of life, etc.). It is crucial to get to know as wide a variety of community members as possible in your efforts to engage the residents. Remember, folks inviting you in need to trust you, and they also want to be sure you are there to help them, in whatever ways *they think* are appropriate. Residents in environmental justice communities are deeply embedded in local structures of power, the issues they deal with are often extraordinarily political, and people are made vulnerable through these structures and issues. A good place to begin engaging is a civic association meeting, where you can clearly identify yourself and why you are there, and then you can start to get to know people. Keep in mind, though, you are an outsider, so take care in your involvement, knowing that the issues you are addressing are very complex within and across communities and that your presence can and will alter the dynamic in discussions of those issues.

If you are welcomed at an initial civic association meeting and feel welcome to attend future meetings, coming and simply (initially) recognizing and giving validation to people's experiences will be a crucial aspect of practicing justice through recognition of community trauma. Civic meetings often have very engaged residents, and this is often the most publicly facing group for outsiders, but realize that a civic association may not reflect or represent all of the voices in a community and that some of the dominant narratives at these meetings about local issues may not be agreed on by others. Over time, it is your obligation to try and meet with as many people as possible and discuss their views on the matters that their community is dealing with. During this time, too, you will likely begin to meet and work with the more visible environmental justice advocates and community leaders that represent these communities.

Second, in Delaware (especially along the Route 9 corridor), there are numerous communities dealing with overlapping issues, that connect them all, yet they all have identities, actors, and histories uniquely their own. This can create some difficulties in addressing environmental concerns, as intra- and inter-community dynamics can play out in conflicts about the best way to proceed, even after a consensus on some issue has been established (which can be very difficult). There are also moments of coordination, cooperation, and agreement, where folks are collectively moving forward on environmental concerns. I say these things with the utmost respect for the variety of individuals and communities in the area, knowing full well that my observations are framed from a more academic perspective; my experiences are extremely limited relative to the generations of people having lived in these areas, and I am a (well-intentioned) outsider.

Third, this is a long-term endeavor. As an academic, I have had to balance my presence in the community between the roles of listener, objective observer and researcher, advocate, friend, and educator, taking years in the making, and I still feel like I have a long journey to go. This blurring of boundaries in my roles always carries a need for what academics call "reflexivity," or thinking deeply about the impact of my presence in the community, and it should be no different for anyone involving themselves in an environmental justice community's effort to address their concerns. As an academic, one particularly effective approach to working with residents is *community-based participatory research* (CBPR), which, ideally, integrates community members and their perspectives on the issues being studied from the very beginning of an academic research project. This ensures a mutual, equitable approach to research and a dual ownership of the process and outcomes. Though they are very difficult to execute and take years, these types of projects have considerable value.

SOME WORKING ENVIRONMENTAL JUSTICE SOLUTIONS AND DIRECTIONS FOR THE FUTURE

Recent news coverage of Southbridge residents' negotiations over a proposed slag facility in their community is a good way to illustrate one initiative that can help an environmental justice community: a *community benefits agreement* (CBA).[56] If a community is zoned for both residential and industrial land uses, then one way to create a more equitable relationship between an industrial corporation and the community is a CBA, which lays out different ways the corporation can give back to the community, such as through jobs, strict environmental controls and health monitoring, and other forms of community investment and amenities. Due to the historical precedent zoning may have established in such communities, a CBA can create a somewhat more harmonious relationship between residents and their industrial neighbors. Sometimes, in communities like Delaware City, the presence of industry is a daily illustration of the balance many residents have to strike between their health and the economic support and jobs the local industry may provide.[57]

Additionally, several urban landscapes along the mid-Atlantic have undertaken greening initiatives to put older (often vacant and contaminated), former industrial properties to use as green spaces, spaces for development, and mixed commercial/residential spaces. Once an old brownfield from an industrial property is cleaned to get rid of its environmental hazards (or those hazards are at least made less severe), mixed-use development or green space could be created for the community to benefit from the land's reuse. The wetland revitalization initiative in South Wilmington is one such example of a contaminated property that will be converted into a city park and stormwater repository,[58] which has the potential to bring positive health and economic impacts to the community of Southbridge. With these initiatives, though, it is also important to ensure affordable housing and other social policies are in place to avoid displacement of the current residents.[59]

There are myriad other ways that environmental justice communities are working with advocates, government, and others to address their concerns, but a full discussion of those initiatives goes beyond the scope of this chapter. Additionally, there are communities in Delaware that I have not focused on in this discussion, including those dealing with groundwater[60] and drinking water[61] contamination issues, while the Coastal Zone Act also remains a subject of debate.[62]

It is worth highlighting here, still, some recent events regarding local Delaware industry that illustrate how environmental advocates stand for many communities to address environmental issues. They pressure industries for stronger environmental protections, more transparency about any potentially harmful environmental side effects of emissions, and communication with and investment in the community. This is particularly the case when an industry has an unexpected, potentially hazardous event, such as a recent flammable gas leak at a manufacturing site in New Castle.[63]

In my view, as we think about a hoped-for future—one of equity, community vitality, and universal protections from environmental harms—we need to take ownership of the fact that much of our environmental injustice stems from the past, in zoning, segregation, discrimination, and industrial growth. We need to continue to work on the methodological and analytical impasses at the intersection of environmental inequality and disparities in human health outcomes by recognizing, acknowledging, and focusing on lessening residential segregation and community isolation, as well as the adversities they create. Further, we need to focus on promoting an enhanced quality of life and health for everyone in the state of Delaware, especially in communities where environmental cleanup will determine the future of those areas. It is only then that we can try and undo the decades of environmental inequity that are rooted in social disparities and geography, and create plans for the future that promote healthy environments in which all groups can live, work, and play.

NOTES

1. This chapter focuses on Northern Delaware, and more specifically on certain communities in New Castle County and in the City of Wilmington. That said, I recognize that many communities in Southern Delaware are dealing with their own environmental struggles. These communities may not always share the same demographic makeup of more urban areas like Northern New Castle County and Wilmington. For an example, see how local groundwater has been impacted by spraying treated wastewater on local fields in a story by Mark Eichmann for WHYY, "Delaware Accuses Chicken Giant of Polluting Water," *WHYY*, November 10, 2017, https://whyy.org/articles/delaware-accuses-chicken-giant-polluting-water/, accessed May 29, 2019.
2. Kristen R. Yount, "The Organizational Contexts of Decisions to Invest in Environmentally Risky Urban Properties," *Journal of Economic Issues*, vol. 31, no. 2 (1997): 367. Quoted in Sangyun Lee and Paul Mohai, "Environmental Justice Implications of Brownfield Redevelopment in the United States," *Society and Natural Resources*, vol. 25, no. 6 (2012): 602.
3. Oliver Milman, "Interview with Robert Bullard: 'Environmental Justice Isn't Just Slang, It's Real,'" *The Guardian*, December 20, 2018, https://www.theguardian.com/commentisfree/2018/dec/20/robert-bullard-interview-environmental-justice-civil-rights-movement.
4. Environmental Protection Agency, quoted in Robert J. Brulle and David N. Pellow, "Environmental Justice: Human Health and Environmental Inequalities," *Annual Review of Public Health* 27 (2006): 104.
5. Brulle and Pellow, "Environmental Justice: Human Health and Environmental Inequalities," 104.
6. Robert D. Bullard, Glenn S. Johnson, Denae W. King, and Angel O. Torres, *Environmental Justice: Milestones and Accomplishments: 1964–2014* (Houston: Barbara Jordan-Mickey Leland School of Public Affairs, Texas Southern University, 2014).
7. Steve Lerner, *Sacrifice Zones: The Front Lines of Toxic Chemical Exposure in the United States* (Cambridge, MA: MIT Press, 2010).
8. Rachel Morello-Frosch, Miriam Zuk, Michael Jerrett, Bhavna Shamasunder, and Amy D. Kyle, "Understanding the Cumulative Impacts of Inequalities in Environmental Health: Implications for Policy," *Health Affairs* 30, no. 5 (2011): 879–87.
9. Rina Li, "Nearly 80% of US Incinerators Located in Marginalized Communities, Report Reveals," *Wastedive*, May 23, 2019, https://www.wastedive.com/news/majority-of-us-incinerators-located-in-marginalized-communities-report-r/555375/.
10. See a discussion of Triana, Alabama, and resident exposure to DDT, as well as cases of Warren County, North Carolina, and PCB exposure, and Louisiana's "cancer alley" in Dorceta E. Taylor, *Toxic Communities: Environmental Racism, Industrial Pollution, and Residential Mobility* (New York: New York University Press, 2014).
11. The *precautionary principle* is an idea that when dealing with the uncertainty of human harm from environmental exposures, caution should be exercised to avoid the exposure to begin with. This is one strategy for performing thorough environmental cleanups when

placing new schools on brownfield sites, for example. On the other hand, uncertainty in the human health consequences of environmental exposures can also lend itself to "business as usual" and continued operations that may expose residents to harmful toxics.

12. Zoe Schlanger, "Race is the Biggest Indicator in the US of Whether You Live Near Toxic Waste," *Quartz*, March 22, 2017, https://qz.com/939612/race-is-the-biggest-indicator-in-the-us-of-whether-you-live-near-toxic-waste/.

13. Jesse Meisenhelter, "How 1930s Discrimination Shaped Inequality in Today's Cities," *National Community Reinvestment Coalition*, March 27, 2018, https://ncrc.org/how-1930s-discrimination-shaped-inequality-in-todays-cities/, accessed June 12, 2019.

14. Bruce Mitchell, "HOLC 'Redlining' Maps: The Persistent Structure of Segregation and Economic Inequality," *National Community Reinvestment Coalition (NCRC)*, March 20, 2018, https://ncrc.org/holc/, accessed June 7, 2019.

15. For an extensive documentation of the environmental justice movement, including a timeline of major events, research reports, and a selected bibliography, see Bullard, Johnson, King, and Torres, *Environmental Justice*.

16. Brian Palmer, "The History of Environmental Justice in Five Minutes," *NRDC*, May 18, 2016, https://www.nrdc.org/stories/history-environmental-justice-five-minutes, accessed June 11, 2019.

17. Matt Reimann, "The EPA Chose This County for a Toxic Dump because its Residents Were 'Few, Black, and Poor,'" *Timeline*, April 2, 2017, https://timeline.com/warren-county-dumping-race-4d8fe8de06cb.

18. Palmer, "The History of Environmental Justice in Five Minutes"; Taylor, *Toxic Communities*, 17.

19. Ibid.

20. Paul Mohai, David Pellow, and J. Timmons Roberts, "Environmental Justice," *Annual Review of Environment and Resources*, vol. 34 (2009): 405–30.

21. Taylor, *Toxic Communities*.

22. Robert D. Bullard, "Solid Waste Sites and the Black Houston Community," *Sociological Inquiry*, vol. 53, no. 2–3 (1983): 273. Quoted in Taylor, *Toxic Communities*, 35.

23. Robert D. Bullard, "Environmental Justice for All," http://www.uky.edu/~tmute2/GEI-Web/password-protect/GEI-readings/BullardEnvironmental%20justice%20for%20all.pdf, accessed June 11, 2019.

24. Robert D. Bullard, *Dumping in Dixie: Race, Class, and Environmental Quality* (Boulder, CO: Westview, 1990).

25. Tanvi Misra, "The Dollar Store Backlash Has Begun," *CityLab*, December 20, 2018, https://www.citylab.com/equity/2018/12/closest-grocery-store-to-me-dollar-store-food-desert-bargain/577777/, accessed June 11, 2019.

26. Alessandra Bergamin, "Dollar Stores Moving to Pull Dangerous Plastics from Shelves," *National Geographic*, May 24, 2019, https://www.nationalgeographic.com/environment/2019/05/dollar-stores-shifting-sales-potentially-dangerous-plastics-phthalates-bpa/.

27. Brulle and Pellow, "Environmental Justice: Human Health and Environmental Inequalities," 110; Robert D. Bullard and William J. Clinton, "Overcoming Racism in Environmental Decision Making," *Environment* 36, no. 4 (1994): 10–44.

28. Gordon Walker, *Environmental Justice: Concepts, Evidence and Politics* (New York: Routledge, 2012), 2.

29. Ibid.

30. Mohai, Pellow, and Roberts, "Environmental Justice," 414.

31. Jeanna Kuang, "Southbridge Tired of Being Treated as Dumping Ground: 'We've Had Enough,' Residents Say as They Seek to Block Industrial Facility," *The News Journal*

(Wilmington, DE), November 8, 2018: A4; Jeanna Kuang, "Facility 'Going to Happen,' Some Say: Southbridge Residents Want to Work with Firm," *The News Journal* (Wilmington, DE), December 19, 2018: A4.

32. Spencer Banzhaf, Lala Ma, and Christopher Timmins, "Environmental Justice: The Economics of Race, Place, and Pollution," *Journal of Economic Perspectives* 33, no. 1 (2019): 185–208.
33. Taylor, *Toxic Communities*, 148.
34. Ibid., 186.
35. Sophia Schmidt, "Residents Living among Industry Take County Survey on Environmental Perceptions, Relocation," *Delaware Public Media*, November 16, 2018 (originally published September 14, 2018), https://www.delawarepublic.org/post/residents-living-among-industry-take-county-survey-environmental-perceptions-relocation.
36. Rachel Morello-Frosch, Manuel Pastor, and James Sadd, "Environmental Justice and the Precautionary Principle," in *Contested Illnesses: Citizens, Science, and Health Social Movements*, eds. Phil Brown, Rachel Morello-Frosch, Stephen Zavestoski, and the Contested Illnesses Research Group (Berkeley: University of California Press, 2012), 64–65.
37. Rachel Morello-Frosch, Miriam Zuk, Michael Jerrett, Bhavna Shamasunder, and Amy D. Kyle, "Understanding the Cumulative Impacts of Inequalities in Environmental Health: Implications for Policy," *Health Affairs* 30, no. 5 (2011): 879–87.
38. Ibid., 881.
39. Mohai, Pellow, and Roberts, "Environmental Justice," 423.
40. Sophia Schmidt, "Public Airs Concerns Following Croda Plant Leak," *Delaware Public Media*, December 20, 2018, https://www.delawarepublic.org/post/public-airs-concerns-following-croda-plant-leak; Sophia Schmidt, "Environmental Justice Advocates Have Questions After Croda Settlement," *Delaware Public Media*, April 2, 2019, https://www.delawarepublic.org/post/environmental-justice-advocates-have-questions-after-croda-settlement.
41. Meredith Newman, "Report: 7 New Castle Communities at Greater Risk for Cancer, Respiratory Illness," *The News Journal* (Wilmington, DE), October 19, 2017.
42. Maddy Lauria, "Advocates Say Delaware is Ignoring Thousands of Vulnerable Residents Breathing Pollution," *The News Journal* (Wilmington, DE), March 17, 2019: A1.
43. Phil Brown, Rachel Morello-Frosch, Stephen Zavestoski, Sabrina McCormick, Brian Mayer, Rebecca Gasior Altman, Crystal Adams, Elizabeth Hoover, and Ruth Simpson, "Embodied Health Movements," in *Contested Illnesses: Citizens, Science, and Health Social Movements*, 16.
44. Schmidt, "Residents Living among Industry Take County Survey on Environmental Perceptions, Relocation."
45. Sophia Schmidt, "State Collects More Data on Dust Problem in Eden Park," *Delaware Public Media*, November 1, 2018, https://www.delawarepublic.org/post/state-collects-more-data-dust-problem-eden-park.
46. Phil Brown, "Popular Epidemiology and Toxic Waste Contamination: Lay and Professional Ways of Knowing," *Journal of Health and Social Behavior* 33, no. 3 (1992): 269.
47. WHYY, "Study Shows Big Difference in Air Quality After Re-Start of Delaware City Refinery," May 20, 2012, https://whyy.org/articles/study-shows-big-difference-in-air-quality-after-re-start-of-delaware-city-refinery/, accessed June 12, 2019.
48. Charles Schmidt, "Beyond Mitigation: Planning for Climate Change Adaptation," *Environmental Health Perspectives*, vol. 117, no. 7 (2009): A307-A309.
49. Ellen M. Douglas, Paul H. Kirshen, Michael Paolisso, Chris Watson, Jack Wiggin, Ashley Enrici, and Matthias Ruth, "Coastal Flooding, Climate Change and Environmental Justice: Identifying Obstacles and Incentives for Adaptation in Two Metropolitan Boston

Massachusetts Communities," *Mitigation and Adaptation Strategies for Global Change* 17, no. 5 (2012): 537–62.

50. The "bathtub model" illustrates what happens when low-lying areas (like Southbridge or New Orleans) are inundated with water. For a discussion of the most recent sea level rise analysis, see Karen B. Roberts, "Delaware Geological Survey, DNREC Update Sea Level Rise Projections for Delaware," *UDaily*, November 27, 2017, https://www.udel.edu/udaily/2017/november/new-sea-level-rise-planning-scenarios/, accessed June 14, 2019.

51. Victor W. Perez and Jennifer Egan, "Knowledge and Concern for Sea-Level Rise in an Urban Environmental Justice Community," *Sociological Forum* 31, no. S1 (2016): 885–907.

52. Bruce Stutz, "A Vulnerable Community Braces for the Impacts of Sea Level Rise," *Yale Environment 360* (Yale School of Forestry & Environmental Studies), January 30, 2017, https://e360.yale.edu/features/a-vulnerable-community-braces-for-the-impacts-of-sea-level-rise.

53. Sophia Schmidt, "South Wilmington Wetland Park Hopes to Start Construction Next Summer," *Delaware Public Media*, June 20, 2018, https://www.delawarepublic.org/post/south-wilmington-wetland-park-hopes-start-construction-next-summer.

54. See the City of Wilmington's Comprehensive 2028 Plan, https://www.wilmingtonde.gov/government/city-departments/planning-and-development/wilmington-2028/comprehensive-plan/full-plan-and-summary-document.

55. Stutz, "A Vulnerable Community Braces for the Impacts of Sea Level Rise."

56. Sophia Schmidt, "Residents Seek Community Benefits Agreement with Slag Grinding Company," *Delaware Public Media*, December 10, 2018, https://www.delawarepublic.org/post/residents-seek-community-benefits-agreement-slag-grinding-company; Jeanna Kuang, "In Wilmington's Southbridge, Talk of New Industry Turns to Jobs, Community Benefits," *Delaware News Journal* (Wilmington, DE), December 17, 2018.

57. Maddy Lauria and Jessica Bies, "After a Fire at Delaware City Refinery, Residents Wary of Their Industrial Neighbor," *Delaware News Journal* (Wilmington, DE), February 4, 2019.

58. Jeanna Kuang, "Wilmington Banks on Wetland Park," *The News Journal* (Wilmington, DE), November 27, 2018: A4.

59. Kenneth A. Gould and Tammy L. Lewis, *Green Gentrification: Urban Sustainability and the Struggle for Environmental Justice* (New York: Routledge, 2017).

60. Scott Goss and Maddy Lauria, "This is Not Going to Cut It," *The News Journal* (Wilmington, DE), December 13, 2017: A4.

61. Karl Baker, "Blades Residents Sue DuPont, Chemours," *The News Journal* (Wilmington, DE), June 5, 2019: A9.

62. William H. Dunn, "Coastal Zone Act Changes Need Local Experts," *The News Journal* (Wilmington, DE), January 28, 2018: A1.

63. Schmidt, "Public Airs Concerns Following Croda Plant Leak"; Schmidt, "Environmental Justice Advocates Have Questions after Croda Settlement."

CONCLUSION

Sustainable Landscapes

ANNA WIK AND SUSAN BARTON

BY NOW, YOU HAVE DEVELOPED a truly cohesive sense of the natural systems that comprise the world around you. Your Delaware Naturalist journey has provided opportunities for you to understand this small state at a variety of scales and in different timelines, beginning with a historic and ecological overview of human impacts on Delaware's landscape and watersheds, including industrial and agricultural changes, urban and suburban development, and issues of environmental justice.

Understanding the geology that created the varied landscapes of Delaware gave you a base from which to examine how water moves through and around the state. Delving into the soil-building processes related to both aquatic and geological activity, we acknowledged the support that this living resource provides for our plant and animal life, as well as for human settlement, building, agriculture, and development. Through field sketching and nature photography you nurtured and broadened your inherent observational skills and identified ways of recording and transmitting information for yourself and others. Foundational explorations of plants, insects, reptiles and amphibians, and birds of the region taught you how to practice identification in the field and at home. An introduction to citizen science highlighted ways that you can engage with the natural world and collect data to further our shared knowledge. You have been introduced to key issues of our time, including climate change and environmental justice.

While continued observation, curiosity, and exploration are key qualities of a Master Naturalist, you may also want to put some of these ideas to the test. The next big question is: what are you going to do with all of the expertise and passion you have cultivated? How can you apply your considerable knowledge, both at home and throughout the state? In this chapter, you will explore how to bring the above topics together and utilize the information you have learned to create and steward sustainable landscapes.

It is in our human nature to create. You may decide to transform a suburban lawn, a wild bit of land, or a newly built lot into a designed space for outdoor enjoyment, to attract pollinators and wildlife, or to reestablish the natural systems that exist on the site. As you now more fully understand, when undertaking such projects, it is important to consider how to approach the given landscape. Each site has specific requirements and an underlying ecosystem.

First and foremost, it is important to hone the observational skills you have been developing through the Master Naturalist program. You have learned to look at insects in all life stages, identify plants (both welcome and invasive), and make note of the geologic history that is just below the Earth's surface. As a designer and steward, it is important to observe and understand what a site can teach you before you start clearing, planting, and putting down bricks to, for instance, make your dream backyard. You can read the cultural history that is present, perhaps indicated by a hedgerow of old trees or a crumbling stone wall. Some experts recommend that we observe a site for a year and a day before making any design decisions about it, but in many cases, we do not have this luxury. If at all possible, take a bit of time to get to know the landscape of the site and the underlying natural systems at work. And consider adopting some of the ecological principles you have learned here and applying them to the landscapes you know best: those surrounding your own homes, and in your own neighborhoods.

UNDERSTANDING ECOSYSTEM SERVICES

As ecologists and landscape designers work to educate the public on the benefits of ecological systems, they have arrived on the term *ecosystem services* to describe "the direct and indirect contributions of ecosystems to human well-being."[1] While healthy ecosystems also sustain communities of life far beyond the human, this term focuses specifically on the benefits these systems provide for human life. In the past, we have been able to count on natural areas in the landscape to provide these services, but as Delaware's forests, farms, and open spaces have been transformed into housing developments and shopping malls, individual (public) natural spaces are no longer large enough to provide the services humans (or many other species) require. As Doug Tallamy and others have pointed out elsewhere in this volume, the ecological renovation that our region requires must happen on private lands.

Simply put, landscape architects and others working with managed landscapes talk about ecosystem services as providing clean water, fresh air, wildlife diversity, and human wellness.

Complex, diverse ecosystems are the basis for all life. These complex systems keep undesirable pests under control, provide habitat and food for spring peepers that loudly announce the early spring, and create habitat and food for the songbirds that greet you each morning and for the foxes whose bright eyes catch the light as they dart across a field at night. It's really quite simple: native plants support native animals, so a greater diversity of native plants means attracting more animals—more butterflies, more birds, and more mammals. And the good news is that each one of us has the capacity to build, or rebuild, our landscape to support this diversity.

Traditionally, home landscapes have contained a limited palette of plants, included large areas of regularly mowed lawn, and provided very little of ecological value. By using more plants (and especially more native plants), planting to conserve energy (by using deciduous trees that shade in the summer and allow sun to heat the home in the winter), and incorporating managed meadows and woods into home landscapes, our planned landscapes can become far more ecologically rich than the current standard, sanitized, and sterile American lawn.

A quick trip to most suburban developments in Delaware reveals a familiar landscape style with just a few trees and a limited variety of foundation plants. There are a high percentage of evergreens, few native plants, large areas of exposed mulch rather than living ground covers, and large expanses of (often chemically sanitized) lawn. There is the perception that "neat and tidy" (but sterile) landscapes like these are easier to maintain than more diverse, open, and "natural" landscapes are. This is the opposite of the truth. In fact, this sort of suburban landscape is ecologically dysfunctional. There are few spaces for people to inhabit, and even fewer places to provide food or habitat for non-humans. Instead, the foundation around the home is "decorated" with plants, and the rest of the property is left as lawn. Why do people continue to landscape their homes this way? Why do they prefer this landscape?

Entire books have been written on this subject, but the roots go back to a need for control over one's domain. Journalist Michael Pollan's famous essay about the "Savannah Syndrome" makes the case that a preference for lawn is built into our DNA; the lawn apparently reminds us of an open grassy landscape resembling the short-grass savannas of Africa on which we evolved and spent our first few million years. Pollan attributes the invention of the American lawn to Frederick Law Olmsted, the designer of Central Park in New York City, who said, "Probably the advantages of civilization can be found illustrated and demonstrated under no other circumstances so completely as in some suburban neighborhoods where each family abode stands 50 or 100 feet or more apart from all others, and at some distance from the public road."[2]

In the nineteenth century, lawns were popular in England, but only for the wealthy. In the United States, Olmsted and others democratized the lawn. In 1870, Frank J. Scott, seeking to make Olmsted's ideas accessible to the middle class, published *The Art of Beautifying Suburban Home Grounds*, the first volume ever devoted to suburban home "embellishment." Scott subordinated all other elements of the landscape to the lawn. Flowers, for example, were permissible, but only on the periphery of the grass: "Let your lawn be your home's velvet robe," he wrote, "and your flowers its not too promiscuous decoration."[3] Early Americans had a desire to tame and control the natural world and, if anything, this impulse has only become more widespread. A neatly mowed lawn provides that perception of "control" that has come to define the domain of the suburban homeowner.

What is wrong with the lawn as the primary vegetation of planned landscapes? It simply does not provide anything like the ecological diversity or other systemic benefits provided by other types of landscaping. It allows little water infiltration, since most water, especially if the lawn is sloped, flows quickly off the surface. It supports almost no wildlife, because grass, as a non-native plant, supports few insects. While grass does photosynthesize and produces some oxygen, its production is far less than that of a complex landscape with multiple layers of vegetation. Lawns remove few particulates from the air, and (obviously) provide no cooling shade. They also offer little help in this era of climate change versus what a more diverse landscape can offer: each person in the United States generates approximately 2.3 tons of carbon dioxide each year, while one acre of healthy forest sequesters about 2.6 tons per year. So, the carbon dioxide produced by one person is offset by one acre of trees. By lowering temperatures and shading buildings in the summer, trees and other plants can reduce buildings' energy use. Less need for fossil fuel also means fewer carbon emissions will be generated.

Supporting ecologically resilient landscapes may strike many readers as intuitive, but researchers have in fact confirmed that humans also need open, outdoor spaces filled with plants and animals to reduce stress, think clearly, increase productivity, and relate to one another more effectively. The stresses and strains of the urban environment have been widely acknowledged. At the end of a long day of work, we often seek "effortless" activities, which contribute to a sense of rest or well-being rather than fatigue. Jogging, swimming, gardening, and playing with one's children are characterized by curiosity, or fascination—similar, in fact, to the pleasure we get from watching a storm or a waterfall, gazing upon a still pond, or listening to the gentle rustle of leaves or a breeze through a meadow. Engagement with the natural world provides a source of restorative experiences that can be active, such as gardening, pulling invasive vines from a forest, or hiking through your local park. Even the conceptual involvement of planning a seasonal garden, mapping out a trail through a natural area, or simply having the knowledge that a park exists around the corner and can be visited in the future can provide some level of restoration. State and national parks or other wildlands offer these benefits, but the urban natural environment can also provide a setting for restorative experiences both physically and conceptually.[4] Natural settings are often believed to instill a sense of peace and serenity, but that does not mean they lack excitement, vibrancy, or sensory richness.

The Japanese have developed (and have now officially formalized) the practice of "forest bathing," which preserves natural landscapes solely to support the psychological and spiritual well-being of its citizens. This is a practice we can emulate in a variety of ways.[5] Some think this need for time in nature traces back to natural selection, when it was clearly to early humans' benefit to observe the natural world closely and discover where berries could be found and in which trees predators hid. Others acknowledge the physical, psychological, and spiritual benefits of spending time outdoors in the company of other beings. The absence of these beings often leaves people feeling what that Native American ecologist Robin Wall Kimmerer has called "species loneliness."[6]

Does that mean that we should completely remove lawn from planned landscapes? Of course not. Grass is the only plant we can routinely walk on, so it provides excellent circulation throughout the landscape. It is also a surface on which we can play, entertain, and gather. In a big city, patches of lawn signal to pedestrians that they have reached a park, a place of respite.

So instead of shunning lawns, we should design them more intelligently. One of the very best ways people can do so is through an understanding of *sustainable landscaping*. This phrase has emerged to refer to landscape design, installation, and management practices that conserve resources, minimize negative environmental impacts, and maximize ecological function while meeting both functional and aesthetic goals.[7] But the trick to scaling this up—especially on land owned by people unfamiliar with ecological principles—is to also make ecological diversity look good. Especially in suburban neighborhoods saddled by decades of ecologically unwise plantings and design, it is vital that newly conceived, ecologically diverse landscapes also be pleasing to look at. New designs must convey the impression that the landscape is being managed intentionally, and not being "abandoned" to natural forces.[8] Only then will this approach spread, and scale up—that is, only then will more and more homeowners choose to incorporate sustainable principles into their own private landscape.

Indeed, if lawns are so important in the landscape, we should design them first for wherever we need to walk, play, or gather. We should then design the rest of the land for something else: landscape beds, forests, and meadows. Landscape beds are easy to incorporate, but they are expensive, requiring purchased plants and maintenance, at least until the plants grow together to prevent weeds from taking over. Forests are loved by most people, but their development takes time, though perhaps not as much as you'd think. A patch of ground behind the greenhouses on the University of Delaware's College of Agriculture and Natural Resources campus in Newark was allowed to develop as forest on its own with no maintenance other than the removal of invasive exotic species. Everything that came up was planted by squirrels, wind, or other natural forces. After ten years, this lovely forest fragment is now a beautiful woodland full of red maple, sweet gum, sycamore, cherries, oaks, black gum, and lots of native shrubs and ground covers. A pine-needle path has been added to provide pedestrian access.

Intentional design, and the intentional "cues" that often go with it, can also enhance the public's acceptance of sustainable and restorative landscaping. This is as true in privately owned neighborhoods as it is in publicly managed areas. As Master Naturalists, we would like to see an uptick in the restoration of our landscape's ecosystem services, but some historical plant and material choices have not been as supportive of wildlife as we might desire. Therefore, we must be strategic about the preservation methods we undertake and consider how education and interpretation may fill in gaps or guide us in simply repairing what could otherwise detract from the overall functionality of the ecosystem.

One of the most important ways we can engage people in the landscape and increase awareness of the beneficial nature of the ecosystem services that sustainable landscapes provide is through interpretation and education. Humans love to learn and discover new things, especially if those things provide positive impacts for us, and if we can take the opportunity to tell the story of the systems at work in a given landscape, then we can prompt a personal connection that could otherwise be lost. Whether the information we share is related to the function of plants, the history of a site, the economic benefits of reduced heating costs, or the ecological benefits of mitigating stormwater runoff, it can spark an interest or engage someone in a topic of which they were unaware. Interpretation can be explicit, through the use of signage or guided tours, or more implicit, by highlighting or making visible aspects of a system that were previously hidden. As stewards of sustainable landscapes, we should make every effort to not let our efforts go unnoticed.

There are stories that are pretty obvious to the unaided eye, and that may only need a brief description or signage to tell. Covered bridges, mills, dams, and garden structures all give clues about previous land use. In addition, there are histories of culturally valuable sites that are equally important, yet harder to pick up on without assistance. Most often the stories we see in these landscapes are those of the owner, the builder, the victor of the skirmish, the person in power. But these landscapes often tell more complicated histories. Did enslaved people or indentured servants maintain the frequently clipped hedges of the European-style *parterre* garden? Which women harvested fruit and vegetables from the kitchen garden? What country did the men who sawed lumber at the mill hail from? Which Native American tribe was on this site before European colonization? Additional research can provide clues to the hidden past and interpretation can make it available for others to understand.

More information about the cultural history of a site gives it more meaning for those who would steward and tend its landscape. Research on restored wetlands has found that structures, signs, and strategically located turf areas help people better understand the stewardship intention and developing beauty of the wetland. Roadside meadow plantings with neatly mowed edges are considered more desirable than meadow treatments containing no mowed edge;[9] and the appearance of maintenance and the inclusion of color increase acceptance by drivers and pedestrians alike,[10] which supports the view that in order for naturalistic plantings to be accepted in urban areas, they must be "visually dramatic."[11]

So it is clear that investing in ecosystem services provides deep and rich human benefits, including nature's capacity to clean water, increase plant and animal diversity, cool the environment, save energy, clean the air, and sequester carbon, and people's ability to enjoy the landscape for the myriad pleasures it can provide. The trouble is, sustainable landscapes are dwindling in the world. So, if we design landscapes that take advantage of natural processes, while providing tangible benefits to the land's human owners, we can then produce ecosystem services in our own backyards and public spaces.

LANDSCAPING FOR ECOSYSTEM SERVICES

The physical environment (soil, light, temperature, air, and moisture) dictates which plants can grow in a given location. The cold hardiness zone map[12] will tell you how far north a plant can grow, but a heat zone map[13] is also important. Plants do not die with one hot day as they might with a freeze, but the accumulation of hot weather for plants adapted to cold conditions will do significant harm. Warm weather can stunt plants, make them more susceptible to insects and disease, and eventually kill plants that thrive in cold weather. Cities are warmer than rural areas and courtyards are warmer than open areas subject to wind. Wind can be especially detrimental to the cold hardiness of broad-leaved evergreens. Broad-leaved evergreens transpire, losing moisture through their leaves. As the wind blows that moisture away, the plant continues to lose water. Leaves eventually turn brown due to lack of water from the combination of cold and wind.

Time is often required to develop the desired effect in sustainable landscape design. Plants are small at the moment of installation and develop depending on localized soil and water conditions. Starting with small plants reduces the transplant shock associated with moving large plants, and also saves money. Starting small allows plants to become established in a site more successfully. The only reason to plant large trees in a landscape installation is for immediate effect, but that may result in a stagnant landscape for several years after planting. It usually takes one year per inch of caliper (tree diameter at breast height) for a tree to become established and start growing normally in a landscape. Therefore, a four-inch caliper tree may look impressive at planting, but four years later, a one- to two-inch caliper tree will catch up in size, with much less expense.

SELECTING PLANTS USING AVAILABLE RESOURCES

Plants for a Livable Delaware (PLD) is a program designed to raise awareness of plants that thrive in Delaware without becoming invasive or suffering from significant disease and insect problems in the landscape. There is no specific list of livable Delaware plants, but they possess characteristics adaptable to varying landscape situations (i.e. drought-resistant, tolerant of poor soils, etc.). The PLD program includes the publication of a series of brochures designed to help people select appropriate plants for landscape planting. The first brochure, "Plants for a Livable Delaware,"[14] outlines ten problem invasive plants and suggests alternatives that perform the same aesthetic and environmental functions in the landscape. For example, English ivy is an effective ground cover, but it has taken over many landscapes and woodlands and is smothering desirable trees and shrubs. When it becomes mature, it fruits, and birds spread ivy seeds far and wide. Some ground cover alternatives include wood aster (*Eurybia divaricatus*), many sedges (*Carex* sp.), goldenstar (*Chrysogonum virginianum*), hay-scented fern (*Dennstaedtia punctilobula*), Allegheny pachysandra (*Pachysandra procumbens*), Virginia creeper (*Pathenoscissus quinquefolia*), Christmas fern (*Polystichum acrostichoides*), barren-strawberry (*Waldsteinia fragearioides*), and yellowroot (*Xanthorhiza simplicissima*).

Another brochure in the series, "Livable Plants for the Home Landscape,"[15] focuses on plant suggestions for specific landscape niches such as meadows, wet areas, dry shade, rain gardens, forest edges, pond/stream edges, sunny slopes, salt and sand conditions, and small gardens and containers. When people are looking for plants, they usually have a specific site in mind, so grouping plant recommendations by site conditions makes sense. One particularly tough site is dry shade. Some grasses and grass-like plants that work well in dry shade are crinkled hairgrass (*Deschampsia flexuosa*), bottlebrush grass (*Elymus hystrix*), and Pennsylvania sedge (*Carex pensylvanica*). While most ferns require moisture, hay-scented fern (*Dennstaedtia punctilobula*), interrupted fern (*Osmunda claytoniana*), and Christmas fern (*Polystichum acrostichoides*) tolerate dry conditions.

Native plants for dry shade include heart-leaved aster (*Symphyotrichum cordifolium*), white wood aster (*Eurybia divaricatus*), hyssop-leaved thoroughwort (*Eupatorium hyssopifolium*), tall white beardtongue (*Penstemon digitalis*), Bowman's root (*Gillenia trifoliata*), wild ginger (*Asarum canadense*), Allegheny pachysandra (*Pachysandra procumbens*), woodland phlox (*Phlox divaricata*), large-flowered merrybells (*Uvularia grandiflora*), hairy alumroot (*Heuchera villosa*), golden ragwort (*Packera aurea*), Jacob's ladder (*Polemonium reptans*), and bluestem goldenrod (*Solidago caesia*). Some nonnative (but also not invasive) plants are great performers in dry shade, including barrenwort (*Epimedium* sp.), Hakonne grass (*Hakonechloa macra* "Aureola"), and Lenton rose (*Helleborus orientalis*). It is harder to find shrubs that tolerate dry shade, but here are a few: pinxterbloom azalea (*Rhododendron periclymenoides*), oakleaf hydrangea (*Hydrangea quercifolia*), and Piedmont rhododendron (*Rhododendron minus*).

A third brochure, "Livable Ecosystems: A Model for Suburbia,"[16] focuses on how to plant and manage different types of landscapes such as rain gardens, windbreaks, meadows, forests, and butterfly gardens.

SELECTING CONSTRUCTION MATERIALS

When considering materials to incorporate into your designed landscape, it is important to recognize the implications of your choices. Our economy, including the landscaping industry, relentlessly promotes products and practices that are "cheaper and faster," and there has been a parallel, exorbitant growth in the global sale of landscaping materials.[17] This in turn has led to moving away from an approach that emphasizes the use of local materials, and lowering transportation costs. Tuscan marble, which is mined in Italy, may be beautiful, but it would be out of place in a naturalistic Piedmont shade garden. Likewise, tropical hardwoods can have great properties, but local woods such as black locust can also provide rot resistance and attractive weathering. If you do choose to use tropical hardwoods or other exotic materials, research their certifications to ensure they have been harvested and processed using sustainable methods.

Some specific questions to ask when sourcing materials are:

1. How is the material acquired, processed, and manufactured? Raw material extraction and primary and secondary processes are used to create the final product that arrives on site. How much energy and habitat loss happens at the place of extraction? What are the environmental impacts of the emissions and wastes generated by processing the raw material to a useable form? What solvents and cleaning fluids containing volatile organic compounds (VOCs) are released in the manufacturing stage?
2. Where is your landscaping material coming from? The Sustainable Sites Initiative (SITES) recommends transporting heavy materials such as bricks, aggregate, and concrete no more than one hundred miles.[18] The energy cost to transport can be more than to it is mine and process materials.
3. What resources are necessary to maintain the material on site? How durable is it? Will it need to be resealed, painted, or cleaned on a regular basis? Does the material off-gas?
4. Finally, what happens to the material at the end of its lifespan? Where does it end up and what does it become?

These lifecycle concerns can guide us to make better decisions about our material selection and how we design our hardscapes. One idea gaining traction recently is designing for *deconstruction* rather than demolition. This practice means that structures are built in such a way that they can (one day) be broken down into reusable parts rather than destroyed and placed in a landfill.

There are a number of projects that the typical homeowner can do to improve the ecosystem services, and thus the "sustainability," of their landscape. Many of these are scalable, meaning they could be done in a small, postage stamp-size urban lot or on a large rural estate.

MANAGING WATER

Delaware is the state with the lowest mean elevation in the country,[19] so all of our state's municipalities are considered vulnerable to coastal flooding or periodic inundation. Managing rainfall from storms is one of the most important functions a person with access to even a small bit of land can provide in their landscape. There are many ways to do this. Here are a few examples at different scales. (For further reference, consult the Delaware Department of Natural Resources and Environmental Control Post Construction Stormwater Best Management Practices Standards & Specifications, February 2019.[20])

SMALL SCALE

Install a rain barrel to capture and store rainwater. One of the most important acts we can do in working with our landscapes is to help slow down that first flush of water that bursts during a rainfall. These bursts often carry pollutants in the form of chemicals, sediments, and pet waste, and run directly over loaded sewer systems and then straight into our waterways. Directing runoff from gutters and roofs directly to a rain barrel is a great way to prevent water from going directly into our stormwater systems. In periods of heavy rain, you can create a chain of rain barrels to store larger amounts of water. Even at the scale of a small apartment balcony, rain barrels can be used to irrigate other gardens when there is no rain.

MEDIUM SCALE

Install a bioretention system, such as a rain garden, or a detention system, such as a wetland swale, to capture, store, and filter rainwater. The term "rain garden" is often used as a catch-all term to describe various methods of stormwater management, but it actually applies to a specific type of garden, which is a *bioretention* system used to manage runoff from small storms. It is upland-based (as opposed to wetland-based), and is typically a shallow, depressed planting bed that uses a combination of well-draining soil and appropriately adapted native or non-invasive plants to retain water, filter it through soil and plant processes, and ultimately recharge the groundwater. In some cases, these gardens slowly release water back into a stormwater conveyance system. Rain gardens must be sited in well-drained soils, ideally halfway between the runoff site (roof, paved surface) and the place the water currently drains. Calculations prepared by a landscape architect or stormwater engineer can help size the rain garden appropriately to manage a storm.

If you live in an area with heavy clay, and want to manage stormwater, you may be better served installing a *biodetention* system, which is designed to capture and slow the flow of water by temporarily holding it before releasing it back into the storm sewer. While your clay soil may ultimately drain, or the carefully selected *facultative* plants (those that can grow in both uplands and wetlands) may evapotranspire some portion of the stormwater, it is unlikely that you will be able to infiltrate all the water on site. Water in biodetention systems, such as constructed wetlands and vegetated roofs, must be carefully directed to slowly release after an initial storm surge is complete.

LARGE SCALE

Install a bioswale or large-scale constructed wetland. If you are trying to manage runoff from a large paved area, you may consider implementing a planted *bioswale*. Bioswales are designed and constructed to mimic natural stream channels. They are linear systems that are designed to slow water entering the storm sewer system and engineered to handle a specific amount of water from larger paved areas, such as parking lots and roadways. These are generally steeper and narrower than a rain garden but use the same bioretention processes as rain gardens to capture, filter, and infiltrate water, as well as direct it into overflow systems.

Constructed wetlands are designed to replicate natural wetland systems in that they are permanently inundated with shallow water levels of one foot or greater. Their primary function is to hold rainwater from large storms, and during this retention time, carried sediments are released and microbes and plant material can uptake and filter pollutants.[21]

With all bioretention and biodetention systems, you should be thoughtful about where you direct overflow. Most municipalities have rules about directing water toward your neighbor's property. Some have rules and regulations about what can be directed off your property.

CONTROLLING WIND

Heating and cooling costs can be intense for the average homeowner. Consider planting a *windbreak*, or barrier, of one or more rows of trees or shrubs to reduce and redirect wind around a property. When the windbreak reduces wind velocity and intensity, a more desirable microclimate is created in the sheltered zone. Trees and shrubs used to create windbreaks also provide wildlife with food sources, shelter, nesting, and protection from the elements. Here are some tips on windbreak planting that can maximize energy benefits:

- Plant deciduous trees to shade east-facing walls and windows from 7 to 11 A.M. and west-facing surfaces from 3 to 7 P.M. during June, July, and August.
- Plant trees with mature heights of at least twenty-five feet approximately ten to twenty feet east and west of your house.

- Trees planted to the southeast, south, or southwest of a building will only shade it in the summer if they extend out over the building's roof. In the winter, when maximum sun is desired, such trees will provide too much shade. Even deciduous trees that have dropped their leaves cast quite a bit of shade in the winter.
- To avoid winter shading, locate trees no closer than 2.5 times their mature height to the south of a building.
- Locate trees on the southeast or southwest of a building at a distance about four times their mature height.
- Plant trees to shade paved areas.
- Shade air conditioners from mid-morning through evening. Prune branches to allow at least several feet clearance around the air conditioning equipment to encourage air flow. Avoid planting shrubs near an air conditioner since they reduce airflow and cooling efficiency.

CREATING PRODUCTIVE FORESTS

Another method you can employ to increase ecosystem services is to intentionally plant trees using the principles of *agroforestry*. Agroforestry views trees as a *functional crop*, and often pairs trees with other existing agricultural systems to maximize benefits and productivity on a given site, while increasing ecosystem services. There are multiple ways of incorporating agroforestry into an existing system, including *alley cropping*, in which trees are planted between rows of other types of crops, and *silvopasture*, in which productive tree species are combined with livestock, who graze beneath the tree canopy, into a single system. These practices can be carried out on a large scale and are worth investigation.

Another type of tree-based system gaining traction across the country is the creation of *food forests*, or edible forest gardens. These designed systems replicate natural ecosystems by incorporating multiple layers of plant material, including canopy and understory trees, a shrub layer, vines, perennials, and bulbs. Plants that comprise a food forest are selected for their productive qualities, whether the product is fruit, nuts, medicine, or pollinator and wildlife support. Each plant has a role to play in the system that provides support to the other plants. For example, plants in the pea family, such as false indigo (*Baptisia australis*), reach deep taproots into the ground to create more air space for other plant roots and fix, or hold, nitrogen, which is a key nutrient for plant growth. Bulbs such as wild onion (*Allium sp.*) provide tasty edible leaves and bulbs for humans while repelling less desirable animals, like rabbits and pest insects, who might cause damage to other plants. Edible forest gardens are akin to orchards, but not in the sense of the traditional monoculture system of orchard planting, which relies heavily on pesticides and fertilizer to encourage yields and viability. Instead, native plants are used, or ornamental non-invasive plants are selected for disease resistance or production quality.[22]

PLANTING FOR POLLINATORS

Pollinators require two essential components in their habitat: somewhere to nest and flowers from which to gather nectar and pollen. Native plants are undoubtedly the best source of food for pollinators, because plants and their pollinators have coevolved. Provide patches of flowers, rather than single plants, to help pollinators find them. Select species that flower throughout the growing season, and avoid modern hybrids, especially double-flowered cultivars that may not produce pollen, nectar, or fragrance to attract pollinators. Use few, if any, pesticides and always select the least toxic option. If you do use a pesticide, try to spray at night, when few pollinators are actively visiting flowers. Include larval hosting plants, like milkweed for monarch butterflies. Leave a few dead limbs on trees, or at least on the ground, to provide nesting sites for native bees. Mix some sea salt or wood ashes into a muddy spot in the landscape to create a salt lick since salt is required by butterflies and bees. Moist animal droppings, urine, and rotting fruit are attractive to butterflies. Try adding sea salt to old fruit and to a moist sponge and see which butterflies come to investigate.

In general, butterflies are less specific with nectar preferences. The plants listed below all supply nectar for a wide range of butterflies. Individual species have more exacting requirements for larval food (see Table 14.1). The following plants are good nectar sources and attract a wide variety of butterflies.

- Asters, *Aster* spp.
- Azaleas, *Rhododendron* spp.
- Bee balm, *Monarda*
- Black-eyed Susan, *Rudbeckia* spp.
- Blue giant hyssop, *Agastache foeniculum*
- Bottlebrush buckeye, *Aesculus parviflora*
- Butterfly weed, *Asclepias tuberosa*
- Bush cinquefolia, *Potentilla fruticosa*
- Clethra, *Clethra* spp.
- Coreopsis, *Coreopsis* spp.
- Gaillardia, *Gaillardia* spp.
- Goldenrod, *Solidago* spp.
- Heliotrope, *Heliotropum* spp.
- Hibiscus, *Hibiscus* spp.
- Ironweed, *Vernonia novaboracensis*
- Joe-Pye weed, *Eupatorium* spp.
- Lantana, *Lantana camara*
- Lavender, *Lavendula* spp.
- Lilac, *Syringa vulgaris*
- Marigold, *Tagetes* spp.
- Milkweed, *Asclepias* spp.
- Mint, *Mentha* spp.
- Pentas, *Pentas lanceolata*
- Phlox, *Phlox* spp.
- Purple coneflower, *Echinacea* spp.
- Rosemary, *Rosmarinus officinalis*
- Shasta daisy, *Leucanthemum cv.*
- Sunflowers, *Helianthus* spp.
- Sweet pea, *Lathyrus odoratus*
- Trumpet honeysuckle, *Lonicera sempervirens*
- Verbena, *Verbena* spp.
- Zinnias, *Zinnia* spp.

TABLE 14.1. MID-ATLANTIC BUTTERFLIES WITH THEIR LARVAL FOOD SOURCE

Butterfly species		Larval food source	
Common name	**Scientific name**	**Common name**	**Scientific name**
Black swallowtail	*Papilio polyxenes*	Carrot	*Daucus carota*
		Dill	*Anethum graveolens*
		Fennel	*Foeniculum vulgare*
		Parsley	*Petroselinum crispum*
Spicebush swallowtail	*Papilio troilus*	Sassafras	*Sassafras albidum*
		Spicebush	*Lindera benzoin*
Tiger swallowtail	*Papilio glaucus*	Ash	*Fraxinus* spp.
		Birch	*Betula* spp.
		Lilac	*Syringa vulgaris*
		Poplar	*Populus* spp.
		Tulip tree	*Liriodendron tulipifera*
		Wild cherry	*Prunus serotina*
Cabbage white	*Pieris rapae*	Garden nasturtium	*Tropaeolum majus*
Great spangled fritillary	*Speyeria cybele*	Violet	*Viola* spp.
Pearl crescent	*Phyciodes tharos*	Asters	*Aster* spp.
Monarch	*Danaus plexippus*	Milkweed	*Asclepias* spp.
Buckeye	*Junonia coenia*	Snapdragon	*Antirrhinum majus*
		Verbena	*Verbena* spp.
Mourning cloak	*Nymphalis antiopa*	River birch	*Betula nigra*
		Elm	*Ulmus* spp.
		Hackberry	*Celtis occidentalis*
		Polar	*Populus* spp.
		Willow	*Salix* spp.
Red-spotted purple	*Limenitis astyanax*	Wild cherry	*Prunus serotine*
		Willow	*Salix* spp.
Painted lady	*Vanessa cardui*	Hollyhock	*Alcea rosea*
		Sunflowers	*Helianthus* spp.

CREATING MEADOWS

In the Midwest, a tallgrass meadow is called a *prairie*, but meadows also grow in the mid-Atlantic in areas that historically have been subjected to periodic disturbance. The New Jersey Pine Barrens even has areas of meadow that are maintained by fire (once as a natural occurrence and now in planned burns).

If you stop mowing your lawn routinely and mow it only once or twice a year, the result will be a meadow. These areas can be highly variable and require management to promote and support desirable species. Their composition depends on soil moisture, sunlight, and surrounding seed sources. Meadows require full sun to thrive, and typically comprise native grasses like little bluestem, Indiangrass, and switchgrass. These grasses will naturally show up in a meadow eventually but can also be seeded in to speed up the process. Moist areas tend to support native flowering perennials and discourage the common invasive plants that can dominate infrequently mowed spaces. You can prevent woody plants from taking over by mowing a meadow at least once per year, preferably in March.

To seed a meadow, consider mixing seed with an organic *carrier* like composted sawdust or yard waste. Start by killing the existing plants with an herbicide or by covering the space with black plastic. Buy a mix of warm-season grass seed with a few species of flowering perennials mixed in. Mix the seed with the organic carrier, and spread the carrier about half an inch to one inch thick over the designated meadow. The carrier provides a moist medium for germination and prevents light from reaching the soil, reducing the germination of common annual weeds crabgrass and foxtail (two highly competitive annuals), which require light to germinate. Thus, the competitive advantage for germination is given to the desired meadow seeds.

MAINTAINING SUSTAINABLE LANDSCAPES

Meadows and rain gardens in managed landscapes look better when they are accompanied by following significant *cues of care*, such as the following. Fill the space in a rain garden with plants rather than with mulch. Mow the edge between a meadow or rain garden and a lawn regularly to provide a clean, well-defined edge. Mow a winding path through the meadow regularly to provide access and show the meadow is planned. Consider planting special warm season grasses or flowering perennials at meadow or rain garden edges to provide color and interest. Mow meadows in early spring to chop up the old leaves and allow new stems and leaves to emerge. Some meadows look better when they are kept shorter by mowing them twice a year. A second mowing in late June will reduce the height of the meadow, keep tall grasses from flopping, and cut down the tall flowering weeds that might look unattractive. While rain gardens can be planted formally or informally, the level of maintenance required is much higher with a formal planting.

THE ROLE OF PATHS

Paths through natural areas can provide access for people to explore and use the landscape. A tallgrass meadow is often viewed as a reservoir of "undesirable" forms of wildlife. But a path allows you to walk through a meadow and observe the grasses and flowering forbs (herbs other than grass) without scratching your legs or attracting ticks. Forests, especially those invaded by invasive vines and shrubs, can be almost impenetrable. A path through the forest invites exploration and can lead to an interesting destination like a rock outcropping with a good view or a collection of Virginia bluebells next to a stream. Paths also provide mystery as they snake out of view of the explorer. Where does the path lead and how far into the forest or meadow will it take me?

Paths are also a signal of human control. Even in a wild landscape, humans often feel more comfortable with the level of order provided by a path. Years ago, I flew over my property in a helicopter. The pilot wasn't particularly interested in the canopy of trees in White Clay Creek Forest, but as soon as we approached my property and he saw the mowed paths through the meadows and the woodchip paths through the woods, he became interested in the landscape below.

MANAGING PESTS WITH INTEGRATED PEST MANAGEMENT

While 95 percent of insects we find in the landscape are not plant pests, there are some insects that cause damage and even death to landscape plants. *Integrated pest management* (IPM) is both a philosophy and a set of prescribed tasks for reducing insect and disease damage in the landscape. The basic premise of integrated pest management is that no single pest control method always provides the best control. Instead, a program of regular monitoring is used to detect problems early and control methods are chosen from biological, physical, and chemical options.

An effective IPM program includes three phases:

1. Plant monitoring
2. Integration of control methods
3. Evaluation of results

To implement IPM, answer the following questions:

- Is a pest present?
- What type of pest is it?
- Where is the pest population located?
- Should the pest be controlled?
- What pest control method should be used?
- When should the pest control be applied?
- Was the pest control effective?

Monitoring is the frequent inspection of plants for the presence of pests and other conditions that could contribute to plant problems. Choose *key plants* to monitor, those that are desirable in a landscape but are most susceptible to insect, disease, and cultural problems. Determine the *key pests*, those that are usually present, usually damaging, and require control on key plants. Monitoring methods include visual inspection facilitated by trapping and beating (a white cloth is placed below a plant and the foliage is tapped to knock insects into the cloth for inspection). *Mechanical control* options include destroying the pests or destroying their habitats. For example, small populations of bagworms can be removed by hand during monitoring. Some galls can be pruned out and destroyed.

The objective of this *cultural control* is to manipulate the environment to make it unsuitable for pests. You can sanitize the growing area by removing and burning infested plants or destroying alternate hosts of rust fungi, such as Eastern red cedar, currants, or barberry. Increase air circulation by pruning trees or provide shade for shade-loving plants to reduce plant stress. Watering more or less, depending on conditions, may discourage a pest population. Proper pruning will reduce entry options for many diseases and insect pests.

Biological controls encourage beneficial parasites, predators, or pathogens with complex plantings in order to suppress pest populations. Antagonistic bacteria will suppress crown gall. Lady beetles have been shown to control aphids on Scot's pine and river birch. Parasitic wasps will reduce pine sawfly on white pine and pine needle scale on Japanese black pine. Predacious mites are available to control spider mites, and entomogenous nematodes attack white grubs, root weevils, and clearwing moth borers. Though it is not currently practical to release large numbers of such beneficial organisms outdoors, strategies should concentrate on promoting existing beneficial organisms by selecting pesticides that are least harmful to beneficial insects and by applying pesticides at times when beneficials would be least affected.

When planning a new landscape or replacing plants in an established planting, choose resistant species and varieties to prevent or suppress pest infestations. Many hybrid rhododendrons are resistant to phytophthora root rot. "Heritage" river birch is resistant to bronze birch borer and can be used in place of clump white birch. Crabapple varieties that show resistance to scab, powdery mildew, and cedar-apple rust have been developed. Serbian spruce is an attractive conifer that is resistant to the adelgids that plague many other conifers. Many native plants have few pest problems, so including natives in your landscape will also reduce undesirable insects and diseases.

Chemical controls can be used when a pest cannot be controlled with another strategy. Once a pest problem has been identified, choose a pesticide that is effective against the pest, compatible with the host plant, and least harmful to beneficial organisms. Spot treating individual pest outbreaks is less harmful and cheaper than cover spraying to prevent problems.

When any control action is taken, record the action, location, pesticide name, application rate, timing, and environmental conditions present. Two weeks after the control action is taken, evaluate the effectiveness of the approach. If the chosen control was not effective, try another option. At the end of the year, evaluate the entire IPM program to determine which control actions were most successful, whether the landscape was maintained at an acceptable "aesthetic injury level," the level below which most people do not perceive a reduction in the attractiveness of the desired planting, and whether pesticide use decreased.

CASE STUDY 1 OF 4

Postindustrial landscape—STAR Campus, Bloom Technology. Postindustrial sites are found throughout the world. When an industry is no longer viable, it may be necessary to abandon its manufacturing sites. Without fail—even, amazingly, in places as degraded as the Chernobyl nuclear site—vegetation always begins to reclaim these spaces. Along New York City's famed High Line, a diverse mix of plants came in and colonized the abandoned rail line, providing inspiration for the now wildly popular park in West Manhattan. An abandoned steel mill near Pittsburgh, the Carrie Furnaces, has been repurposed as an event space. Ten different signs there lead visitors through the Iron Garden and explain how nature is reclaiming the land. Minimal maintenance, primarily mowing paths, has made the site accessible.

FIGURE 14.1. Graphic from design charette for Bloom Technology. (Courtesy of Susan Barton.)

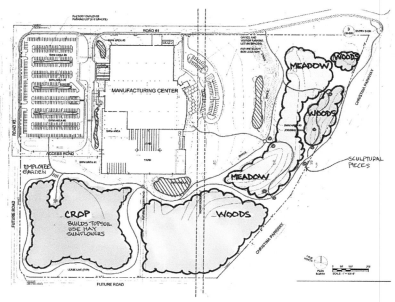

In 2009, the University of Delaware officially acquired the Chrysler property in Newark. The site, now called the STAR Campus, has been developed in partnership with commercial businesses. The first business to lease the land and build on the site was Bloom Energy, headquartered in San Jose, California. They were planning to build their new fuel cell manufacturing plant on fifty acres of the former Chrysler site. The plan included rain gardens in the parking lot to manage stormwater but was otherwise traditional, with minimal landscaping and plans for mowed lawn on the bulk of the property. The university's College of Agriculture and Natural Resources was able to work with the Bloom CEO, construction manager, and future maintenance manager to discuss alternate landscape plans, and held a design charrette with members of the Plant and Soil Sciences Department and Bloom. The maintenance manager was NOT interested in mowing grass on fifty acres, so the group was able to come up with a variety of different landscape solutions, including creating a hay field for horse feed, a developing woodland, and a seeded meadow, all connected with a series of walking paths.

CASE STUDY 2 OF 4

Forest fragment (FRAME—Forest Fragments in Managed Ecosystems) (Tara Trammell, Assistant Professor, University of Delaware, and Vince D'Amico, U.S. Forest Service Research Entomologist). In the United States, the hundred or so counties in states along the Boston-to-Washington, D.C., corridor (the "Megalopolis") are more densely populated than any other region of the U.S. Despite this, more than 250,000 forests larger than one hectare grow in this matrix of human development. Most (98 percent) are small forested green spaces under fifty hectares, but they can range in size up to large forests covering thousands of hectares. Forests in developed areas, no matter their size, make contributions for local residents, and our research reveals many areas where the closest forest for thousands or even millions of citizens is under ten hectares. These small forests, despite their importance, are often subject to local human influences such as invasive species spread. Non-native invasive species are one of the greatest threats to biodiversity and ecosystem conservation worldwide,[23] and it has long been recognized that plant invaders alter ecosystem structure (like species composition) and function (like nutrient cycling).[24]

Furthermore, invasive plants can affect disease transmission via complex interactions with plants and animals.[25] To understand the variety of influences on these forest ecosystems, it is important to study many small forests across a range of non-native plant invasion and development intensities. Thus, researchers from the USDA Forest Service and the University of Delaware established a long-term urban forest network, the FRAME (FoRests Among Managed Ecosystems), to monitor plants, birds, insects, and many other organisms, to estimate ecosystem services (like carbon storage and sequestration), and to study ecosystem functioning (like nutrient cycling).

FIGURE 14.2. Invasive plants can overtake a woodland landscape. (Photo courtesy of Vince D'Amico.)

The researchers are conducting non-native invasive plant removal experiments to establish the potential best management practices for invasive plant eradication efforts. Forest restoration is crucial for sustaining healthy forest ecosystems that are vulnerable to human influences and are essential for providing ecosystem services to urban and suburban residents.

CASE STUDY 3 OF 4

Residential landscape—Applecross Development. In 2011, five faculty in the University of Delaware College of Agriculture and Natural Resources received a National Integrated Water Quality Program (NIWQP) Award to study water quality and ecosystem services resulting from landscape management best practices that enhance vegetation in urbanizing watersheds. Part of this project included establishing a sustainable landscape on a residential site in Delaware. A home in the Applecross development in Wilmington was selected due to the sustainable features already present in the development's public spaces, as well as its proximity to Winterthur Museum and Gardens, another study site. A one-acre lot was planted to reduce lawn by half, manage stormwater through plantings, include primarily native plants, incorporate a meadow and reforestation area, and engage people in the landscape. The final design was a compilation of several student projects and included a six thousand-square-foot meadow, a three thousand-square-foot reforestation area, and about eleven thousand square feet of landscape beds (as seen in Figures 1.4 B–D). Details about the project can be found at https://sites.udel.edu/sld/.

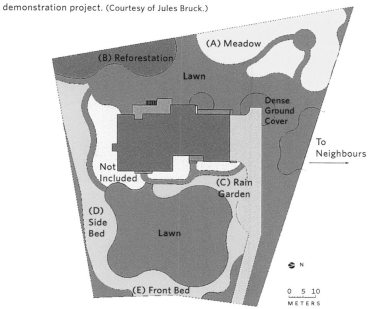

FIGURE 14.3. Proposed design for the Applecross demonstration project. (Courtesy of Jules Bruck.)

FIGURES 14.4 A–D. Applecross demonstration project: (A) residential landscape before modification; (B) meadow; (C) forest; (D) landscape bed. (Photos courtesy of Susan Barton.)

FIGURES 14.5A-B.
Box scroll garden, Winterthur, 1920 and today. (Courtesy of Chris Strand.)

CASE STUDY 4 OF 4

Cultural landscape (Chris Strand, Brown Harrington Director of Garden & Estate, Winterthur Museum, Garden and Library). A cultural landscape is an area associated with an activity, a historic event, or people that possesses cultural or historic value. This bounded, geographic area generally includes cultural resources, such as evidence of human use within the landscape (disturbance, roads, ponds, earthworks), architecture, and hardscape elements (paths, steps, fences, benches, sculpture). It also includes natural resources, such as plants, domestic animals, and wildlife.

The garden and estate at Winterthur Museum, Garden & Library is an example of a cultural landscape with a long history of use that includes field crops, a dairy, hay, a sawmill, nineteenth- and twentieth-century gardens, and a museum. In 1969, with the death of H. F. du Pont, the garden and wider estate became the responsibility of the nonprofit corporation managing Winterthur Museum. The board of trustees established the preservation of the garden as a primary goal by ratifying a board policy for the garden. It states, among other things, that "the original design of the garden and grounds has been substantially retained over the years. It is Winterthur's

FIGURES 14.6A-B.
Route 52, circa 1917 and today.
(Courtesy of Chris Strand.)

policy to conserve this design for posterity, while recognizing the elements of change inherent in living organisms."[26]

This preservation of the Winterthur landscape (rather than rehabilitation, reconstruction, or restoration) has followed a careful process of documentation and record keeping, evaluation of the "design intent" of areas of the garden and wider property, development of management plans, site-specific restoration and replanting, and restoration of objects and hardscape elements.

CARRYING THE MESSAGE

In this era of rapid development, climate change, and population growth, it is more imperative than ever to develop and maintain sustainable landscapes. Both natural and designed spaces have the ability to provide clean water, fresh air, wildlife diversity, and human wellness—in other words, ecosystem services. There are many ways to participate in the creation and stewardship of sustainable landscapes. First and foremost, you can implement best practices at home. In all cases, follow the tenets of Earth care, people care, fair share: make decisions that support the Earth and other humans, and be responsible with resources. Plant native or non-invasive ornamentals suited to your site conditions, consider stormwater management, respect the cultural and natural history of a site, and use integrated pest management.

In addition to improving your home landscape, there are infinite opportunities to make a difference in your community and the state of Delaware. Volunteer your time and expertise to local and regional organizations invested in making positive changes in the environment. Collaborative action can make quick work of an otherwise daunting job and provides opportunities to bolster ties within your community and engage with like-minded folk. In addition, not-for-profit organizations across the region fundraise to support other community-based initiatives financially. Giving donations to these groups is another way you can support and foster the development of sustainable landscapes.

And remember: the skill set you have developed in the Master Naturalist program needs to be shared! Consider how can you take this message and educate others about it, whether they be students or other community members. What area has been most exciting to you in your studies? Often, we are best equipped to teach about the ideas we are most passionate about. Find a specific passion, be it native pollinator support, agroforestry, or invasive plant removal, and throw yourself into this cause. Your passion can rally others to join your effort and build an amazing network of advocates for sustainable landscapes across the state.

NOTES

1. TEEB, *The Economics of Ecosystems and Biodiversity: An Interim Report*, http://doc.teebweb.org/wp-content/uploads/Study%20and%20Reports/Additional%20Reports/Interim%20report/TEEB%20Interim%20Report_English.pdf.
2. K. T. Jackson, *Crabgrass Frontier: The Suburbanization of the United States* (New York: Oxford University Press, 1985), 45.
3. Ibid.
4. R. Kaplan, "The Impact of Urban Nature: A Theoretical Analysis," *Urban Ecology* 8 (1984): 189–97.
5. C. R. Hall, and M. W. Dickson, "Economic, Environmental, and Health/Well-Being Benefits Associated with Green Industry Products and Services: A Review," *Journal of Environmental Horticulture* 29(2) (2011): 96–103.
6. R. W. Kimmerer, *Braiding Sweetgrass* (Minneapolis: Milkweed Editions, 2013), 208.
7. Sustainable Sites Initiative, "Frequently Asked Questions," http://www.sustainablesites.org/faqs; D. Welker, and D. Green. *Sustainable Landscaping* (Washington, DC: U.S. Environmental Protection Agency and Smithsonian Institution's Horticultural Services Division, 2003).
8. D. E. Hands, and R. D. Brown, "Enhancing Visual Preference of Ecological Rehabilitation Sites," *Landscape and Urban Planning* 58(1), 2002, 57–70; J. I. Nassauer, "Messy Ecosystems, Orderly Frames," *Landscape Journal* 14 (1995): 161–70.
9. S. Barton, "Enhancing Delaware Highways: A Natural Vegetation Project" (PhD Diss., University of Delaware, 2005).

10. A. Lucey, "Influencing Public Perception of Sustainable Roadside Vegetation Management Strategies" (M.S. thesis, University of Delaware, 2010).
11. James D. Hitchmough, "New Approaches to Ecologically Based, Designed Urban Plant Communities in Britain: Do These Have Any Relevance in the United States?" *Cities and the Environment (CATE)* 1, no. 2 (2008): 10.
12. United States Department of Agriculture, "USDA Plant Hardiness Zone Map," https://planthardiness.ars.usda.gov/PHZMWeb/, accessed September 15, 2019.
13. American Horticultural Society, "Heat Hardiness Map," https://www.ahsgardening.org/gardening-resources/gardening-maps/heat-zone-map, accessed January 1, 2017.
14. Susan Barton, and Gary Schwetz, "Plants for a Livable Delaware," University of Delaware Cooperative Extension, http://s3.amazonaws.com/udextension/lawngarden/files/2012/06/PLD.pdf, accessed September 15, 2019.
15. Susan Barton, Sarah Deacle, Gary Schwetz, and Douglas Tallamy, "Livable Plants for the Home Landscape," University of Delaware Cooperative Extension, http://s3.amazonaws.com/udextension/lawngarden/files/2012/06/lowres18spreads.pdf, accessed September 15, 2019.
16. Susan Barton, "Livable Ecosystems," University of Delaware Cooperative Extension, http://s3.amazonaws.com/udextension/lawngarden/files/2012/06/live_eco_final.pdf, accessed September 15, 2019.
17. Meg Calkins, "Site Design: Materials and Resources," in *The Sustainable Sites Handbook: A Complete Guide to the Principles, Strategies, and Best Practices for Sustainable Landscapes*, ed. Meg Calkins (Hoboken, NJ: John Wiley & Sons, Inc., 2012), 322.
18. The Sustainable Sites Initiative (SITES) is a comprehensive system for creating sustainable and resilient land development projects. The program was developed through a collaborative, interdisciplinary effort of the American Society of Landscape Architects, the Lady Bird Johnson Wildflower Center at the University of Texas at Austin, and the United States Botanic Garden. Administered by Green Business Certification Inc. (GBCI), SITES offers a comprehensive rating system designed to distinguish sustainable landscapes, measure their performance, and elevate their value. SITES certification is for development projects located on sites with or without buildings—ranging from national parks to corporate campuses, streetscapes to homes, and more.
19. University of Delaware Research Online Magazine. "Sea Level Rise: America's Flattest Site Prepares for the Future," https://www1.udel.edu/researchmagazine/issue/vol4_no1/slr_intro.html, accessed on September 30, 2019.
20. DNREC Division of Watershed Stewardship, *Delaware Post Construction Stormwater BMP Standards and Specifications* 2019.
21. Ibid.
22. USDA, *Agroforestry Strategic Framework: Fiscal Years 2019–2024*, https://www.usda.gov/topics/forestry/agroforestry.
23. Anthony Ricciardi, "Are Modern Biological Invasions an Unprecedented Form of Global Change?" *Conservation Biology* 21, no. 2 (2007): 329–36.
24. Vincent D'Amico, Tara L. E. Trammell, Jeffrey J. Buler, Jake Bowman, Zachary Ladin, Solny Adalsteinsson, and W. Gregory Shriver, "In the FRAME: Monitoring the Novel Assemblages and Emerging Ecology of Urban Forest Fragments," *Frontiers to Urban Ecology*. In preparation.
25. Solny A Adalsteinsson, Vincent D'Amico, W. Gregory Shriver, Dustin Brisson, and Jeffrey J. Buler, "Scale-Dependent Effects of Nonnative Plant Invasion on Host-Seeking Tick Abundance," *Ecosphere* 7, no. 3 (2016): e01317.
26. The Henry Francis Du Pont Winterthur Museum, Inc., Board of Trustees Policy and Guidelines for Winterthur's Garden and Grounds, revised November 2008.

CONTRIBUTORS

SUSAN BARTON, PHD, is an extension specialist and professor in the Plant and Soil Sciences Department at the University of Delaware. She has worked closely for the past twenty years with the Delaware Department of Transportation to research and implement new roadside vegetation management strategies. She has also worked with partners to develop the Plants for a Livable Delaware Program, designed to provide alternatives to known invasive plant species and to promote sustainable landscaping. She teaches Plants and Human Culture, Farm to Table, Field Sketching of Landscape Subjects, the Landscape Architecture symposium, and coordinates the Landscape Horticulture Internship. She also works closely with the nursery and landscape industry, writing newsletters, organizing short courses, and conducting horticulture industry expositions with the Delaware Nursery and Landscape Association. Barton received the Nursery Extension Award in 1995 from the American Nursery and Landscape Association and the Ratledge Award for service from the University of Delaware in 2007.

DR. JULES BRUCK is Professor and Director of Landscape Architecture at the University of Delaware, where she teaches courses in creativity, design process, field sketching, and planting design. She is a registered landscape architect, and serves on the Delaware Board of Landscape Architecture. In April 2018, she co-founded the Coastal Resilience Design Studio and helped to launch the Delaware Resilience Awareness Program and the Coastal Observer online application for citizen scientists interested in documenting changes to the coastal environment. Her current research interests are in coastal resilience, green infrastructure, and public perception of sustainable landscape practices such as designing for ecosystem services. Dr. Bruck has a PhD in Agricultural Education from Texas A&M University.

JON COX is a National Geographic Explorer and an assistant professor in the Department of Art and Design at the University of Delaware, with an MFA in Photography and a BS in Entomology and Plant Pathology. He also serves as a board member of the Dorobo Fund for Tanzania and is president of the Amazon Center for Environmental Education and Research. Cox co-authored a six-year collaborative documentary book project with hunter-gatherers in Tanzania titled *Hadzabe, By the Light of a Million Fires* (African Books Collective, 2013). He was a pioneer in the field of digital photography, served as the adventure photographer/writer for *Digital Camera Magazine*, and authored two Amphoto digital photography books. Cox is a co-recipient of a National Geographic—Genographic Legacy Fund Grant to support a collaborative cultural mapping initiative with the Ese'Eja hunter-gatherers living in the Amazonia basin of Peru. A book and traveling exhibition to accompany this project titled *The Ese'Eja People of the Amazon: Connected by a Thread* is currently on tour. Cox is currently working on cultural mapping initiatives with the Lenape Indian Tribe of Delaware funded by National Geographic, the Delaware Humanities Forum, the University of Delaware's Partnership for Arts and Culture, and the Interdisciplinary Humanities Research Center.

MCKAY JENKINS is the Cornelius Tilghman Professor of English, Journalism, and Environmental Humanities at the University of Delaware. He is the author of many books and articles about environmental issues, including *Food Fight: GMOs and the Future of the American Diet* (Avery, 2017); *ContamiNation: My Quest to Survive in a Toxic World* (Avery, 2016); and (with E. G. Vallianatos) *Poison Spring: The Secret History of Pollution and the EPA* (Bloomsbury, 2014). Jenkins holds an MS in Journalism from Columbia's Graduate School of Journalism, and a PhD in English from Princeton. A former staff writer for the *Atlanta Constitution*, he has also written for *Outside, Orion, The New Republic,* and many other publications. He is a recipient of both the University of Delaware's Excellence in Teaching Award and the College of Arts and Sciences Excellence in Teaching Award.

GERALD MCADAMS KAUFFMAN is Director of the University of Delaware Water Resources Center, one of the fifty-four National Institutes for Water Resources (NIWR) supported by the United States Geological Survey at landgrant universities in the fifty states, District of Columbia, and three island territories of Guam, Puerto Rico, and U.S. Virgin Islands. Kauffman holds faculty appointments in the Biden School of Public Policy and Administration, Geography Department, and Department of Civil and Environmental Engineering. He wrote a book on water resources engineering with Dr. Jonathan Brant from the University of Wyoming entitled *Water Resources and Environmental Engineering* (Professional Publications, Inc., 2011), and is currently writing a book on sustainable watershed management to be published with John Wiley and Sons. Kauffman also serves as Delaware's first water master, appointed by the governor and General Assembly in accordance with the Water Supply Coordinating Council Act of 2000. He likes bicycling and cross-country skiing, especially in the White Clay Creek National Wild and Scenic River near Newark, Delaware.

TOM MCKENNA is a hydrogeologist at the Delaware Geological Survey and Associate Professor in the Department of Geological Sciences at the University of Delaware. He earned his PhD in Geology from the University of Texas after having earned a Master's at the University of South Carolina and a Bachelor's degree at Stockton University, where Professor Mike Hozik inspired him and his friends to be geologists. McKenna currently lives in Newark,

Delaware, and is married to a geologist (Kim). They have a daughter who is also a geologist (Jillian).

VICTOR PEREZ is an Associate Professor of Sociology at the University of Delaware, with specializations in environmental justice, health and illness, and the sociology of risk. A unifying theme throughout his career is the entwined configuration of health, risk, and society, with a focus on environmental and health issues through constructionist and social justice lenses. Currently, his research focuses on migration/relocation due to climate change impacts and environmental burdens, planned/adaptive relocation of environmentally burdened communities, green/environmental gentrification, and intensive zoning. He also volunteers in South Wilmington and serves as a core member of the South Wilmington Planning Network.

IAN STEWART spent a happy childhood exploring the woods and seashore of his native Northeast England, where he developed a lifelong interest in nature and especially birds. He completed a PhD on the breeding biology of birds at the University of Leicester, followed by research positions at the University of Kentucky and the University of Delaware. He currently works as an ornithologist at the Delaware Nature Society, where he conducts research into how the society's land management practices are affecting biodiversity, with a particular focus on birds. He also leads bird walks and nature outreach programs including public bird banding sessions. He lives with his wife and two young sons in Kennett Square, Pennsylvania.

DOUG TALLAMY is a professor in the Department of Entomology and Wildlife Ecology at the University of Delaware, where he has authored ninety-five research publications and has taught insect-related courses for forty years. Chief among his research goals is to better understand the many ways that insects interact with plants and how such interactions determine the diversity of animal communities. His book *Bringing Nature Home: How Native Plants Sustain Wildlife in Our Gardens* (Timber Press, 2007) was awarded the 2008 Silver Medal by the Garden Writers' Association. *The Living Landscape*, co-authored with Rick Darke, was published by Timber Press in 2014. Doug's latest book *Nature's Best Hope* was released by Timber Press in February 2020. Among his awards are the Garden Club of America Margaret Douglas Medal for Conservation, the Tom Dodd, Jr. Award of Excellence, the 2018 American Horticultural Society B. Y. Morrison Communication Award, and the 2019 Cynthia Westcott Scientific Writing Award.

TARA TRAMMELL is a research ecologist and the John Bartram Assistant Professor of Urban Forestry at the University of Delaware. She leads the Urban Ecology Lab in the Plant and Soil Sciences Department, which conducts research on understanding human impacts on plants and soils in urban environments. In the Urban Ecology Lab, a variety of field, lab, and modeling techniques are used to understand controls on plant species composition and diversity and to assess changes in plant-soil carbon and nitrogen cycling in urban ecosystems. Research projects include examination of management impacts on lawn carbon and nitrogen cycling, and of non-native plant invasion and sea level rise impacts on small forests. Trammell's research program and publication record of over thirty peer-reviewed journal articles and eight invited book chapters focuses on understanding how global environmental change, such as urbanization and plant invasion, alters terrestrial ecosystems embedded within urban, suburban, and exurban landscapes.

JENNIFER VOLK is the Associate Director of University of Delaware Cooperative Extension and is also an Environmental Quality Extension Specialist. She received both her undergraduate degree in Chemistry and graduate degree in Marine Studies from the University of Delaware. For nine years following graduate school, Volk worked in the Delaware Department of Natural Resources and Environmental Control's Watershed Assessment Section as an Environmental Scientist, where she worked on water quality issues and watershed management plans. At the University of Delaware, she continues to work on environmental issues in the state and the region and has broadened her focus to include climate change impacts and responses. She serves as the university's liaison to the Northeast Climate Hub's University Partnership Network, where she collaborates with other research and extension professionals in the region on climate-related initiatives.

JOCELYN WARDRUP is a soil and wetland scientist experienced in the mid-Atlantic region. After receiving a Bachelor's degree in Geography, she obtained a position with a soil scientist and fell in love with variations of soils, their morphology, classification, and interpretations. She returned to the University of Delaware for a Bachelor of Science in Environmental Soil Science, earning that degree in 2011. She has eight years of experience working with the private, public, and nonprofit sectors in Delaware, Maryland, Pennsylvania, New Jersey, Virginia, West Virginia, and Ohio. Wardrup holds Certified Professional Soil Scientist and Professional Wetland Scientist certifications. She has experience in soil suitability evaluations for wastewater treatment, basements, and stormwater management, wetlands delineations and permitting, forest stand delineations, and wetlands monitoring and assessment. After hiking the Appalachian Trail in 2017 and 2018, she returned to the University of Delaware for a Master's in Plant and Soil Science focusing on a new aspect of soil science: digital soil mapping. A lifelong Delawarean, she encourages everyone to examine and explore the important layers of the Earth that lie hidden just beneath our feet.

AMY (WENDT) WHITE is co-author of the field guide *Amphibians and Reptiles of Delmarva* (Tidewater Publishers, 2002; rev. 2007) with her husband Jim (see below). She grew up in Delaware and holds an undergraduate degree in Geology from the College of William and Mary and a graduate degree in Environmental Engineering from the University of Delaware. Working for the Delaware Nature Society, she coordinated the Delaware Stream Watch program for the first five years of that program, and she also worked as a project engineer with Malcolm Pirnie, Inc., investigating soil and groundwater contamination. White is currently a teacher-naturalist at the Delaware Nature Society.

JIM WHITE, a native Delawarean, graduated from the University of Delaware with a degree in Entomology and Applied Ecology. He was one of the primary investigators for the five-year Delaware Herpetological Survey (1986–91), which gathered distributional information throughout the state, and is an active contributor to the current five-year Delaware Herpetological Atlas. White has been with the Delaware Nature Society for thirty-six years, currently as Senior Fellow for Land and Biodiversity Management. He also teaches a course in herpetology at the University of Delaware.

The Delaware Naturalist Handbook is typeset in Chronicle (2002) and Whitney (1996), both created by Jonathan Hoefler of Hoefler & Co. in New York. The Scotch Roman roots of the serifed text typeface Chronicle date back to 1809, the birth year of the great English naturalist Charles Darwin. The sans-serif Whitney numerals and italics have been especially adapted for this volume by Robert Wiser to be compatible with the size and slant of the Chronicle type.